Ice Cream

JOIN US ON THE INTERNET
WWW: http://www.thomson.com
EMAIL: findit@kiosk.thomson.com

Visit Chapman & Hall's Internet Resource Center for information on our new publications, links to useful sites on the World Wide Web and an opportunity to join our e-mail mailing list. Point your browser to:
http://www.chaphall.com/chaphall.html or
http://www.chaphall.com/chaphall/foodsci.html for Food Sciences

A service of I(T)P

Ice Cream

FIFTH EDITION

Robert T. Marshall
Department of Food Science and Human Nurition
University of Missouri, Columbia

W. S. Arbuckle
Department of Dairy Science
University of Maryland, College Park

CHAPMAN & HALL

I(T)P® International Thomson Publishing

New York • Albany • Bonn • Boston • Cincinnati • Detroit • London • Madrid • Melbourne
Mexico City • Pacific Grove • Paris • San Francisco • Singapore • Tokyo • Toronto • Washington

Copyright © 1996 by Chapman & Hall

Printed in the United States of America

For more information, contact:

Chapman & Hall
115 Fifth Avenue
New York, NY 10003

Chapman & Hall
2-6 Boundary Row
London SE1 8HN
England

Thomas Nelson Australia
102 Dodds Street
South Melbourne, 3205
Victoria, Australia

Chapman & Hall GmbH
Postfach 100 263
D-69442 Weinheim
Germany

Nelson Canada
1120 Birchmount Road
Scarborough, Ontario
Canada M1K 5G4

International Thomson Publishing Asia
221 Henderson Road #05-10
Henderson Building
Singapore 0315

International Thomson Editores
Campos Eliseos 385, Piso 7
Col. Polanco
11560 Mexico D. F.
Mexico

International Thomson Publishing - Japan
Hirakawacho-cho Kyowa Building, 3F
1-2-1 Hirakawacho-cho
Chiyoda-ku, 102 Tokyo
Japan

All rights reserved. No part of this book covered by the copyright hereon may be reproduced or used in any form or by any means —graphic, electronic, or mechanical, including photocopying, recording, taping, or information storage and retrieval systems —without the written permission of the publisher.

2 3 4 5 6 7 8 9 10 XXX 01 00 99 98 97

Library of Congress Cataloging-in-Publication Data

Marshall, Robert T., 1932-
   Ice Cream / Robert T. Marshall and W. S. Arbuckle.-- 5th ed.
     p.   cm.
   Fourth ed. / W. S. Arbuckle. Westport Conn. : AVI Pub. Co., c1986.
   Contents: Includes bibliographical references and index.
   ISBN 0-412-99491-7 (hardcover : alk. paper)
   1. Ice Cream Industry.   I. Arbuckle, W. S. (Wendell Sherwood),
1911-1987. II. Title.
TX795.M295   1996   96-3282
637' .4--dc20   CIP

To order this or any other Chapman & Hall book, please contact **International Thomson Publishing, 7625 Empire Drive, Florence, KY 41042.** Phone: (606) 525-6600 or 1-800-842-3636. Fax: (606) 525-7778. e-mail: order@chaphall.com.

For a complete listing of Chapman & Hall's titles, send your request to **Chapman & Hall, Dept. BC, 115 Fifth Avenue, New York, NY 10003.**

Contents

Preface	xi

1

The Ice Cream Industry	**1**
Definition and Composition	*1*
Historical Background	*3*
Production and Consumption	*6*
References	*9*

2

Energy Value and Nutrients of Ice Cream	**10**
Energy Value and Nutrients	*10*
Caloric Content of Ice Cream and Related Products	*11*
Protein Content of Ice Cream	*13*
Milkfat Content	*13*
Carbohydrates in Ice Cream	*14*
Minerals in Ice Cream	*15*
Vitamins in Ice Cream	*16*
Palatability and Digestibility of Ice Cream	*16*
References	*17*

3

Classifications of Ice Cream and Related Products	**18**
Commercial Grouping of Ice Cream and Related Products	*18*
Flavor Categories for Ice Cream	*20*
References	*21*

4

Composition and Properties	**22**
Complexities of Composition	*25*
Characteristics of a Satisfactory Composition	*26*
The Role of the Constituents	*26*

Importance of Flavor	*33*
The Balanced Mix	*33*
References	*44*

5

Ice Cream Ingredients — 45

Optional Ingredients	*45*
Composition of Milk	*45*
Milk Products Used in Ice Cream	*51*
Sweeteners	*58*
References	*69*

6

Stabilizers and Emulsifiers — 71

Uses of Stabilizers	*71*
Kinds of Stabilizers	*72*
Characteristics of Individual Stabilizer Ingredients	*72*
Ice Cream Improvers	*75*
Emulsifiers	*75*
Industrial Usage	*79*
References	*80*

7

Flavoring and Coloring Materials — 81

Flavors for Frozen Desserts	*82*
Vanilla	*83*
Chocolate and Cocoa	*87*
Fruits in Frozen Desserts	*91*
Procedures and Recipes	*96*
Nuts	*100*
Spices and Salt	*101*
Color in Frozen Desserts	*101*
Flavoring Lowfat and Nonfat Ice Cream	*102*
References	*103*

8

Calculation of Ice Cream Mixes — 104

The Importance of Calculations	*104*
Mathematical Processes Most Frequently Used	*104*
Methods of Calculating Mixes	*105*
Standardizing Milk and Cream	*106*
Calculating Mixes with the Serum Point Method	*108*
Mix Decisions	*113*
Simple Mixes	*113*
Complex Mixes	*117*

CONTENTS vii

Use of Computers in Ice Cream Production	*127*
References	*131*

9

Mix Processing — **139**

Preparation of the Mix	*139*
Pasteurization of the Mix	*147*
Homogenization	*151*
Aging Mixes	*158*
Packaging Mixes for Sale	*159*
Flavoring Mixes	*160*
References	*163*

10

The Freezing Process — **164**

Prefreezing Tests	*164*
Freezing Operations	*164*
Changes That Take Place During the Freezing Process	*166*
Refrigeration Needed to Freeze Ice Cream	*172*
Calculating Refrigeration Requirements	*175*
Types of Freezers	*177*
The Batch Freezer	*195*
References	*198*

11

Packaging, Labeling, Hardening and Shipping — **200**

Considering the Package	*200*
Labeling	*201*
The Packaging Operation	*205*
The Hardening Process	*211*
Handling, Storing and Shipping	*216*
Quality is the Goal	*220*
References	*220*

12

Soft-Frozen Dairy Foods and Special Formulas — **222**

Soft-Serve Products	*222*
Freezers for Soft-Serve and Shakers	*222*
Cleaning and Sanitizing Soft-Serve Freezers	*226*
The Heat Treatment Freezer	*228*
Soft-Serve Desserts Containing Particulates	*229*
Soft Serve Mix Composition	*229*

13

Sherbets and Ices — 234

The Composition of Sherbets and Ices	234
Preparation of Ices	236
Preparation of Sherbets	237
Freezing Ices and Sherbets	239
Defects	239
References	240

14

Fancy Molded Ice Creams, Novelties, and Specials — 241

Production Systems	242
Specialty Equipment	242
Ice, Fudge and Cream Stick Items	249
Ice Cream Bars	250
Chocolate Coatings	250
Other Special Products	253

15

Defects, Scoring and Grading — 258

Flavor Defects	258
Sweetner System	261
Body and Texture Defects	262
Color	267
Package	267
Melting Quality	267
Evaluating Frozen Desserts	269
Scoring Methods	270
Ice Cream Clinics	272
References	275

16

Cleaning, Sanitizing, and Microbiological Quality — 276

Cleaning Equipment	278
Functions of Detergents in the Dairy	278
Major Detergent Components and Their Functions	279
Principles of Cleaning	280
Sanitization of Equipment	286
Sanitary Environment	287
Hygienic Personnel	288
Tests of the Finished Procuct	289
Summary	290
References	291

17

Refrigeration — 292

Methods of Refrigeration	292
Types of Refrigerants Commonly Used	293
Principles of Mechanical Refrigeration	294
Operating Precautions	296
Defrosting Methods	297
Methods of Cooling	298
Terms Used in Refrigeration	300

18

Sales Outlets — 301

Wholesaling	302
Retailing	302
Merchandising	302
Drive-In Store	303
The Soda Fountain	303
Personnel	305
Training for Fountain Service	306
Basic Soda Fountain Preparations	311
Specials for Seasons and Memorable Occasions	312
References	313

19

Formulas and Industry Standards — 314

Plain Ice Cream	314
Candy or Confection Ice Cream	315
Chocolate Ice Cream	315
Fruit Ice Creams	316
Frozen Yogurt	317
Nut Ice Creams	318
Puddings	319
Parfait	320
Mousse	320
Frappé	320
Sorbets	320
Punch	321
Granite	321
Sherbets	321
Soufflé	321
Lacto	321
Fruit Salad	321
Fancy Molded Ice Cream	322
Frosted Malted	322
Specials	322
Homemade Ice Cream	323
Low- or Reduced-Lactose Ice Cream	324

Ice Creams of Lowered Fat Content *325*
References *327*

Appendices

A. Historical Chronology of Ice Cream Industry **328**

B. Miscellaneous Tables **331**

C. Frozen Desserts Plant Inspection Form **336**

Index **339**

Preface

Ice cream, favorite food of millions. Rich and creamy, sweet and flavorful, pleasing to the palate and the eyes. How rewarding it is to learn about and write about ice cream. The many forms and flavors of these dispersions of tiny air bubbles, ice crystals, fat globules and casein micelles in thick, sweet-flavored syrup have the potential to provide the human sensory receptors with the highest levels of enjoyment. The study of the creation, production, and distribution of ice cream is important to many people. This is why my forebears, Professors W. S. Arbuckle and J. H. Frandsen, wrote the earlier editions of *Ice Cream*. I am most pleased to have the opportunity to write the fifth edition. It is a complete revision of previous editions.

Emphases have been placed on newest technologies and developments in science that affect the ice cream industry. Additionally, many changes took place in the realm of governmental control since the last edition, and these changes have been put in perspective for the reader. The contributions of the previous authors continue to play an important role in the organization and content of the book. However, with technology continuously moving forward and in the interest of clarity and brevity, I have elected to omit many of the older references from the book and suggest that those who wish to search the older literature consult a copy of the earlier editions.

Subjects that have been added or expanded include nonfat and lowfat ice cream products, fat replacers, the new labeling regulations, changes in the standard of identity, programmable freezers and aerators. Other subjects that have been amplified include flavorings, fruits, soft-serve freezing and dispensing, production of novelties, pasteurization controls, packaging, hardening, storage-retrieval systems, and cleaning and sanitizing.

About eighty illustrations have been added or completely revised to bring increased understanding of the scope and depth of the subject of frozen dessert manufacture to the reader.

While recognizing that computers are being used to do much of the calculations of mixes, I have included the well-done section on calculations (Chapter 8) that has served so many so well. However, the section has been made more concise, and the chapter on calculations of some unusual mixes has been eliminated. Furthermore, cost and percentage overrun are incorporated in the material on mix calculations rather than being given chapter status.

It is important to acknowledge the invaluable assistance of the ice cream industry in this writing effort. The Dairy and Food Industries Supply Association supported me in a sabbatical leave during 1993 so that I could visit and study the industry personally. Much gratitude is owed to those manufacturers and suppliers who hosted and enlightened me during that significant six months. The International Ice Cream Association has continually supported my efforts in ways too numerous to mention. Opportunities to work with the National Ice Cream and Yogurt Retailers Association have also been instructive. Support from my colleagues in the American Dairy Science Association, the International Association of Milk, Food and Environmental Sanitarians and the Institute of Food Technologists has been invaluable.

The Arbuckle Endowment at the University of Missouri, started by Wendel and Ruth Arbuckle and enhanced by gifts from industry, started me on the way to this project. Professors Arbuckle and Frandsen provided the first-hand knowledge of the developing and maturing ice cream industry. Now it is a mature industry. Yet it continues to change with each new day. It is being concentrated more and more into larger plant operations under ownership of large firms, some of them international in scope. However, it is also an industry in which there is still room for the enterprising person to begin a small business. Suppliers to the industry are increasingly providing technical assistance to manufacturers, enabling small firms to gain knowledge that places them in competition with firms that have their own research units.

Ice cream never improves after it leaves the freezer. Therefore, the closer the freezer is to the consumer, the better the product stands to be. The wide variety of ingredients and flavors available to the manufacturer and the specialized equipment there is to make frozen desserts make it possible to differentiate a firm's products from all others. For these reasons, entry into the market is possible for many. This makes it highly relevant for a host of people to know the principles of the science and technology of ice cream manufacture. To those curious persons who find this treatise valuable, I extend my regards as well as the invitation to contribute their knowledge and experience in every effort to make this wonderful industry even more precious to the people who enjoy eating ice cream and related products.

1
The Ice Cream Industry

> The current fashionable thesis is that the main asset of a large company is its human skills; the collection of individuals that work for it. It follows that companies likely to be most successful will be those who train their people best and nurture their skills.
> —Hamish McRae, British economic commentator

DEFINITION AND COMPOSITION

Ice cream is a frozen mixture of a combination of components of milk, sweeteners, stabilizers, emulsifiers and flavoring. Other ingredients such as egg products, colorings, and starch hydrolysates may be added also. This mixture, called a mix, is pasteurized and homogenized before freezing. Freezing involves rapid removal of heat while agitating vigorously to incorporate air, thus imparting the desirable smoothness and softness of the frozen product.

The broad term *frozen desserts* refers to ice cream and related products. Specific products include ice cream and its lower fat varieties, frozen custard, mellorine (vegetable fat frozen dessert), sherbet, water ice, and frozen confections. Some of these desserts are served in either the soft frozen or hard frozen form.

In the United States ice cream is defined by federal standards as containing not less than 10% milkfat and 20% total milk solids (TMS), as weighing not less than 4.5 lb/gal, containing not less than 1.6 lb of food solids per gallon, and containing not more than 0.5% stabilizer. (Unless specified otherwise, percentages for ice cream are stated by weight.) However, by adding a descriptive modifer before the name ice cream, manufacturers can market reduced fat, light or lite, lowfat or nonfat ice cream. These products are described in Chapter 4. Bulky flavored ice creams are those containing large amounts of flavorings such as chocolate or strawberries. Percentages of fat and TMS in bulky flavored products are permitted to be as low as 8 and 16, respectively.

The composition of ice cream varies widely depending on the intended market. Typical composition of a product labeled vanilla ice cream is 11% milkfat, 11% nonfat milk solids (NMS), 12% sugar, 5% corn syrup solids, and 0.3% stabilizer/emulsifier. This makes the total solids approximately 39%. Products having the words flavored or artificially flavored along with the name of the flavoring on the label tend to contain the minimal amounts of solids from milk and to contain higher amounts of whey solids and corn syrup solids than do products having the name of the flavoring only. The range in composition may be 0–20% fat, 8–15% NMS, 13–20% sweetener, 0–1.0% stabilizer-emulsifier, and 32–43% total solids (TS).

The physical structure of ice cream is a complicated physicochemical system. The three phases of the system are liquid, solid, and gas. Air cells and ice crystals are dispersed in a continuous liquid phase. The liquid phase also contains solidified fat, colloidal milk proteins, insoluble milk salts, lactose crystals in some cases, colloidal stabilizers, and sugars and soluble salts in solution.

Ice cream is a palatable, nutritious, healthful, and relatively inexpensive food. One serving of ice cream of average composition, 4 fluid ounces, weighs about 70 g, and supplies about 130 calories, 3 g protein. 100 mg calcium, 70 mg phosphorus, 250 international units (IU) vitamin A, 120 µg riboflavin, and 30 µg thiamine.

Frozen custard (also known as French ice cream) has the formula of ice cream except that it must contain at least 1.4% egg yolk solids.

Sherbet contains a small amount of milk solids plus sugar, fruit juices, fruit flavorings, and stabilizer. It must contain 1–2% milkfat and 2–5% total milk solids. A sherbet must weigh not less than 6 lb/gal and have not less than 0.35% acidity calculated as lactic acid. It must contain 2% citrus fruit, 6% berry fruits, or 10% other fruits. The acidity requirement is waived for nonfruit flavors, which do not have to contain any fruit or fruit juice.

Water ice is similar to sherbet but does not contain milk solids or egg ingredients other than egg white. Otherwise, the standards are the same as for sherbets.

Frozen confections include stick novelties, ice cream sandwiches, and other special single serving items. Quiescently frozen dairy confections are sold in the form of individual servings on a stick. They must contain not less than 13% milk solids and 33% TS. Overrun, the increase in volume of product over the volume of mix due to incorporation of air, must not exceed 10%. Quiescently frozen confection is similar to quiescently frozen dairy confection except that it must contain at least 17% TS. It may or may not contain milk solids.

Mellorine-type products are similar to ice cream except that the milkfat has been replaced by a suitable vegetable oil or animal fat. Sources of vegetable oils include soybean, cottonseed, canola, corn, and coconut. Mellorine-type frozen desserts must contain not less than 6% fat from one or more specified sources and not less than 3.5% protein of biological value at least equivalent to that of milk protein. Vitamin A must be added in such quantity as to provide 40 IU per gram of fat.

Parevine-type products are similar to ice cream except that no dairy ingredients are used, these having been replaced by safe and suitable ingredients.

Frozen yogurt is a cultured frozen dessert containing the same ingredients as ice cream. It contains not less than 3.25% milkfat and not less than 8.25% NMS. The lowfat and nonfat forms contain 0.5–2% and less than 0.5%, respectively. The frozen product weighs not less than 5 lb/gal. There are curently no federal standards for frozen yogurt.

Freezer-made milk shakes contain 3.25–6% milkfat and not less than 10% NMS. Freezer-made shakes are similar but contain less than 3.25% milkfat.

1 THE ICE CREAM INDUSTRY

Ice cream made with goat's milk is prepared from the same types of ingredients and with similar formulas as that prepared with cow's milk, but it must be labeled as made from goat's milk.

Federal standards for frozen desserts are published in the Code of Federal Regulations, Title 21, Part 135. This is referenced elsewhere as 21 CFR 135. Some other frozen desserts not defined by federal standards may be defined by some state standards.

HISTORICAL BACKGROUND

Although the ice cream industry was largely developed in the United States, the product was introduced to the United States from Europe. Ice cream probably evolved from the iced beverages and water ices that were popular in Europe during medieval times. We do know that wines and fruit juices with honey were cooled with ice and snow brought from the mountains of the Apennines to the court of the Roman emperor Nero in the first century AD. Unfortunately, no definite description exists, except that snow and ice were used to cool and possibly to freeze sweet desserts. It is conceivable that this idea originated in ancient Egypt or Babylon, where sweetmeats and other dainties may have been iced.

In the thirteenth century Marco Polo returned to Italy from his famous journey to the Orient and brought recipes for water ices said to have been used in Asia for thousands of years. The art of making these products then moved to France, Germany, and England during the next few centuries. By 1560 an Italian had written of a food ". . . made of milk sweetened with honey and frozen . . ." In 1660 the Cafe Procope was founded in Paris by an Italian named Cotelli, and water ices were manufactured and sold. An anonymous 84-page manuscript entitled *L'art de Faire des Glaces* (The Art of Preparing Ice Cream) was written about 1700. In 1768 a 240-page treatise, *The Art of Making Frozen Desserts*, appeared in Paris. It gave formulas for "food fit for the gods" and offered theological and philosophical explanations for phenomena such as the freezing of water.

Ice cream probably came to the United States with the early English colonists. The first written evidence of ice cream in America was in a letter of May 17, 1744, by a guest of proprietary governor William Bladen of Maryland. The letter stated that "a dessert no less curious, among the rarities of which it was composed, was some fine ice cream which with the strawberries and milk, eat most deliciously" (Arbuckle 1981). George Washington spent about $200 for ice cream in New York during the summer of 1790. But the masses had to wait for developments of ice harvesting, insulated ice houses, and the hand-cranked ice cream freezer (invented in 1846 by Nancy Johnson and patented in 1848 by W. G. Young). In 1851 the first wholesale ice cream industry in the United States was established in Baltimore, Maryland, by Jacob Fussell. Plants were established soon thereafter in St. Louis, New York, Washington, Chicago, and Cincinnati. Two of the most important contributions to development of the industry were the perfection of mechanical refrigeration (1878) and the invention of the direct expansion ice cream freezer (1913).

The development of condensed and dry milk, the introduction of the pasteurizer and homogenizer, and improved freezers and other processing equipment accompanied a slow growth in the industry until after 1900. The ice cream soda was introduced in 1879. The ice cream cone was first produced by an Italian emigrant, Italo Marchiony, in 1896 in New York City. He was granted a patent for his special mould in 1903. A Syrian waffle concessionaire, E. A. Hamwi, started rolling waffles into the shape of a cone, and an ice cream vendor in the adjoining booth used them as ice cream containers at the 1904 St. Louis World's Fair.

Annual production of ice cream in 1905 was only 5 million gallons. At that time there was no national trade organization, and only one college offered instruction in ice cream manufacture. Most products were being made without much guidance in quality or content. In 1905 Thomas D. Cutler founded *The Ice Cream Trade Journal,* the predecessor to *Ice Cream Field* and that to *Dairy Field*. In that same year Emery Thompson, manager of the ice cream and soda fountain department in a New York City department store, developed the gravity-fed batch ice cream freezer. The invention enabled nearly continuous production. The company remains under family ownership and continues to supply freezers to industry.

In recognition of 90 years of publication of their trade journal for the ice cream and dairy industry, *Dairy Field* reviewed growth of the ice cream industry in their January 1995 issue. Many of the following citations are from that issue.

In the 1900–1919 era the United States experienced rapid industrialization and urbanization. Ice cream standards were adopted by some states, and many dairy organizations were formed. New York City was the first municipality to inspect dairy farms for sanitary milk production practices, and the first dairy show was held in Chicago in 1906. Chicago led the way toward safe ice cream and dairy foods by adopting the first compulsory pasteurization regulation in 1909; however, it left the loophole that the requirement applied to milk from herds that had not been tested for tuberculosis. In 1917 the newly organized *Association of Ice Cream Supply Men*, the forerunner of today's *Dairy and Food Industries Supply Association*, held its first trade exposition in Boston.

In 1921 Mojonnier Brothers introduced the first ice cream packaging machine, and in 1923 the Nizer Cabinet Company introduced the first automatic electric freezer. The Eskimo Pie, Good Humor ice cream bar, and Popsicle were all invented around 1920. Christen Nelson invented the I-Scream bar in 1919. After Nelson took Russell Stover as his partner, Stover named the bar the Eskimo Pie. The Popsicle was first called the Epsicle in honor of its inventor, Epperson, a concessionaire of an amusement park. The idea came to him, legend has it, when he left a glass of lemonade containing a spoon in an open window on a cold night. By morning the lemonade was frozen. He immersed the glass in water and removed the frozen mass. He patented the invention in 1924. The ingenuity of Harry Burt and the prompting by his daughter led to the Good Humor Ice Cream Sucker, a chocolate-coated ice cream bar on a stick.

The 1930's saw ice cream firms expanding after the great depression. The Beatrice Creamery Company was the first firm to advertise in a national

1 THE ICE CREAM INDUSTRY

consumer magazine, *The Saturday Evening Post*. United Airlines became the first airline to serve ice cream on airplanes in 1937.

The Second World War challenged the industry in the early 1940's. Dictator Benito Mussolini of Italy thought ice cream "too American" and banned its sale in Italy. Dairy Queen was started in 1940, and ice cream sandwiches were sold on the streets of New York about that time. 1945 was an important year, because Bill Rabin developed a mechanized process for producing ice cream sandwiches. Additionally Irving Robbins and Burton Baskin opened the Snowbird Ice Cream Store in Glendale, CA, offering 21 flavors. Nine years later they adopted the "31" Baskins-Robbins logo and embarked on a nationwide franchise program. In 1947 Lady Borden ice cream made its debut, and half-gallon packages appeared in stores in 1948.

Television began to influence marketing in the 1950's. Attention was focused on packaging, automated equipment, and increased efficiency. The Carnation Company introduced the Ice Cream Bon Bon. Soft-serve drive-ins flourished with the names Dairy Queen, Tastee Freeze, and Carvel leading the way.

Testing of nuclear bombs in the 1960's made people anxious over the possibility that radionuclides might get into dairy products. Still, new developments spurred the industry. Ben Cohen and Jerry Greenfield met in 1963 while in the seventh grade. Fifteen years later they opened their first Ben and Jerry's Homemade ice cream shop in a renovated gasoline station in Vermont. In the mid-1960's Good Humor reported it had tested more than 3,000 ice cream flavors. The biggest flops were Chile Con Carne, Kumquat, Licorice, and Prune.

In the 1970s two major fast food chains moved new ice cream products into their stores. McDonald's began testing McSundaes, soft-serve vanilla ice cream topped with strawberry, hot fudge, pineapple, or caramel topping. Dairy Queen began selling soft-serve yogurt in their stores in 1977.

The 1980's saw continued amalgamation of the industry with Pillsbury purchasing Haagen-Daz Company (1983) and Nestle' S. A. purchasing Carnation (1984). This consolidation continues as the twenty-first century approaches. The acquisition in 1993 of the Sealtest and Breyers lines from Kraft Foods by the British firm Unilever made the resultant ice cream firm, Good Humor Breyers, the largest in the United States and continued the trend toward globalization of the ice cream industry. During this time, the Baskin-Robbins International Division established manufacturing plants in several countries to augment the supplies of ice cream shipped from the United States to their many stores in other nations.

The many areas of progress reviewed in the preceding discussion, as well as many others to numerous to cite, have been made possible largely because of advances in transportation, general availability of electricity and consequently of refrigeration, improved packaging, and improved qualities of ingredients, not the least of which are those made from milk.

Modern automated, high-volume operations provide a plentiful supply of ice cream in a wide variety of fat contents, flavors, packages, and prices. Novelty manufacturers produce thousands of items of many different types per minute. Specialty producers, often at the retail store level, produce ice cream cakes, pies, and molded items. The ice cream trade is big business, and it serves to satisfy one of the most ardent cravings for food among humans.

PRODUCTION AND CONSUMPTION

The leading nation in the world in ice cream production and consumption is the United States with annual per capita consumption of all ice cream, frozen yogurt, and ice products at about 23.5 quarts (188 4-fl-oz servings; Figure 1.1). For the next five countries, average per capita consumption figures in 4-fl-oz servings follow: New Zealand, 151 servings; Australia, 130 servings; Belgium/Luxembourg, 126 servings; Sweden, 120 servings; and Canada, 108 servings.

Monthly production figures for the United States indicate that ice cream consumption is seasonal; however, it is much less so than in former years (Figure 1.2). Whereas in 1921 production in July was twice the monthly average for the year, this statistic dropped to 1.6 times the monthly average by 1941. Further leveling of production followed, and for the last 35 years production in July has ranged from 1.2 to 1.3 times the monthly average for the respective years. Lowest production occurs in November through January with average production at 75–80% of the monthly average for the year.

California leads all states in the United States in the production of frozen desserts with over 172 million gallons per year. The remaining top four states and the millions of gallons produced annually in 1993 were: Pennsylvania, 107.1; Ohio, 80.2; Texas, 70.1; and Indiana, 62.7. The top five states produced about 35% of the U.S. production. The number of plants producing ice cream

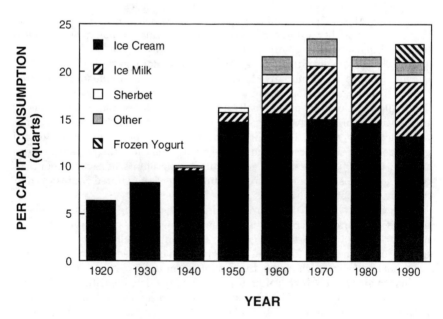

Figure 1.1. Per capita consumption of frozen dessert products in the United States for each ten years since 1920.

1 THE ICE CREAM INDUSTRY

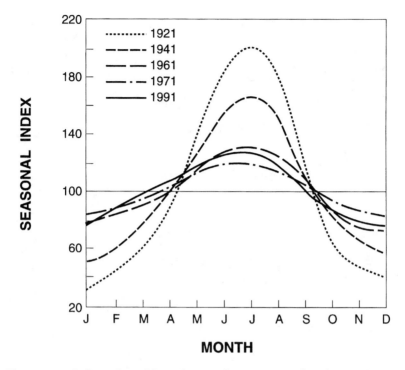

Figure 1.2. Indices of monthly production of ice cream in selected years since 1921.

dropped from 1628 in 1970 to 949 in 1980 to 713 in 1990 and to fewer than 500 in 1995. Of course, there are many thousands of retail stores and food service establishments that freeze ice cream, yogurt, sherbets, sorbets, and ices. However, most of the mix frozen by these firms is made in large factories that deliver directly to the ice cream retailer's door. An interesting statistic shows how the production of vegetable fat frozen desserts has practically disappeared: 1,792 ice cream plants produced mellorine in 1968. By 1988 only 33 plants were reported making the product. Subsequently the data ceased to be reported.

The distribution of products and records of production in the United States for selected years since 1859 are shown in Table 1.1.

The history of frozen desserts shows that mankind has made great efforts to produce and consume these highly enjoyable foods. Those who first consumed them were the elite of society. Today's offerings to consumers are truly "fit for royalty" and prices of some items limit their regular consumption to wealthy persons. But ice cream is also made for the masses. The average retail price of one-half gallon of ice cream is about $2.65, and that equates to about 14 min of labor at the average wage in the United States (IICA. THE LATEST SCOOP, 1994 edition). When one considers that a half gallon

Table 1.1. Ice Cream and Related Products Total and Per Capita Production (Hard and Soft) 1879–1994

Year	Ice Cream Total 1,000 Gallons	Ice Cream Per Capita Quarts	Ice Milk[a] Total 1,000 Gallons	Ice Milk[a] Per Capita Quarts	Frozen Yogurt[b] Total 1,000 Gallons	Frozen Yogurt[b] Per Capita Quarts	Sherbert Total 1,000 Gallons	Sherbert Per Capita Quarts	Water Ices[c] Total 1,000 Gallons	Water Ices[c] Per Capita Quarts	Other Frozen Dairy Products[d] Total 1,000 Gallons	Other Frozen Dairy Products[d] Per Capita Quarts	Total Frozen Products Total[e] 1,000 Gallons	Total Frozen Products Per Capita Quarts
1879	144	0.01											144	0.01
1889	851	0.06											851	0.06
1899	5,021	0.27											5,021	0.27
1909	29,637	1.31											29,637	1.31
1919	152,982	5.86											152,982	5.86
1920	171,248	6.43											171,248	6.43
1930	255,439	8.30											255,439	8.30
1940	318,088	9.64	10,457	0.32			8,089	0.24			2,910	0.09	339,544	10.29
1950	554,351	14.66	36,870	0.98			17,018	0.45			8,230	0.22	634,768	16.79
1960	699,605	15.55	145,177	3.23			40,734	0.91	18,299	0.48	50,127	1.11	969,004	21.54
1970	761,732	14.95	286,663	5.63			48,887	0.96	33,361	0.74	58,597	1.15	1,193,144	23.42
1980	829,798	14.61	293,384	5.17			45,187	0.80	37,265	0.73	23,468	0.41	1,225,223	21.58
1985	901,449	15.16	301,312	5.07			48,206	0.81	33,386	0.59	60,165	1.01	1,360,974	22.88
1986	923,597	15.38	314,673	5.24			49,679	0.83	49,842	0.84	43,443	0.72	1,382,329	23.02
1987	928,356	15.33	327,561	5.41			49,998	0.82	50,937	0.85	48,835	0.81	1,401,848	23.15
1988	882,079	14.43	354,831	5.80			52,175	0.85	47,098	0.78	48,892	0.80	1,388,429	22.70
1989	831,159	13.47	376,507	6.10	82,454	1.34	52,660	0.85	50,362	0.82	43,324	0.70	1,435,274	23.26
1990	823,610	13.21	352,271	5.65	117,577	1.89	50,278	0.81	49,168	0.80	32,389	0.52	1,426,829	22.89
1991	862,638	13.68	341,793	5.42	147,137[b]	2.33	47,379	0.75	50,704	0.81	35,997	0.57	1,490,935	23.64
1992	866,110	13.58	328,185	5.15	134,067	2.10	49,940	0.78	55,991	0.89	51,759	0.81	1,482,969	23.25
1993	866,248	13.42	325,346	5.04	149,933	2.32	50,813	0.79	52,908	0.83	65,963	1.02	1,516,432	23.49
1994	876,434	13.47	359,895	5.53	150,795	2.32	54,771	0.84	63,873	0.98	60,172	0.92	1,565,964	24.06

Source: International Ice Cream Association from data published by the U.S. Department of Agriculture.
[a]Includes freezer-made milkshakes.
[b]Frozen yogurt production data were not collected by USDA prior to 1989. USDA increased the overrun factor for frozen yogurt in 1991, which may have resulted in an overstatement of production for that year. IICA estimates total frozen yogurt production in 1991 was closer to 130 million gallons.
[c]Also includes sorbet, frozen juice bars, and gelatin pops.
[d]Through 1989 includes mellorine-type and other vegetable fat products, pudding pops, tofu-based products, and miscellaneous frozen dairy desserts. Beginning in 1990, includes nonfat frozen desserts, pudding pops, tofu-based products, and miscellaneous frozen desserts but excludes mellorine and mellorine-type products. USDA discontinued reporting mellorine production in 1990 since the majority of mellorine mix is shipped out of the U.S.
[e]Data from 1919 indicate only the estimated trend of production. Frozen yogurt included in total beginning in 1989. Mellorine no longer included as of 1990. USDA no longer reports mellorine production.

contains 16 servings of 4 fl oz, this makes the cost of a serving average less than $0.20, a remarkable buy for the nutrition and appetite satiation that it brings.

REFERENCES

Arbuckle, W. S. 1981. *The Little Ice Cream Book.* Library of Congress Catalog no.: 81-90341. p. 63.

2
Energy Value and Nutrients of Ice Cream

ENERGY VALUE AND NUTRIENTS

The property whereby a food produces heat and energy within the body may be expressed in terms of energy value. The unit customarily used by nutritionists for measuring human energy needs and expenditures and the energy value of foods is the kilocalorie, which is the amount of heat required to raise the temperature of 1 kg of water 1°C. A calorie is the amount of heat required to warm 1 g of water 1°C.

Digestion is the disintegration of food into simple nutrients in the gastrointestinal tract to prepare them for absorption. Metabolism consists of the chemical changes that nutrients undergo from the time they are absorbed into the body until they appear as excretory products. It includes the distribution of the absorbed food, the building (anabolism) and breaking down (catabolism) of tissues, and the absorption and release of energy. All nutrients are equally important to the extent they are needed in a particular diet.

The energy value and nutrients of ice cream depend upon the food value of the ingredients from which it is made (see Table 2.1). The milk products that go into the mix contain the constituents of milk, but in different amounts. On a weight basis ice cream contains three to four times as much fat, and about

Table 2.1. Composition of Plain Ice Cream (per 100-g Edible Portion)

Constituent	Good average ice cream	For different fat content				Ice cream cones	Water ice
		~5%	10%	~12%	~16%		
Water (%)	61.7	66.7	63.2	62.1	62.8	8.9	66.9
Food energy (cal)	196.7	152.0	193.0	207.0	222.0	377.0	78.0
Protein (%)	4.1	4.8	4.5	4.0	2.6	10.0	0.4
Fat (%)	12.0	5.1	10.6	12.5	16.1	2.4	Trace
Total carbohydrate (%)	20.7	22.4	20.8	20.6	18.0	77.9	32.6
Weight per 100-cal portion (g)	50.8	65.6	51.7	48.3	45.0	26.5	128.4

12–16% more protein than does milk. In addition, it may contain other food products, such as fruit, nuts, eggs, candies, and sugar, and these may enhance its nutritive value. Ice cream contains about four times as much carbohydrate as milk. The milk solids in ice cream are usually subjected to higher heat treatments than are those of pasteurized milk; they also are subjected to lower temperatures in the freezing process; and they are stored longer before consumption. Like milk, ice cream is not a good source of iron and some of the trace minerals.

Ice cream is an excellent source of food energy. The fact that the constituents of ice cream are almost completely assimilated makes ice cream an especially desirable food for growing children and for persons who need to put on weight. For the same reason, its controlled use finds a place in the diet of persons who need to reduce or who do not wish to gain weight.

CALORIC CONTENT OF ICE CREAM AND RELATED PRODUCTS

Ice cream manufacturers may need information regarding the nutritional value of their products, especially when they do nutritional labeling. School and hospital officials and dietitians often need nutritional data when recommending diets. The wide variation in the composition of ice cream and related products makes it impractical to provide nutritional data that will apply to all products. It is, however, possible to calculate for practical use the food energy value of a given product if the composition is known.

The total caloric value of ice cream depends on (1) the percentage of carbohydrates including lactose, added sweeteners, many bulking agents, and sugars that may be present in fruit or flavoring; (2) the percentage of protein including that from milk, whey protein-based fat replacers, nuts, eggs, or stabilizer; and (3) the percentage of fat from any source including cream, emulsifier, egg, cocoa, or nut fat.

When the caloric contents of fats, proteins, and carbohydrates are measured in a calorimeter, their yields in cal/g are 9.45, 5.65, and 4.10, respectively. Not all the digestible food material is assimilated by the body. On average 5% of the fat, 8% of the proteins, and 2% of the carbohydrates are not absorbed. The amount of energy normally expected to be derived from milk per gram of fat, protein and carbohydrate is as follows: fat, 8.79 cal; protein, 4.27 cal; carbohydrates, 3.87 cal. These values are the amounts of energy released from the food nutrients as heat units or calories, and are referred to as physiological fuel values. In every day usage these numbers are rounded to 9, 4, and 4, respectively. Neither minerals nor vitamins furnish appreciable amounts of energy. The type of sugar has little relation to the fuel value derived from it, and all sugars have about the same energy value.

The caloric value of 100 g of vanilla ice cream containing 12.5% fat, 11% NMS, 15% sugar, and 0.3% gelatin (a protein) may be calculated as follows, assuming for the NMS that lactose content is about 52% and protein content is about 36%:

Fat	12.5 × 9 = 112.50
Carbohydrates	[15 + (11 × 0.52)] × 4 = 82.88
Protein	[(11 × 0.36) + 0.30] × 4 = 17.04
	212.42 Total

The caloric value of a serving of ice cream varies with the composition of the mix and the weight of mix per gallon of finished ice cream. The serving size of ice cream for nutritional labeling purposes has been set at one-half cup or 4 fl oz (Federal Register, Vol. 58. No. 158, August 18, 1993, p. 44053). The serving size is 85 g for frozen flavored and sweetened ice, pops, and frozen fruit juices. Sundaes have a one-cup serving size.

Obviously, the weight of a serving of ice cream and the composition are the major variables. Ice cream is sold by volume with a minimum weight specified. Weight can be estimated by calculating the density of the mix and multiplying by the volume of mix per serving. Density is a function of overrun and composition. It varies from about 1.06 to 1.15 g/ml, and the typical 4-fl-oz serving contains between 63 and 68 g of mix when the overrun is 100%. If the overrun is 50%, the same 4-fl-oz weighs 94 to 102 g. As an example, let us calculate the calories in a high solids, high fat mix that contains 16% fat, 10% NMS, and 17% sweeteners on a weight basis and is frozen at 100% overrun.

1. Calculate mix density, where densities of constituents are fat = 0.93 g/ml, nonfat solids =1.58 g/cc, and water = 1 g/ml.

$$\text{Density} = \frac{100}{(16/0.93) + (27/1.58) + 57} = 1.095 \text{ g/ml}$$

2. Calculate the weight of mix in 4 fl oz (118.3 ml) when overrun is 100% (one-half of the volume is air). The mix constitutes 2 × 29.58 ml = 59.16 ml

and: 59.16 ml × 1.095 g/ml = 64.78 g

3. Calculate calories:

Carbohydrates	[17 + (10 × 0.52)] 4 =	88.8
Protein	(10 × 0.36) 4 =	14.4
Fat	16 × 9 =	144.0
	Total	= 247.2

Using the same approach, the calories per serving have been calculated for a variety of ice creams at high and low overrun values (Table 2.2).

Caloric values of one serving of representative types of ice creams vary widely as shown on nutritional labels (Table 2.3). Consumers often ask what impact consumption of ice cream may have on their weight. Data in Tables 2.2 and 2.3 show that one serving of ice cream contributes from less than 5% to more than 10% of the calories of a 2,000 calories daily intake, which is generally deemed appropriate for adults whose activity levels are limited. As in all aspects of healthy eating, variety and moderation are important practices.

2 ENERGY VALUE AND NUTRIENTS OF ICE CREAM

Table 2.2. Calories per Serving of Ice Creams of Widely Varying Fat Content and Overrun

Fat	NMS Percentages	Sweeteners[a]	Density (g/ml)	Weight (g/serving)	Overrun (%)	Calories (4 fl oz)
16	10	17	1.095	64.8	100	160
16	10	17	1.095	97.2	50	240
10	10	17	1.042	61.6	100	119
10	10	17	1.042	92.4	50	179
5	13	17	1.120	66.2	100	105
5	13	17	1.120	99.3	50	158
0	13	22	1.150	67.9	100	91
0	13	22	1.150	102.0	50	136

[a]Includes maltodextrins, polydextrose, and corn syrups that may constitute parts of lowfat and nonfat ice creams.

Table 2.3. Average Values Claimed on Nutrition Labels by Category of Vanilla Ice Cream

Category	Calories	Fat (g)	Protein (g)	Carbohydrate (g)	Sugar (g)	Calcium (g)	n*
Nonfat	90	0	3.5	19	13	10	3
Lowfat	100	2.3	3.0	18	15	9	6
Light	110	3.3	3.0	17.2	15	9	5
Reduced fat**	100	4.0	3.0	13	3	8	1
Regular	135	7.2	2.3	15	13	8	15
Premium	165	9.8	2.4	17	14	7	5

*n=number of samples per category
**The single product was "sugar-free"

PROTEIN CONTENT OF ICE CREAM

Ice cream has a high concentration of NMS, which is 34–36% milk protein. The milk proteins contained in ice cream are of excellent biological value, because they contain all the essential amino acids. Milk proteins are important sources of tryptophan and are especially rich in lysine. Proteins are essential in animal life as components of protoplasm of each living cell. Milk proteins are not only known to be complete, but the assimilation of ingested milk proteins is 5–6% more nearly complete than for other proteins in general.

Protein content is calculated from determination of the nitrogen content in the food. Early analysis of proteins showed that they have close to 16% nitrogen. The general practice then was to multiply the nitrogen content by the 6.25 conversion factor for the protein content. The accepted value for milk protein is 6.38.

MILKFAT CONTENT

Milkfat consists mainly of triglycerides of fatty acids, 95.8% on a weight basis (Bitman and Wood 1990). Glycerides are compounds in which one, two, or three fatty acid molecules are linked by ester bonds with the trihydric alcohol, glycerol. Mono-, di-, and triglycerides contain one, two, and three fatty acids, respectively. Milkfat is highly complex, containing almost 400 fatty acids (Jensen et al., 1991). It is unique among fats and oils in that it contains 11.8 and 4.6 moles of butyric (4-carbon) and caproic (6-carbon) acids, respectively, per 100 moles of total fatty acids. All of the butyric and 93% of the caproic acids are esterified to the third carbon (sn-3 position) of the glycerol molecule. Milkfat also contains 2.25% diacylglycerols, 1.11% phospholipids (nine different ones), 0.46% cholesterol, 0.28% free fatty acids, and 0.08% monoacylglycerols. Although milkfat is relatively low in polyunsaturated fatty acids (about 4.5%), it contains about 27% monounsaturated fatty acids (Jensen et al. 1991).

Milkfat content in ice cream is usually determined by extracting and weighing the ether-soluble fraction. Interest in milkfat is centered on its nutritional and functional attributes. It supplies energy, essential fatty acids, fat-soluble vitamins, saturated and unsaturated fatty acids, and sterols including cholesterol. It functions to provide unique flavor, to carry fat-soluble flavors, to lubricate the mouth, and to affect the structure, thus the texture, of frozen desserts.

CARBOHYDRATES IN ICE CREAM

Carbohydrates include starch, dextrin, cellulose, sugars, pectins, gums, and related substances. Carbohydrates serve as a source of heat and energy in the body. They are broken down to simple sugars under the action of specific enzymes secreted into the digestive tract and the principal end product is glucose. Sugars of several kinds may be used in the manufacture of ice cream. The commonly used sugar is sucrose, a disaccharide. It may come from either cane or beet as these are identical in composition. Corn sugar, now used extensively, is predominantly glucose (dextrose) or is converted to maltose or fructose (levulose). The sugars of most fruits are sucrose, fructose, and glucose. Invert sugar, a mixture of equal amounts of the monosaccharides fructose and glucose, is used at times.

Lactose, milk sugar, is a disaccharide that constitutes over one-third of the solid matter in milk and approximately 20% of the carbohydrate in ice cream. Lactose is unique in that it is found only in milk, whereas other types of sugars are fairly widely distributed in nature. Adults of Asian and African descent may produce insufficient lactase enzyme (β-D-galactosidase) to fully hydrolyze the lactose in a full serving of ice cream. Lactose cannot be absorbed through the intestine wall unless it has been split from the disaccharide to the monosaccharide form. This condition may result in physical discomfort due to bloating and, in extreme cases, to diarrhea. These symptoms arise when the lactose moves into the large intestine, where it raises the osmotic pressure causing water to migrate into the intestine (hence the diarrhea). Furthermore, the lactose is a substrate for coliform bacteria of the colon. They ferment it, producing liberal quantities of acid and gas, the latter causing the victim to have a bloated feeling.

Four approaches can be used to reduce the possibility of an individual experiencing this problem. First, the consumer may select ice creams high in fat.

The higher the fat content of the ice cream, the lower the nonfat solids content and, consequently, the lower the lactose content. High-fat ice creams tend to be the superpremium types, and the source of NMS in such ice creams is usually limited to skim milk solids. Skim milk solids contain about 50% less lactose than do whey solids. The latter may replace up to one-fourth of the NMS in ice cream, and whey solids, being low in cost, are often used to the extent permitted to replace skim milk solids in economy ice creams.

A second approach is to consume ice cream at the end of a meal. This ensures relatively slow flow of lactose through the digestive system and reduces the load on the enzyme that is present. It also presents a more dilute solution of lactose to the colonic bacteria.

The third approach is to consume frozen yogurt. This product, properly prepared, carries living yogurt bacteria that have already fermented part of the lactose in the skim milk solids used to make the yogurt. To the extent these bacteria remain alive to the time of eating they supply lactase to the human intestine. Experiments with human volunteers demonstrated the effectiveness of this approach in relieving symptoms of lactose malabsorption in susceptible persons (Tamine and Robinson 1985). The author and his graduate students (Sheu and Marshall 1993; Hong 1995) found that freezing killed up to 90% of the yogurt bacteria. However, when they used naturally encapsulated yogurt bacteria only about 50% of them died on freezing. When they enrobed the bacteria in calcium alginate, up to 90% of them survived freezing.

The fourth approach to alleviating lactose malabsorption is to eliminate lactose from the frozen dessert. This can be done by hydrolyzing the lactose with purified β-D-galactosidase before the product is frozen. The enzyme is relatively expensive and several hours are needed for the process. Furthermore, since two molecules are produced for each lactose molecule split, the freezing point of the mix may be lowered excessively. Another approach is to remove lactose from skim milk by ultrafiltration. Removal of 50% of the lactose by ultrafiltration followed by enzymatic hydrolysis of the remainder would provide concentrated skim milk solids with about the same freezing point as concentrated skim milk of the same solids content.

For calculating energy values of milk carbohydrates, the coefficient of digestibility, heat of combustion, and ingested-nutrients factors are usually given as 98%, 3.95 cal/g, and 3.87 cal/g, respectively.

MINERALS IN ICE CREAM

Certain inorganic elements are essential for growth and performance. Those needed in substantial amounts, calcium, phosphorus, magnesium, sodium, potassium, and sulfur, are termed major minerals or macronutrients. Those needed in small amounts, copper, cobalt, iodine, manganese, zinc, fluorine, molybdenum, and selenium, are termed trace minerals. The inorganic nutrients are interrelated and should be in particular proportions in the diet. Calcium and phosphorus are of vital concern since they are very important nutritionally, especially for building strong bones and teeth, and for unique functionality in dairy foods. About 85% of the phosphorus in the human body is combined with calcium in bones and teeth. Milk and its products, including ice cream, are among the richest sources of calcium. The calcium and phosphorus content of ice cream derive almost entirely from the nonfat milk solids and are therefore found in proportion to the content of NMS. Since NMS content can range from about 6–14%, and calcium averages about 13.8 mg/g NMS, 70-g servings of ice cream with these extremes of concentration would contribute from 58–135 mg of calcium to the diet. The midpoint between these numbers, 96.5 mg, is 8–12% of the recommended daily allowance for calcium in the human diet (RDA for children is 800 mg and for most adults is 1,200 mg, National Academy of Sciences, 1993). Similarly, 1 g of NMS contains about 10.7 mg of phosphorus, and 70 g of the ice creams formulated at the extremes of NMS content would contain about 45–105 mg of phosphorus. Since the RDA for phosphorus is 1,200 mg per day for persons ages 11–24 and 800 mg per day for others except for infants, it is obvious that ice cream can be a significant source of phosphorus. One serving can furnish about 10% of the RDA.

Fortunately, milk contains little copper or iron, the two minerals that catalyze oxidation. Since ice cream is often stored for weeks to months, it is imperative that contamination of any of the ingredients with these two minerals be prevented. This is a major reason why manufacturers exclude copper from dairy equipment.

VITAMINS IN ICE CREAM

Like milk, ice cream is an important source of several vitamins, the content depending primarily on the amount of milk solids and the weight of a serving. The fat-soluble vitamins, A, D, E, and K, are contained mainly in the fat and are absent in unfortified nonfat products. Milkfat is a good source of vitamin A. Manufacturers are required to add vitamin A to lowfat and nonfat ice creams. The content of water-soluble vitamins is proportional to the concentration of NMS in plain ice creams. The highest concentration is expected in nonfat ice cream and the lowest concentration in high-fat ice cream. Fruits and nuts also contribute some of these vitamins. Ice cream is considered a good source of riboflavin. As with calcium, 70 g of ice cream contributes about 10% of the RDA of riboflavin, i.e., about 18 µg/g of NMS and from 75–175 µg/70 g of ice cream. Ice cream can also furnish significant amounts of thiamine, pyridoxine, and pantothenic acid.

PALATABILITY AND DIGESTIBILITY OF ICE CREAM

The high palatability of ice cream is an important factor in the choice of it as a food. Chewing is not required with most flavors, and the smooth velvety texture soothes the palate. Its coldness makes it especially desirable during hot weather. Digestibility is generally high. The exception can be with the lactose malabsorbing person. Still, it was recently shown (Johnson et al. 1993) that even most lactose malabsorbers can consume with no significant symptoms the equivalent of a cup of milk a day. This is about the amount of lactose in a serving of ice cream. Thus, ice cream is an ideal food for times when other foods do not appeal. No other food contributes so much food value in as attractive and appealing form or is so universally liked and distributed as is ice cream.

REFERENCES

Bitman, J., and D. L. Wood. 1990. Changes in the milk fat phospholipids during lactation. *J. Dairy Sci.* 73:1208.

Hong, S. H. 1995. Enhancing survival of lactic acid bacteria in ice cream by natural encapsulation and gene transfer. Ph. D. dissertation. University of Missouri, Columbia. August.

Jensen, R. G., A. M. Ferris, and C. J. Lammi-Keefe. 1991. The composition of milk fat. *J. Dairy Sci.* 74:3228–3243.

Johnson, A. O., J. G. Semenya, M. S. Buchowski, C. O. Enowonwu, and N. S. Scrimshaw. 1993. Adaptation of lactose maldigesters to continued milk intakes. Am. J. Clin. Nutr. 58:879.

Sheu, T. Y., and R. T. Marshall. 1993. Microentrapment of lactobacilli in calcium alginate gels. *J. Food Sci.* 54:557–561.

Tamine, A. Y., and R. K. Robinson. 1985. The nutrient value of yogurt. In *Yogurt Science and Technology*. Pergamon Press Ltd. Heading Hill Hall, Oxford, England. pp. 365–373.

3
Classifications of Ice Cream and Related Products

Classifications of the various products considered to be related to ice cream has been done in many different ways. Early workers (Washburn 1910; Mortensen 1911; Frandsen and Markham 1915) divided them into two to ten groups, primarily depending on whether the product did or did not contain eggs. Mortensen considered commercial practices in suggesting the ten groups. Turnbow et al. (1946) defined 20 commonly agreed on terms used to categorize different frozen desserts.

Standards of Identity have been issued for ice cream and frozen custard, goat's milk ice cream, mellorine, sherbets, and water ices (21 CFR 135). Additionally, regulations published by FDA (21 CFR 101) established pursuant to the 1993 National Labeling and Education Act (NLEA) have made possible the sale of a wide variety of ice cream products under four new names, viz., reduced fat, light, lowfat, and nonfat ice creams. To avoid conflict with these new regulations the Standard of Identity for ice milk was revoked by the U.S. Food and Drug Administration effective September 14, 1995. This standard had identified as ice milk those products containing 2–7% milkfat and 12–15% total milk solids, and being sweetened, flavored, and frozen like ice cream. The products were sold in both the soft-frozen and hard-frozen forms. They are now sold under the new names as affected by the fat content primarily.

Within the product ice cream there are numerous variations of formula, dairy ingredients, sweeteners, stabilizers and emulsifiers, flavors, fruits, nuts, colors, methods of freezing, sizes, shapes, techniques for dispensing into packages, and other variables that make possible the creation of a wide variety of products. Many of these are described in the following presentation.

COMMERCIAL GROUPING OF ICE CREAM AND RELATED PRODUCTS

Ice cream. A product containing at least 10% milkfat, 20% total milk solids, safe and suitable sweeteners, and defined optional stabilizing, flavoring, and

3 CLASSIFICATIONS OF ICE CREAM AND RELATED PRODUCTS

dairy derived ingredients. The finished ice cream weighs at least 4.5 lb/gal and contains at least 1.6 lb of food solids per gallon. If the milkfat is increased above 10%, the NMS may be decreased to the same extent to as low as 6%. Whey solids, including modified whey products, may be added to ice cream to replace up to 25% of the NMS. Hydrolyzed milk proteins may be added at a concentration not to exceed 3% when the mix already contains at least 20% TMS.

Plain ice cream. Ice cream in which the total amount of color and flavoring ingredients is less than 5% of the volume of the unfrozen ice cream. Examples are vanilla, coffee, maple, and caramel ice creams.

Bulky flavored. Ice cream in which the total volume of the coloring and flavoring ingredients is at least 5% of the volume of the unfrozen product, or it can be ice cream containing visible particles of products such as cocoa, nuts, fruit, candy, or confections. A reduction in the minimal fat content is permitted.

Frozen custard, French ice cream, French custard ice cream. One or more of the optional egg ingredients are used in such quantity that the total weight of egg yolk solids is not less than 1.4% of the weight of the finished frozen product or less than 1.12% for bulky flavored products.

Reduced fat. Ice cream made with 25% less fat than the reference ice cream.

Light or Lite. Ice cream made with 50% less fat or 1/3 fewer calories than the reference ice cream, provided that in the case of the caloric reduction less than 50% of the calories are derived from fat.

Lowfat. Ice cream containing not more than 3 g of milkfat per serving.

Nonfat. Ice cream containing less than 0.5 g of fat per serving.

Mellorine. A food similar to ice cream but having the milkfat replaced in whole or part with vegetable or animal fat. It contains, by FDA Standard of Identity (21 CFR 135.130), not less than 6% fat and 2.7% protein, the protein having a protein efficiency not less than that of milk protein. Provision is made for reduction of the fat and protein by added bulky flavoring agents. Vitamin A must be present in a quantity that will ensure that 40 IU are available for each gram of fat in mellorine. Nonnutritive sweeteners are not permitted in mellorine.

Fruit. Ice cream containing fruit, with or without additional fruit flavoring or color. The fruit, such as strawberry, may be fresh, frozen, canned, or preserved.

Nut. Ice cream containing nutmeats, such as walnuts, almonds, pecans, and pistachio, with or without added color or flavoring.

Confection. Ice cream with appropriate flavorings plus particles of candy such as peppermint, butter crunch, or chocolate chip.

Bisque. Ice cream containing appropriate flavorings and particles of grape-nuts, macaroons, ginger snaps, sponge cake, or other bakery products.

Puddings. High fat ice cream containing generous amounts of mixed fruits, nutmeats, and raisins, with or without liquor, spices, or eggs. Examples are nesselrode and plum puddings.

Variegated (also called ripple). Ice cream into which a syrup such as chocolate or strawberry has been distributed so as to produce a marbled effect in the hardened product. A prime example is MIZZOU TIGER STRIPE, gold-colored French vanilla ice cream variegated with a thick chocolate syrup made from heavily "Dutched" cocoa. The design that shows in a perfect dip is that of a tiger skin.

Neopolitan. Two or more distinct flavors in the same package.

Fanciful name. Ice cream that, because it contain a combination of flavoring ingredients, is best described with a name that "stirs the fancy" of the potential consumer (see the list of fanciful names in Chapter 7).

Fancy molded. Ice cream or frozen yogurt molded into shapes of fruit or other attractive or festive forms. The group includes brick ice cream in one, two, or more layers, or with fancy centers.

Cake roll. Layered ice cream on moist cake, rolled like a jelly roll.

Spumoni. A combination of vanilla ice cream, chocolate mousse or chocolate ice cream, cherries, and tutti-frutti ice cream, or whipped cream combined with fruits arranged in a spumoni cup and hardened.

Aufait. Two or more layers of ice cream with pectinized fruits or preserves spread thinly between the layers; or the fruits may be stirred gently into the ice cream as it comes from the freezer to give a marbled appearance.

Soft-serve. Ice cream and related products sold as drawn from the freezer without hardening.

Frozen yogurt or lacto. A nonstandardized product quite similar to ice cream that is low in fat. It must contain yogurt bacteria of the species *Lactobacillus bulgaricus* and *Streptococcus thermophilus*. Some regulatory bodies set a minimum standard for acidity at 0.3% to 0.5% calculated as lactic acid. Others prefer to set a standard for numbers of viable bacteria in the product at some time after manufacture.

Mousse. Whipped cream plus sugar, color, and flavoring and frozen without further agitation. Condensed milk may be added to improve consistency.

Gelatin cube. Ice cream in which fruit-flavored gelatin cubes are substituted for fruit.

Fruit sherbet. A product made of fruit juices, sugar, stabilizer, and small amounts of milkfat and nonfat milk solids. It contains at least 0.35% acidity. A mixture of 4 parts water ice mix with 1 part ice cream mix can constitute a sherbet mix.

Soufflé. Sherbet containing egg yolk or whole eggs.

Water ice. Also known as *ice,* the product is made of fruit juice, sugar, and stabilizer, with or without additional fruit acid, flavoring, color, or water and frozen with or without agitation. No dairy products are contained.

Frappé. An ice made with a mixture of fruit juices, frozen to a slushy consistency, and served as a drink.

Granite. Water ice frozen with very little agitation.

FLAVOR CATEGORIES FOR ICE CREAM

The code of Federal Regulations (21 CFR 101.22) provides specific labeling instructions for flavorings and colorings. Natural and artificial flavors are defined. A "natural flavor" is the essential oil, oleoresin, esssence or extractive, protein hydrolysate, distillate, or any product of roasting, heating, or enzymolysis that contains the flavoring constituents derived from a spice, fruit or fruit juice, vegetable or vegetable juice, edible yeast, herb, bark, bud, root, leaf or similar material, meat, seafood, eggs, dairy products, or fermentation products thereof, whose significant function in food is flavoring rather than nutritional. Artificial flavors are not derived from these materials.

Foods that contain artificial flavors, artificial colors, or preservatives must be labeled as necessary to render the statement likely to be read by the ordinary person under conditions of purchase and use of the food.

In the labeling of frozen desserts to indicate flavors it is necessary to designate whether the flavoring(s) is/are natural, artificial, or a mixture of both. The name of the flavoring must appear on the principal display panel in letters at least one-half as high as the name of the food.

The International Ice Cream Association developed the following guidelines to fulfill the labeling standards of the FDA for all flavors of ice cream.

Category I—Those products that contain no artificial flavor. The label reads the name of the flavor followed by ice cream, e.g., **vanilla ice cream**.

Category II—Those products that contain both natural and artificial flavor, but the natural flavor predominates in quantity. The label reads the name of the flavor followed by the word flavored and ice cream, e.g., **vanilla flavored ice cream**.

Category III—those products that are flavored exclusively with artificial flavor or with a combination of a natural and artificial flavor in which the artificial predominates. The label reads artificially flavored, the name of the flavor, and ice cream, e. g., **artificially flavored vanilla ice cream**.

Flavor suppliers are required to certify that flavors supplied are in fact natural, artificial, or mixtures of the two. FDA is responsible for verifying that certifications are accurate. Other aspects of labeling are covered in the packaging section of Chapter 11.

REFERENCES

Frandsen, J. H., and E. A. Markham. 1915. *The Manufacture of Ice Cream and Ices.* Orange Judd Publ. Co., New York.
Mortensen, M. 1911. Classification of ice cream and related frozen products. Iowa AES Bull. 123.
Turnbow, G. D., P. H. Tracy, and L. A. Raffetto. 1946. *The Ice Cream Industry,* 2nd ed. Wiley, New York.
Washburn, R. M. 1910. Principles and practices of ice cream making. Vermont AES Bull. 155.

4
Composition and Properties

Ice cream is composed of a mixture of milk products, sweetening materials, stabilizers, flavors, or egg products, which are referred to as ingredients. The wide variety of ingredients that may be used to produce different kinds of ice cream is apparent from the classifications discussed in Chapter 3. Furthermore, any one kind of ice cream may be made by combining the ingredients in different proportions. However, the effects of these ingredients upon the finished product are due to the constituents of the ingredients.

An ice cream mix is the unfrozen blend of all the ingredients of ice cream with the exception of air and flavoring materials. The composition of ice cream is usually expressed as a percentage of its constituents, e.g., percentage of milkfat, nonfat milk solids (NMS), sugar, egg yolk solids, stabilizer, and total solids.

The composition of ice cream varies in different localities and in different markets. The best ice cream composition for a manufacturer to produce is often difficult to establish. After consideration is given to legal requirements, quality of product desired, raw materials available, plant processes, trade demands, competition, and cost, there is a choice of a minimum, average, or high milk solids composition of the product as well as the ratio of fat to NMS. Some firms may choose to manufacture only one of these products, others two, and still others several.

It may be inadvisable for a small manufacturer to produce more than one brand of ice cream. If product of only one composition is manufactured, it is important that it be the best product possible.

In ice cream, the percentage of milkfat varies more than any other constituent. The milkfat content may vary from 0–24%, depending upon such factors as regulations, grade, price, and competition. As the fat content of ice cream is increased, the NMS must be decreased so as to avoid "sandiness" (i.e., the crystallization of milk sugar or lactose in the finished ice cream). Table 4.1 suggests compositions that avoid sandiness and permit recognition of particular local preferences as to sugar content or fat content of commercial ice creams and related products. These local preferences and the quality of the ingredients, as well as the technique of manufacture, are fully as important as the composition in determining the best ice cream for that locality. Table 4.2 gives Federal composition requirements for frozen desserts.

Table 4.1. Approximate Composition (%) of Commercial Ice Cream and Related Frozen Desserts.

Product	Milkfat	NMS	Sweeteners[a]	Stabilizers[b] and emulsifiers	Approximate TS
Nonfat ice cream (hard)[c]	<0.8	12–14	18–22	1.0	35–37
Lowfat ice cream (hard)[c]	2–4	12–14	18–21	0.8	35–38
Light ice cream (hard)[c]	5–6	11–12	18–20	0.5	35–38
Reduced fat ice cream (hard)[c]	7–9	10–11	18–19	0.4	36–39
Soft-serve ice cream	3–4	12–14	13–16	0.4	29–31
Economy ice cream	10.0	10.0–11.0	15.0	0.30	35.0–37.0
	12.0	9.0–10.0	13.0–16.0	0.20–0.40	
Trade brand ice cream	12.0	11.0	15.0	0.30	37.5–39.0
	14.0	8.0–9.0	13.0–16.0	0.20–0.40	
Deluxe ice cream (premium-super premium)	16.0	7.0–8.0	13.0–16.0	0.20–0.40	40.0–41.0
	18.0–20.0	6.0–7.5	16.0–17.0	0.0–0.20	42.0–45.0
	20.0	5.0–6.0	14.0–17.0	0.25	46.0
Mellorine	6.0–10.0	2.7 (Protein)	14.0–17.0	0.40	36.0–38.0
Frozen yogurt	3.25–6.0	8.25–13.0	15.0–17.0	0.50	30.0–33.0
	0.5–2.0	8.25–13.0	15.0–17.0	0.60	29.0–32.0
	<0.5	8.25–14.0	15.0–17.0	0.60	28.0–31.0
Sherbert	1.0–3.0	1.0–3.0	26.0–35.0	0.40–0.50	28.0–36.0
Ice	—	—	26.0–35.0	0.40–0.50	26.0–35.0

[a]Includes sucrose, glucose, fructose, corn syrup solids, maltodextrins, polydextrose, and other bulking agents, some of which contribute little or no sweetness.
[b]Includes cellulose gum and cellulose gel.
[c]Terms for specific fat content claims are defined in 21 CFR 101.62.

Table 4.2. Federal Standards of Composition for Frozen Desserts

Product	Minimum fat (%)	Maximum fat (%)	Minimum protein (%)	Minimum TMS[a] (%)	Weight (lb/gal)	Minimum TS (lb/gal)	Minimum acidity (%)	Egg yolk solids (%)
Ice cream[b,c]								
Plain flavor	10	—	—	20.0	4.5	1.6	—	<1.4
Bulky flavor	8	—	—	16.0	4.5	1.6	—	<1.12
Sherbert	1	1–2	—	2–5	6.0	—	0.35	—
Water ice	—	—	—	—	6.0	—	0.35	—
Mellorine-type product[d]								
Plain flavor	6	—	2.7	—	4.5	1.6	—	—
Bulky flavor	4.8	—	2.2	—	4.5	1.6	—	—

Source: 21 CFR 101 and 135 (1995).
[a]TMS: total milk solids.
[b]Frozen custard, French ice cream, and French custard ice cream have the same solids requirements, except they must contain not less than 1.4% egg yolk solids (plain flavor) or not less than 1.12% egg yolk solids (bulky flavors).
[c]May be made in reduced fat, light, lowfat, or nonfat forms that weigh as little as 4.0 lb/gal and are not nutritionally inferior to ice cream.
[d]Vegetable or animal fats other than milkfat may be used.

COMPLEXITIES OF COMPOSITION

Although the best of ice cream can be made from cream, concentrated or condensed milk, sugar, stablizer, and high quality flavoring, ice cream manufacturers have found it highly desirable to use a variety of ingredients. To learn about the extent of variation in composition, the author and co-workers did a market survey in 1995, the results of which appear below. A total of 24 different ingredients appeared on the ingredient labels of the 38 samples examined. The lists were composed by taking the most frequently used ingredients for each class of product. Therefore, the lists are unlikely to match exactly the ingredients on a single manufacturer's container.

It is important to remember that water is a major portion of ingredients such as milk and that milk solids in ice cream are typically nearly twice as concentrated as in milk; furthermore, sweeteners and other ingredients dilute out the milk solids. Therefore, the simplified listing on the ingredient labels should imply to the ice cream expert that at least one source of concentrated or dried milk solids is included in the formula.

Ingredient labels for representative vanilla ice creams follow with the ingredients listed in the order of highest to lowest use concentration.

Superpremium: skim milk, cream, sugar, egg yolk, vanilla extract.

Premium: milk, cream, sugar, corn syrup, mono/diglycerides, guar gum, carrageenan, annatto color, locust bean gum, cellulose gum, vanilla extract.

Regular: milk, nonfat milk, cream, whey, sugar, corn syrup, mono/diglycerides, guar gum, carrageenan, locust bean gum, cellulose gum, polysorbate 80, vanilla extract, vanillin, annatto color.

Reduced fat, sugar free: milk, cream, polydextrose, maltodextrin, egg yolk, guar gum, locust bean gum, carrageenan, vanilla extract, vitamin A.

Light: skim milk, cream, sugar, corn syrup, whey, guar gum, carrageenan, mono/diglycerides, polysorbate 80, annatto color, vanilla extract, vitamin A.

Lowfat: nonfat milk, cream, sugar, corn syrup, whey, polydextrose, maltodextrin, guar gum, cellulose gum, locust bean gum, carrageenan, mono/diglycerides, annatto color, vanilla extract, vitamin A.

Nonfat: nonfat milk, sugar, corn syrup, maltodextrin, polydextrose, mono/diglycerides, microcrystalline cellulose, guar gum, cellulose gum, carrageenan, vanilla extract, vitamin A.

Fat free premium: skim milk, corn syrup, sugar, pectin, vanilla, vitamin A.

Fat free, sugar free: nonfat milk, polydextrose, maltodextrin, sorbitol, milk protein isolate, egg yolk, cellulose gum, cellulose gel, locust bean gum, carrageenan, mono/diglycerides, polysorbate 80, vanilla extract, aspartame, vitamin A.

In a companion survey of ingredient labels of chocolate ice creams the only two ingredients in addition to those found in the vanilla ice creams were cocoa and chocolate liquor. The latter was found only in premium ice cream. Chocolate liquor would not, of course, be found in fat free products, because the cocoa butter content would furnish too much fat to permit a label claim of less than 0.5 g of fat per serving (Chapters 3 and 13). There also was a higher incidence of the use of whey in chocolate than in vanilla ice cream, probably because the potential for the whey flavor to be detected by consumers is much less with a highly flavored than with a delicately flavored product.

The survey disclosed that manufacturers used a wide variety of label statements to show sources of milk solids in products with lowered fat content. Although skim milk and nonfat milk are interchangeable words, some manufacturers appeared to consider skim milk to represent the liquid form and nonfat milk to represent the dried form of the product, so they listed both on the label.

Among the ice creams with full fat content, the choices of label terms to indicate sources of milk solids were: skim milk, milk and cream, 37%; nonfat milk (or skim milk) and milkfat, 30%; milk and cream, 20%; milk and nonfat milk, 10%; and skim milk and cream, 3%. Since it would not be possible to supply enough milkfat with milk and nonfat milk to make a full fat ice cream, the products so labeled should be considered mislabeled.

CHARACTERISTICS OF A SATISFACTORY COMPOSITION

Some of the characteristics that merit consideration are cost, handling properties (including mix viscosity, freezing point, and whipping rate of the mix), flavor, body and texture, food value, color, and general palatability of the finished product. In developing a formula to fulfill the needs of any particular situation, numerous factors must be considered. These include customer demands for flavor, body, texture, and color characteristics of the finished product; for example, natural flavor or flavor fortified with artificial flavoring; smooth, chewy-to-heavy, or coarser texture. Other factors might include meeting composition standards; the nature of the competition; type of manufacturing operation; source, availability, quality, and cost of ingredients; and production capacity.

Although the methods of processing and freezing influence the characteristics of the mix and of the finished product, the effects of constituents supplied by the ingredients are also important. Therefore, each constituent (fat, NMS, sweetener, egg products, stabilizers, emulsifiers, total solids, salts, optional ingredients, flavors, and colors) contributes to the characteristics of the ice cream.

THE ROLE OF THE CONSTITUENTS

Basic Ingredients

When commercial ice cream was first being introduced in this country, the ingredients were cream, fluid milk, sugar, and stabilizer. Later, condensed milk, nonfat dry milk, and butter became popular ice cream ingredients. Technological developments and changes in marketing and economic conditions have since encouraged the development and use of many other ingredients.

A wide range of ingredients for ice cream is now available from numerous sources. These ingredients may be grouped as dairy products and nondairy products. Dairy products furnish the basic ingredients of milkfat and NMS, which have essential roles in ice cream. Some dairy products provide fat, some provide NMS, others supply both fat and NMS, and still others, e.g., whey solids, supply bulk to the mix. The nondairy products include sweeteners, stabilizers and emulsifiers, egg products, fruits, nuts, flavors, special products, and water. The basic ingredients in frozen dairy foods are milkfat, NMS, sweeteners, stabilizers and emulsifiers, flavorings, and water. The functional properties imparted by these basic ingredients from which the mix is formulated are quite varied.

Milkfat

Milkfat is an ingredient of major importance in ice cream. The use of the correct percentages is essential not only to balance the mix properly, but also to satisfy legal standards. Studies show consistently that fat globules concentrate at the surfaces of air cells during the freezing of ice cream. This accounts, in part, for milkfat imparting a rich characteristic to the flavor. Increasing the fat content of ice cream decreases the sizes of ice crystals by interrupting the space in which they have to form. Therefore, other means of restricting ice crystal size and growth are highly important in ice creams made low in fat. Milkfat, because it is not dissolved, does not lower the freezing point. It tends to retard the rate of whipping. High fat content may limit consumption, will impart a high caloric value, and may increase the cost. The fat content of commercial ice cream is usually 10–12%. When the fat percentage is lower than 10%, a descriptor (reduced fat, light, lowfat, and nonfat or fat free) must be used in the label. The best source of milkfat is fresh cream. Other sources are frozen cream, plastic cream, butter, butter oil, anhydrous milkfat, fractionated milkfat and concentrated milk blends. Milkfat is associated with a small amount of phospholipids of which lecithin contributes importantly to its properties. Doan and Keeney (1965) stated that milkfat contributes a subtle flavor quality, is a good carrier and synergist for added flavor compounds, and promotes desirable tactual qualities. The characteristics noted by these workers account for the superior flavor qualities of frozen dairy foods made from fresh milkfat compared with those made from vegetable fat.

Nonfat Milk Solids

NMS are the solids of skim milk, and consist of protein (37%), milk sugar (lactose; 55%), and minerals (8%). These solids are high in food value, inexpensive, and while not adding high flavor notes to the ice cream, enhance its palatability. Lactose adds slightly to the sweet taste, largely produced by added sugars, and the minerals have a slightly salty taste, which rounds out the flavor of the finished product. The proteins in NMS help to make the ice cream more compact and smooth, and thus tend to prevent a weak body and coarse texture. Therefore, as much NMS as can be added is desirable—except that an excess of NMS may result in a salty, overcooked, or condensed-milk flavor and increase the risk of lactose crystallization during storage. NMS increase viscosity and resistance to melting, but also lower the freezing point. Variations over the usual range of concentration have no pronounced effect on whipping ability, but variations in the quality of NMS do have an important influence on it. NMS content is varied inversely with fat content to maintain the proper mix balance and to ensure good body, texture, and storage properties. However, a high concentration of lactose, which may crystallize under certain conditions, can cause sandiness of texture. Because of the many factors that may affect lactose crystallization, it is difficult to give a statistically certain limit to the percentage of NMS that should be used in an ice cream mix. However, as a rule of thumb, the NMS should be no more than 15.6–18.5% of the total solids in the mix, depending on whether the turnover will be slow or rapid and how much heat shock may occur.

To estimate the range of the maximum NMS content of a mix, subtract from 100 the sum of the percentages of all the other solids in the mix, and then divide by a factor of 6.4 (15.6%) and then again separately by 5.4 (18.5%). For example, for a mix with 10% fat, 15% sweetener solids, and 0.3% stabilizers, the highest percentage NMS for expected rapid turnover would be

$$\frac{100 - (10.0 + 15.0 + 0.3)}{5.4} = \frac{74.7}{5.4} = 13.83$$

For the same mix, the highest percentage NMS for expected slow turnover would be

$$\frac{100 - (10.0 + 15.0 + 0.3)}{6.4} = \frac{74.7}{6.4} = 11.67$$

Sweeteners

Sweetening Value. Sweetening value means the sweetening effect of added sugars and is expressed as the weight of sucrose necessary to give an equivalent sweet taste. For many years sucrose was the only sweetening agent added to ice cream; consequently, it has been used as a standard in comparing the sweetening effect of other sugars. However, there has been an increasing tendency to obtain the desired sweetness by blending sucrose with other sugars. This tendency has been due to the improvement in quality of other more economically priced sweeteners and to a desire to increase the total solids of some ice creams without exceeding the limit of desirable sweetness. Optimal sweetness can be obtained only by using some sucrose in the blend. The percentage of the sweetening agent that can be obtained from other sources is influenced mainly by (1) the desired concentration of sugar in the mix, (2) the total solids content of the mix, (3) the effect on the properties of the mix, such as freezing point, viscosity, and whipping ability, (4) the concentration in the sweetener of substances other than sugar (e.g., the undesired flavor of honey or the undesired color of molasses), and (5) the relative inherent sweetening power of the sweeteners other than sucrose.

For most ice cream formulations the sweeteners can be either sucrose (cane or beet sugar) alone or sucrose in combination with some product of hydrolyzed corn starch. The sugar may be used in dry or liquid form.

Although it is commonly agreed that the best ice cream is made from sucrose, approximately 45% of the sucrose can be replaced by corn sweeteners for economy, handling, or storage reasons. However, corn products with low to medium dextrose equivalents (DE) may impart off-flavors and should constitute no more than about one-third of the total sweetener solids. Many sugar blends are commercially available. Low-conversion corn syrup solids (low in DE) increase the solids content without imparting excess sweetness. Blends of sucrose and medium- or high-conversion corn syrup solids have also been used advantageously.

The main function of sugar is to increase the acceptance of the product by making it sweet and by enhancing the pleasing creamy flavor and the delicate fruit flavors. Lack of sweetness produces a flat taste; too much tends to overshadow desirable flavors. The total amount of sweetness expressed as sucrose may vary from 12–20%, while 14–16% is usually most desirable. Sweeteners increase the viscosity and the total solids (TS) of the mix. This improves the body and texture of the ice cream, provided the TS content does not exceed about 42% or the sugar content does not exceed about 16%. Above these limits, the ice cream tends to become soggy and sticky. Sweeteners, being in solution, depress the freezing point of the mix. This results in slower freezing, thus, a lower temperature is needed for proper hardening. In addition to their effect on the quality of the ice cream, sweeteners, except the nonnutritive or low caloric types, are usually the lowest cost source of TS in the mix.

Egg Yolk Solids

Egg yolk solids are high in food value but usually increase the cost of ice cream. They impart a characteristic delicate flavor, which aids in obtaining a desirable blending of other flavors, but even slight off-flavors in egg products are easily noticeable in the ice cream.

They have a pronounced effect in improving the body and texture, have almost no effect on the freezing point, and increase the viscosity. Egg yolk solids, regardless of their source, improve whipping ability, presumably due to lecithin existing in a lecithin-protein complex. They are especially desirable in mixes of low TS concentration and in mixes for which the fat is obtained from butter, butter oil, or vegetable fat.

Stabilizers

Stabilizers are used to prevent the formation of objectionable large ice crystals in ice cream and are used in such small amounts as to have a negligible influence on food value and flavor. They are of three general types: (1) gelatin stabilizers, which come from animal sources (such as calf skin, pork skin, and bones); (2) vegetable stabilizers, such as sodium alginate, carrageenan, agar-agar, and CMC (sodium carboxymethylcellulose), and (3) gums, such as guar, locust bean, tragacanth, karaya, and oat. All stabilizers have a high water-holding capacity, which is effective in smoothing the texture and giving body to the finished product. They increase viscosity, have no effect on the freezing point, and, with a few exceptions, tend to limit whipping ability. Their most important function is to prevent growth of ice crystals as temperatures fluctuate during storage.

The amount of stabilizer to use varies with its properties, with the solids content of the mix, with the type of processing equipment, and with other factors. The amount used in regular ice creams may be in the range 0–0.5%, but generally is from 0.2–0.3%. When fat content is reduced, stabilizer content is increased with cellulosic products being added.

Stabilizers extensively used in frozen dairy foods include guar gum, locust bean gum (carob bean gum), carboxymethyl cellulose (CMC), carrageenan (Irish moss extract), sodium and propylene glycol alginates, gelatin, and pectin. The alginates have an immediate stabilizing effect upon addition to the mix because they interact with calcium. CMC produces a chewy characteristic in the finished product. Gelatin produces a thin mix and requires an aging period. Pectin is used in combination with the gums as a stabilizer for sherbets and ices.

Stabilizers (1) improve smoothness of body, (2) aid in preventing ice crystal formation in storage, (3) give uniformity of product, (4) give desired resistance to melting, and (5) improve handling properties. The problems that come from using excessive amounts of stabilizers include (1) undesirable melting characteristics and (2) soggy or heavy body. Commercial stabilizer products are usually blends of several stabilizing materials in the proportions necessary to give the desired characteristics to the frozen product.

Emulsifiers

Emulsifiers are used in the manufacture of ice cream to produce a finished product with a smoother texture and stiffer body and to reduce whipping time. The use of emulsifiers results in air cells that are smaller and more evenly distributed throughout the internal structure of the ice cream. While egg yolk solids may produce similar results, their effect is not so pronounced.

The emulsifying ingredients commonly used in the ice cream industry are monoglycerides and diglycerides (commonly denoted as mono/diglycerides or mono- and diglycerides) composed of glycerol and selected fatty acids. The total amount of emulsifiers by weight may not exceed 0.2%. The use of two polyoxyethylene-type emulsifiers—sorbitan tristearate and monooleate—is permitted in frozen dairy foods, but only up to 0.1%. The excessive use of emulsifiers may result in slow melting, and body and texture defects.

Total Solids

TS replace water in the mix, thereby increasing the nutritive value and viscosity and improving the body and texture of the ice cream. This is especially true when the increase in TS is due to added carbohydrates, sweet cream buttermilk solids, or eggs. Egg yolk solids, like sweet cream buttermilk solids, improve the whipping ability and shorten the freezing time. Increasing the percentage of TS decreases the percentage of frozen water and frequently permits a higher overrun while maintaining the minimum of 1.6 lb of food solids per gallon of ice cream. A heavy, soggy product results when the TS

4 COMPOSITION AND PROPERTIES

content is too high, i.e., above 40–42%. Furthermore, displacement of water with TS results in a warmer product.

Water and Air

Water and air are important constituents of ice cream, but their effects can be overlooked easily. Water is the continuous phase. It is present as a liquid, a solid, and as a mixture of the two physical states. The air is dispersed through the fat-in-serum emulsion. The interface between the water and air is stabilized by a thin film of unfrozen material and by partially churned fat globules (Figure 4.1). The interfaces of the fat are covered by a layer of fat-emulsifying agent.

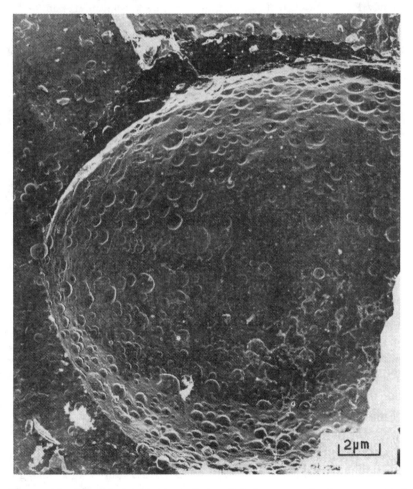

Figure 4.1. The interior of a typical air cell in ice cream showing numerous fat globules in the lamella of the air cell. (Courtesy K. G. Berger, Lyons Central Laboratories, London.)

Water in the ice cream mix comes from fluid dairy products and syrups, or from added water. The water from milk, having passed through the mammary gland, may be expected to be clean. The water from other sources, must come from a supply where purification is ensured.

In the manufacture of ice cream, the overrun, or the increase in volume of ice cream over the volume of mix used, is produced by incorporation of air. The amount of air in ice cream is important because it influences quality and profits and is involved in meeting legal standards. Maintaining a uniform amount of air is essential in controlling both quality and quantity. Air filters are used on continuous freezers to remove particulates from air entering the freezer.

Studies have been conducted on gases other than air in ice cream. Researchers have developed a method of injecting into the mix during the freezing process liquid nitrogen (N_2) at atmospheric pressure and at a temperature well below that of milk. Others have described a process and apparatus for injecting a mixture of nontoxic, inert gas into ice cream mix. Still others have added finely shredded solid carbon dioxide (CO_2) to ice cream during the manufacturing process to replace the air. They claim an improved product.

Optional Ingredients

Several special ingredients are used for their desirable effects in the preparation of the mix and on the finished products. Ordinary salt is sometimes used in ice cream. This is usually unnecessary, except in certain flavors such as custards and nut ice creams. Some believe that a small amount of salt (~<0.1%) improves the flavor of ice cream. Perhaps this is a carryover from earlier times when ice cream formulations contained a lower percentage of NMS and thus less natural milk salts. In any case, a salty flavor should be avoided.

The caseinate derivatives, especially sodium caseinate, are effective in increasing the whipping rate and overrun; however, they may so stabilize the emulsion that insufficient fat is destabilized during freezing. In this case the product may lack dryness and stiffness.

The mineral salts, including citrates and phosphates, and calcium and magnesium salts, affect mix and finished-product qualities. Mineral salts are usually used in limited amounts, and affect handling properties and appearance of the product. The citrates and phosphates are good casein solvents and increase the hydration of the casein. They also impart stability to the mix during heating and processing.

Calcium salts may decrease the stability of the protein, but they also give the mix and finished product a creamy, rich appearance. Calcium sulfate in small quantities affects the dryness and stiffness of the frozen product as it is drawn from the freezer.

Specially prepared low-lactose milk solids are available for increasing the total milk solids (TMS) without causing an excessive lactose content. These products contain the milk protein and mineral salts and impart their nutritive value and properties to the product.

IMPORTANCE OF FLAVOR

Flavor is generally considered the most important characteristic of ice cream. It is easily confused with taste, which includes the "feel sensation" of body and texture as well as the true flavor. The flavor of ice cream is the result of blending the flavors of all the ingredients, some of which may not be sufficiently pronounced to be recognizable, although each contributes to the final effect. This makes it difficult to predict the effect of a certain ingredient upon the flavor of the ice cream. Furthermore, the desirability of a particular flavor, or more properly "blend of flavors," depends upon the individual doing the tasting. The subjectivity involved in flavors explains why a certain blend of flavors is less popular to some persons than to others.

Flavor has two important characteristics: type and intensity. Flavors that are delicate and mild are easily blended and tend not to become tiresome even when very intense, while harsh flavors soon become tiresome even in low concentrations. As a general rule, therefore, delicate flavors are preferable to harsh ones; but in any case the flavor should be only intense enough to be easily recognized and delicately pleasing to the taste.

THE BALANCED MIX

A balanced mix is one in which the proportions of the ingredients will produce a satisfactory finished product—a frozen dessert in which the defects, if any, cannot be further corrected by any change in the composition or ingredients of the mix.

Defects such as rancid flavor, feed flavor, or uneven color cannot be corrected by changing the concentration of the constituents. Therefore, they do not indicate a poorly balanced mix. However, other defects, such as (1) lack of flavor—insufficient concentration of flavoring, (2) lack of richness—insufficient concentration of fat, (3) sandiness—too high concentration of lactose, or (4) weak body—low total solids or low stabilizer, may be corrected by changing the composition of the mix. These defects do, therefore, indicate that the mix is incorrectly balanced.

Conditions That Limit the Balancing of a Mix

Balancing is done to give desirable results under certain limited conditions of processing and handling the mix or of handling the finished ice cream. For example, a mix may be properly balanced for a finished ice cream that is to have a rapid turnover, but the components might cause sandiness if the ice cream were to be stored for an extended time. Another mix may be properly balanced for freezing in a batch freezer but not in a continuous freezer. A mix may be thrown out of balance by changing the source of the constituents. For example, if the fat in the mix is obtained from butter, the mix may need added egg yolk solids to improve its whipping ability and to give it the proper balance, but if the mix is made with sweet cream, the egg yolk solids would not be necessary. A knowledge of the role of each constituent together with its advan-

Table 4.3. Advantages and Limitations of Selected Ice Cream Constituents

Constituent	Advantages	Limitations
Milkfat	Increases richness of flavor and smoothness of texture, lubricates and insulates the mouth, adds body	Relatively high cost Hinders whipping May limit consumption due to high calories and satiating effect
Nonfat milk solids (NMS)	Improves texture, builds body, allows higher overrun	High amount may cause sandy texture and cooked or salty flavor
Sugar	Lowers freezing point Displaces water Improves flavor/texture Low hardening temperature	Excess sweetness possible Lowers whippability Increases freezing time
Corn syrup solids	Low cost solids Low relative sweetness Bind and replace water	Impart off-flavor when overused
Stabilizers	Make texture smooth Provide body	Excess firmness may occur High melt resistance
Egg yolk solids	Improve whippability Impart custard flavor	Foamy melted product
Total solids (TS)	Smoother texture Firmer body Higher nutrient content Lessen excess coldness	Heavy, soggy or sticky body Reduce coldness
Flavoring	Increases acceptability	Intensities and harshness may be unacceptable
Coloring	Improves attractiveness Aids flavor identification	Allergic reactions of some people to yellow no. 5 or 6

tages and limitations is necessary in selecting a desired composition and in properly balancing a mix. Usually an ice cream mix that is properly balanced for average commercial conditions will have between 36–42% TS and between 20–26% TMS (obtained by adding the percentage of fat to the percentage of NMS). This does not apply to a mix for ice cream with lowered fat content, a sherbet, an or ice. (Calculations for balancing mixes are given in Chapter 8.) For easy reference the advantages and limitations of ice cream's constituents are summarized in Table 4.3.

Mix Properties

The ice cream mix represents a complex colloidal system, of which many of the properties have not been fully investigated. In the mix, some of the substances occur in true solution (the sugars, including lactose, and the salts), others are colloidally suspended (casein, stabilizers, insoluble sweetener solids, and some of the calcium and magnesium phosphates), and the fat globules are in coarse dispersion. Although the whey proteins are dissolved, they have little effect on the freezing point.

The substances in true solution are small molecules or ions and have a strong affinity for water. The substances in colloidal suspension typically have particles with an opposite electrical charge to that of the solvent, and the mutual attraction keeps them together in suspension. The like electric charges on the particles keep them apart, which helps to maintain the suspension, as do collisions between particles in suspension with those of the solvent.

4 COMPOSITION AND PROPERTIES

Occasionally, the substances in suspension may not have sufficient attraction for the solvent and there may not be sufficient viscosity to keep them suspended. Different substances also have differing affinities for water. Some particles, the hydrophobic colloids, have so little affinity for water that if there is no charge on the particle precipitation occurs. On the other hand, substances with high degrees of affinity for water, the hydrophilic colloids, may remain in suspension even with no electrical charge. Substances in coarse dispersion or suspension do not stay uniformly dispersed, but settle or rise depending on their density relative to the suspending medium.

Colloidal suspensions are sensitive to changes in pH and salt concentration and, since ice cream is so complex, many factors have an impact on mix properties.

Ice cream mixes consist of (1) an aqueous phase (58–68% water) in which carbohydrates, whey proteins, and minerals are dissolved, (2) casein and the substances associated with it are colloidally suspended, and (3) milkfat is emulsified. This complex emulsion can withstand the stress of freezing, mechanical agitation, and concentration. Aeration is remarkable considering the inherent instability of the fat globules, casein micelles, and lactose under these conditions.

Mix properties of practical importance include stability, density, acidity, surface tension, interfacial tension, viscosity, absorption, freezing point, and whipping rate.

Mix Stability

Mix stability refers to the resistance to separation of the milk proteins in colloidal suspension and the milkfat in emulsion. Instability results in separation of (1) protein particles as coagulated or precipitated material, (2) whey from melted mix, or (3) syrup in the mix upon aging.

Homogenization, mix acidity, dehydrating salts, ratio of fat to TMS, heat-treatment, freezing, aging time, and the extent to which the water in the mix is bound all affect mix stability.

Hydration. The most stable mix particles are both charged and hydrated (hydrophilic suspension, Figure 4.2a); the next most stable are those that are

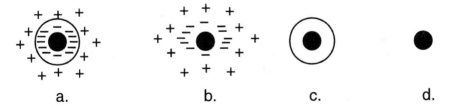

Figure 4.2. (a) Hydrophilic suspension with particle charged and hydrated; (b) hydrophilic or hydrophobic suspension with particle charged but not hydrated; (c) hydrophilic suspension with particle hydrated but not charged; (d) suspension with particle neither charged nor hydrated.

hydrated but not charged (hydrophilic or hydrophobic suspension, Figure 4.2b); less stable are particles that are hydrated but have no charge (a hydrophilic suspension, Figure 4.2c); least stable is a suspension in which the particles are neither hydrated nor carrying a charge (Figure 4.2d). When particles are neither hydrated nor charged, mixes are usually unstable.

Factors affecting the hydration of milk proteins include temperature, previous heat-treatment, salts, extremes of acidity or alkalinity, and homogenization.

Colloidal substances in ice cream mixes become more strongly hydrated at low temperatures, but increased affinity of the protein for water is produced through unfolding of protein molecules as a result of heat-treatment. Calcium salts may be expected to cause a greater depressing effect on the hydration of proteins than salts of sodium or potassium. The hydrating effect of citrates or phosphates for casein is shown in the following double decomposition reaction:

calcium caseinate + sodium citrate $\rightleftarrows$ sodium caseinate + calcium citrate
calcium caseinate + sodium phosphate $\rightleftarrows$ sodium caseinate
+ calcium phosphate

Any charge that shifts the reaction to sodium caseinate would be expected to increase the hydration of the casein. (Hydration is at a maximum at pH 6.2 to 6.4) Two-stage homogenization may be expected to increase slightly the bound water over that achieved with single-stage homogenization.

Emulsion Stability. Mix stability is dependent on emulsion (fat) and colloid (protein) stability. Ice cream is homogenized to reduce the relatively large fat globules to fine particles with a high degree of dispersion. The fat globules in the homogenized mix are surrounded by an interfacial layer, which may be hydrated and is thick with respect to the dimensions of the fat globules. The interfacial layer consists mainly of complexed milk protein and has been reported to be about 0.3 μm thick. An increase in internal liquid cohesion decreases churning of the fat in the freezer and increases the resistance to foam destruction when the mix is extended into the lamellae during shipping of the frozen mix.

The state of dispersion of milkfat in ice cream depends on the forces that tend to drive the fat globules apart—the emulsifying effects of the homogenizing valve and the mutual repulsion of the globules due to their electric charges. The forces that tend to bring the globules together are the collisions of the globules as they emerge from the valve of the homogenizer, the Brownian movement of the very small globules, the cohesiveness of the adsorption layers surrounding the globules, the interfacial tension between globules in close proximity, and the concentrating effect of freezing on the proximity of fat globules to each other in the serum.

High temperatures increase the electric charge on the fat globules, thus decreasing clumping, and citrates and phosphates increase the negative charge, thereby also decreasing clumping. Low temperatures and calcium salts increase the positive charge and thus cause or increase clumping.

In ice cream it is desirable to retain, as far as possible, the original fine dispersion of the fat, and the extent to which and form in which milkfat globules may aggregate are closely linked to the behavior of the fat globule surface. Making the fat globule surface partly or totally hydrophobic is the first stage in "demulsification" of the milkfat.

The second stage takes place when the fat globules are in a partially solidified state, containing the crystalline and liquid fractions of the fat in an optimal ratio. If clumping is to occur, a certain amount of the liquid fat has to be released from the fat globules. The liquid fat remains attached to the outer surface of the globule membrane, and, when two or more globules approach each other closely enough, the liquid fat fuses.

At low temperatures, when only small quantities of liquid fat are present, little clumping takes place. Freezing alters the fat globule membrane, and, during subsequent thawing, the adjacent globules join together in clumps, because liquid fat is spread between the outer surface of the membrane and the plasma, which renders the fat hydrophobic.

Optical methods with special immersion media are used to identify the fat globules in the internal structure of ice cream. Fat globules can be detected dispersed in the unfrozen material, around the air cells individually, and in chainlike arrangements. There is a relationship between the degree to which the fat globules are grouped in chainlike arrangements and clusters (which is affected by the freezing and whipping process of the ice cream mix) and different percentages of fat and overrun content.

One of the factors affecting the stability of fat in ice cream is the process of freezing. The fat globules begin to agglomerate due to the agitation and the concentration effect of freezing; when observed with the microscope, the agglomerates begin to look like bunches of grapes. The rate of agglomeration and coalescence is a function primarily of the degree of agitation, but it is affected also by such factors as protein stability, melting point of the fat, temperature of the freezer, the types of emulsifier, stabilizer, and sugars, and the salt content.

Dryness in ice cream is directly correlated with emulsion instability, and the greatest dryness and stiffness is obtained in ice cream in which the maximum amount of fat clumping has taken place short of churning. As ice cream is manufactured under lower freezing temperatures and with longer agitation periods, a greater degree of fat destablilization is exhibited.

Dryness and stiffness are primarily due to the agglomeration of the milkfat globules. This agglomeration of fat results in slowing of the rate of melt down. Therefore, fat agglomeration during freezing is beneficial, particularly in the case of the continuous freezer.

If agglomeration is carried too far, as might be the case with the extended agitation received in the soft-serve freezer, the result will be churning, with the production of visible granules of butter. The negative charges carried by the fat globules, which cause them to repel each other, are lost or overcome during the agitation. An ideal ice cream would be one in which all of the fat is agglomerated but in which none has churned out as visible butter granules. Ice cream in this condition would possess optimal properties of texture, body, dryness, and stiffness as well as an improved apparent richness.

The agitation of relatively cold and stiff ice cream mix increases the rate of churning, contrary to the belief held by some that a long initial freezing time leads to increased churning. In fact, at the higher temperatures associated with long initial freezing, churning takes place much more slowly.

Certain emulsifiers tend to destabilize the milkfat, thus accelerating churning. Excess destabilization of fat is indicated when there is a poor melt down, i.e., the ice cream does not melt to a smooth consistency but retains much of its original shape or structure even after it has been exposed to room temperature.

Results of a survey have shown that ice cream sandwiches manufactured under conditions designed to extrude a dry product have approximately three times the amount of destabilized fat as ice cream frozen under conditions where dryness is not of particular importance.

Density of Mixes

The specific gravity or density of ice cream mix varies with composition. The specific gravity may be measured by a hydrometer and the density by weighing a known volume of mix at a known temperature on a gravimetric balance. We can also calculate the density for a mix at 60°F (15°C) by the formula

$$\frac{100}{[\%\text{fat}/0.93] + [\%\text{sugar}, \%\text{NMS}, \%\text{stabilizer}/1.58] + \%\text{water}}$$

Wolff (1982) used a value of 1.601 for the nonfat solids, which gives a value approximately 0.0026 higher than the 1.58 factor. Investigations indicate that the specific gravity of a mix may vary from 1.0544 to 1.1232.

Acidity of Mixes

The normal titratable acidity of mixes varies with the percentage of NMS contained and may be calculated by multiplying the percentage of NMS by the factor 0.017. Thus, a mix containing 11% NMS would have a normal titratable acidity of 0.187%. The normal pH of ice cream mix is about 6.3.[1] The acidity and pH are related to the composition of the mix—an increase in NMS raises acidity and lowers the pH. The percent titratable acidity and pH values for mixes of various NMS content are given in Table 4.4.

Table 4.4. Titratable Acidity[a] and pH Values for Ice Cream Mixes Containing 7–13% NMS

NMS (%)	Approximate acidity (%)	Approximate pH
7	0.119	6.40
8	0.136	6.35
9	0.153	6.33
10	0.170	6.32
11	0.187	6.31
12	0.204	6.30
13	0.221	6.28

[a] As lactic acid.

---

[1] A neutral substance (i.e., neither acidic nor alkaline) would have a value of 7.0, with decreasing values indicating increasing acidity.

If fresh milk components of excellent quality are used, the mix can be expected to have a normal acidity. The apparent or natural acidity of ice cream mix is caused by the milk proteins, mineral salts (mostly phosphates and citrates), and dissolved CO_2. Developed acidity is caused by the production of lactic acid by bacterial fermentation of the lactose in dairy products. When the mix or ice cream acidity is above normal, developed acidity was probably present in the dairy products used in the mix. A high acidity is undesirable as it contributes to excess mix viscosity, decreased whipping rate, inferior flavor, and a less stable mix. The latter may contribute to "cook on" during processing and pasteurization, because heat and acidity accelerate the denaturation of proteins.

Influence of Mineral Salts

Various mineral salts have been used to help control churning and separation of the fat in the mix during the freezing process and to increase the stiffness and smoothness of the finished ice cream. Studies on the effects of sodium and magnesium phosphates, calcium and magnesium oxides, and sodium bicarbonate have shown that they tend to improve the body, texture, and general characteristics of the finished product.

Calcium and magnesium oxides or carbonates are recommended in preference to the sodium products because sodium's strong wetting effects counteract and nullify most of the beneficial effect the additive might have on the protein. Sodium citrate and disodium phosphate are effective protein stabilizers.

Disodium phosphate, sodium tetrapyrophosphate, sodium hexametaphosphate, and sodium citrate have been studied for their possible effect on controlling the churning defect of soft-serve ice cream. The mineral salts were used at concentrations of 0.1 and 0.2%, and the products were compared with a control that did not contain added salt. The compounds were blended with the stabilizer before mix preparation. Without exception the ice cream samples containing one of the salts showed less fat destabilization than the control sample drawn after the same length of time in the freezer. These buffering salts most likely influenced fat emulsion stability through some mechanism involving the milk proteins. The effects of these salts on acidity were very small.

Mix Viscosity

Rheology is a branch of physics concerned with the composition and structure of flowing and deformable materials. Published data on the rheological characteristics of ice cream mix and ice cream are mainly tabulations of research results that express the effect of various mix constituents on product properties. The flow properties are expressed in terms of viscosity, and considerable attention has been given to factors affecting mix viscosity.

Viscosity, the resistance of a liquid to flow, is the internal friction that tends to resist the sliding of one part of the fluid over another. The unit of viscosity is the poise, which is the force in dynes/cm^2 required to maintain a relative velocity of 1 cm/sec between two parallel plates 1 cm apart. Viscosity may be expressed in absolute or relative values. The absolute unit of measurement commonly used is the centipoise, 1/100 poise. The absolute viscosity of water at 68°F (20°C) is 1.005 cP. Ice cream mix has both apparent viscosity, i.e., the measured viscosity progressively decreases as the shear rate increases, and true viscosity, the resistance to flow that remains after the apparent viscosity disappears. The true viscosity of ice cream mix may range from 50–300 cP. The higher the viscosity of a mix, the greater the power required to freeze that mix. A certain level of viscosity seems essential for proper whipping and for retention of air. The viscosity of a mix is affected by:

- Composition—viscosity is influenced more by the fat and the stabilizer than by the other constituents.
- Kind and quality of ingredients—those carrying the fat are especially important. Also, heat and salts (such as calcium, sodium, citrates, phosphates) greatly affect the viscosity due to their effect on the casein and other proteins.
- Processing and handling of the mix—the steps in processing that have the greatest effect are pasteurization, homogenization, and aging.
- Concentration—TS content.
- Temperature.

Although much has been written about the causes and effects of viscosity, there has been no final answer to the question of how much is desirable in ice cream mixes and how it can be accurately measured. A high viscosity was believed essential at one time, but for fast freezing (rapid whipping) in modern equipment a lower viscosity seems desirable. In general, as the viscosity increases, the resistance to melting and the smoothness of texture increases, but the rate of whipping decreases. Viscosity is now considered a characteristic that frequently accompanies rather than causes desirable whipping, body, and texture. Therefore, the mix should be properly balanced (in regard to composition, concentration, and quality of ingredients) and then properly processed to produce the desired whipping ability, body, and texture. Under these conditions a desirable viscosity is ensured. Viscosity values of ice cream mix are valuable in indicating whether any factors may be influencing the mix unduly.

Viscosity may be measured in three ways: (1) by time required to flow under a fixed pressure through a pipet or specially constructed tube; (2) by measuring the force required to move one of two parallel plates or coaxial cylinders between which a layer of the liquid sample is placed; or (3) by measuring the fall of a ball through a column of liquid.

Data indicate that ice cream should be considered a viscous system rather than a plastic. The absolute viscosity of ice cream at 100% overrun is 20 billion cP at –8.0°C. Lowering the temperature a few degrees may double or triple this value.

Surface Tension

Surface tension is a force resulting from an attraction between surface molecules of a liquid that gives surfaces filmlike characteristics. The greater the attraction between the molecules, the higher the surface tension value, and the less the attraction between the molecules of the liquid, the lower the surface tension value.

The unit of measurement of surface tension is the dyne. The du Nouy apparatus is commonly used to determine the surface tension of ice cream mixes.

This apparatus measures the force required to pull a platinum ring with a circumference of 4 cm free from the surface of the liquid. The normal range of surface tension values for ice cream mix is 48–53 dynes.

Increasing the surface tension above that of the freshly processed mix made from fresh ingredients is difficult; however, the surface tension may be readily decreased by the addition of emulsifiers. Mixes with low surface tension values caused by the addition of emulsifier have shown excessive rates of whipping, fluffy and short body characteristics, and high susceptibility to shrinkage.

Interfacial Tension and Absorption

Interfacial tension is the force involved at the interface between two liquids, a liquid and a gas, or a liquid and a solid. The properties of interfaces are determined by the kind, orientation, and number of molecules residing in them. Surface tension and interfacial tension are related, because the conditions that produce stress on the surface of a liquid may also produce stress in the bounding surface at an interface. Surface tension and interfacial tension values vary inversely.

Adsorption involves dissolved substances at the interface. Substances adsorbed at the interface form an adsorption film and lower the interfacial tension. Substances accumulate at a surface in the order of their ability to lower the interfacial tension. Less effective depressants are excluded. Thus, emulsifiers, especially those with a low hydrophilic lipophilic balance (HLB), displace proteins adsorbed to fat globules. In case of dilution, the concentration of the adsorption film decreases except when the adsorption is of a nonreversible character.

Whereas smoothness of texture is enhanced by homogenization of ice cream mixes, the fat globule membrane formed during homogenization differs markedly from the natural fat globule membrane. Its structure depends primarily on the surface-active materials in the mix during homogenization. These materials, in order of their potential to adsorb to the newly formed fat globules, are emulsifiers, especially the "poly" types, phospholipids, whey proteins, and casein. This order of adsorption is determined by the propensities of these components to lower the interfacial tension, the whey protein being more surface active than casein. In the conventional mix, containing 10% milkfat and 10% nonfat milk solids but no added emulsifier, a major component of the fat globule membrane is casein, and the membrane is much thicker than when whey proteins comprise a large part of the mix. Furthermore, emulsifiers, by displacing protein from the fat globule surface, decrease the thickness of the fat globule membrane. Goff et al. (1989) recently demonstrated that when whey proteins are adsorbed on fat globules in preference to casein, the resulting emulsion is more readily destabilized during freezing and the ice cream has a more dry appearance. However, the high viscosity produced by the whey proteins, especially after pasteurization at temperatures above 70°F, tends to limit the amount of whey protein that can be added.

Goff and Jordan (1989) showed that the destabilizing power of size types of commercial emulsifiers depended upon the interfacial tension between the serum and lipid phases of the mixes. Polyoxyethylene sorbitan monooleate

(Tween 80) was the most powerful destabilizing agent. Reduction of the amount of protein adsorbed to the fat globule surfaces increased the susceptibility of the fat globules to coalescence induced by the shear forces and ice crystals in the freezer.

Schmidt and Smith (1989) demonstrated that the amount of fat destabilized was higher when the mix was unhomogenized or was homogenized at 35 kg/cm^2 (500 psi) than when it was homogenized at 282 kg/cm^2 (4,000 psi). Interestingly, the unhomogenized mix with no added emulsifier was more highly destabilized than was the reference mix that contained an emulsifier (0.075%; 4 parts mono/diglycerides and 1 part polysorbate 80) and was homogenized conventionally (140 and 35 kg/cm^2, 2,000 and 500 psi). No differences were shown in stiffness or overrun, but mixes homogenized at very high pressures had the highest glossiness as well as the least fat destabilized. Few differences in sensory qualities were found between samples given different homogenization treatments and the reference sample.

Freezing Point

The freezing point of ice cream is dependent on the concentration of the soluble constituents and varies with the composition. The freezing temperature can be calculated with considerable accuracy and can be determined in the laboratory with a cryoscope.

An average mix containing 12% fat, 11% NMS, 15% sugar, 0.3% stabilizer, and 61.7% water has a freezing point of approximately 27.5°F (–2.5°C). The freezing point of mixes with high sugar and NMS content may range downward to 26.5°F (–3°C), while for mixes with high fat, low NMS, or low sugar content it may range upward to 29.5°F (–1.4°C).

The initial freezing point of the average ice cream mix is approximately 27 to 28°F (–2.8 to –2.2°C) and essentially reflects the freezing point lowering due to the sweetener content of the mix. When latent heat is removed from water and ice crystals are formed, a new freezing point is established for the remaining solution since it has become more concentrated with respect to the soluble constituents. A typical freezing curve for ice cream shows the percentage of water frozen at various temperatures (Figure 4.3).

The calculated freezing points of ice cream mixes and the quantities of ice separated during the freezing process are in close agreement with experimental values.

Whipping Rate

To be most enjoyable on eating, most frozen desserts must contain air that is whipped in as minute bubbles. The rate of incorporation of these bubbles and their individual stabilities determine the overall whipping rate.

The major factors at work during the freezing process that affect whipping rate are effective agitation in the presence of a controlled volume of air and concomitant freezing of the mix.

It is vital that the mix contain surface-active components that will quickly

4 COMPOSITION AND PROPERTIES

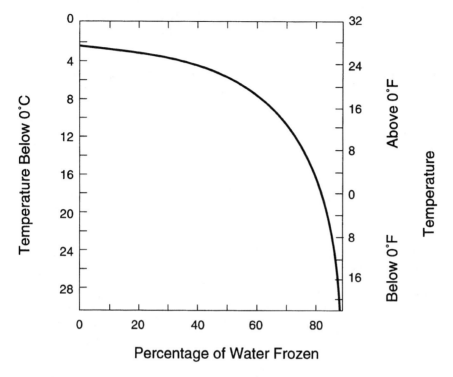

Figure 4.3. Typical freezing point curve for ice cream showing the percentages of water frozen at various temperatures. (Courtesy Doan and Keeney 1965.)

migrate to the surfaces of the formed air cells to stabilize them. This function is performed by proteins, phospholipids, and added emulsifiers.

It is also important that fat globules and ice crystals not mechanically interrupt and weaken the lamellae of the air cells. Therefore, as freezing starts, it is required that fat globules be small and well dispersed. However, to prevent collapse of the foam, especially during storage, and to produce dryness and stiffness, it is vital that fat globules be partially destabilized, i.e., churned.

The size, number, and physical condition of fat globules in an ice cream mix determine the rate of whipping and the stability of the whipped product. Small fat globules and limited clumping enhance whipping. Nonfat mixes whip more rapidly than those containing fat, but when frozen, they possess a foam structure that is susceptible to shrinkage. Partial churning of the fat of ice cream during freezing produces a bridging structure that provides resistance to shrinkage. Mixes made with butter, butter oil, anhydrous milkfat, or frozen cream have poorly dispersed fat and, consequently, poor whipping ability. Egg yolk solids and buttermilk solids from sweet cream improve whipping ability, presumably because of lecithin existing as a lecithin-protein complex. Emulsifiers, especially the "poly" types, also improve whipping ability. Usual variations in concentration of NMS have no pronounced effect on whipping, but qualitative variations in NMS are important. Sodium caseinate improves whipping properties and affects air cell and ice crystal distribution to an extent hardly expected of any other commonly used ice cream constituent. Sugar decreases the whipping ability except when added after homogenizing and then it increases whipping ability. Finally, the design and operation of the freezer determine whether the maximum whipping ability of a given mix is obtained.

Rate of whipping is determined by measuring the overrun at 1-min intervals while the mix is being frozen in a batch freezer. Normally, within 3.5 min after the freezing process is started, the mix has begun to freeze and within 7 min an overrun of 90% is obtained. In mixes possessing a high whipping rate, 90% overrun may be reached within 5 min. Mixes requiring 8 min or more to reach 90% overrun are considered to have a low whipping rate.

REFERENCES

Doan, F. J., and P. G. Keeney. 1965. Frozen dairy products. In *Fundamentals of Dairy Chemistry*. B. H. Webb and A. H. Johnson (eds.). AVI Publ. Co., Westport, CT.

Goff, H. D., and W. K. Jordan. 1989. Action of emulsifiers in promoting destabilization during the manufacture of ice cream. *J. Dairy Sci.* 72:18–29.

Goff, H. D., J. E. Kinsella, and W. K. Jordan. 1989. Influence of various milk protein isolates on ice cream emulsion stability. *J. Dairy Sci.* 72:385–397.

Schmidt, K. A. and D. E. Smith. 1989. Effects of homogenization on sensory characteristics of vanilla ice cream. *J. Dairy Sci.* 71:46–51.

Wolff, I. A. 1982. Handbook of Processing and Utilization in Agriculture, Vol. 1, pp. 323–325. CRC Press, Boca Raton, FL.

5
Ice Cream Ingredients

Essential for the manufacture of ice cream of the highest quality are ingredients of excellent quality, a mix that is formulated and balanced to provide proper function of each component, and intelligent processing, freezing, and hardening of the product. However, the selection of excellent ingredients is, without doubt, the most important factor in successful manufacture of frozen desserts. The clean, fresh, creamy flavor desired in ice cream can be repeatedly secured only by the use of ingredients that have been carefully produced and handled. In general, the more an ingredient has been handled or processed and the longer it has been stored, the less desirable is its flavor and function in the ice cream.

OPTIONAL INGREDIENTS

Frozen desserts can be made with a wide variety of ingredients, and they can be grouped by type as shown in Table 5.1. Federal definitions and standards for frozen desserts (Code of Federal Regulations Title 21, Part 135) specify optional dairy ingredients, optional caseinates, safe and suitable sweeteners, and other ingredients.

COMPOSITION OF MILK

Because milk is the source of dairy ingredients in ice cream, it is important to understand milk's composition and properties. Milk is composed of water, milkfat, and nonfat milk solids (NMS). The latter two components compose the total solids (TS) of milk. NMS are the solids of skim milk and include lactose, proteins, minerals, water-soluble vitamins and enzymes plus some minor constituents. These solids are also referred to as skim milk solids and serum solids (SS).

The composition of milk (Table 5.2) is influenced by numerous factors. For example, as fat content changes, so does NMS content and in the same direction. The magnitude of the change is dependent primarily on the breed of cow with Holstein cows having a change of approximately 0.5% NMS with each 1% change in fat content. Milk is produced in the mammary gland from constituents

Table 5.1. Essential and Optional Ingredients for Ice Cream

Dairy ingredients
Cream
Plastic cream
Anhydrous milkfat
Butter oil
Butter
Milk
Concentrated milk
Evaporated milk
Sweetened condensed milk
Superheated condensed milk
Dried milk, dried cream
Skim milk, same forms as milk
Sweet cream buttermilk
Condensed or dried sweet cream buttermilk
Concentrated skim milk that has been treated with Ca(OH)→ and Na_2HPO_4
Whey and those modified whey products (e.g., reduced lactose whey, reduced minerals whey, and whey may be added in the protein concentrate) that have been determined by FDA to be generally recognized as safe (GRAS). Whey and whey products contribute no more than 25% by weight of the total NMS in the food.
Hydrolyzed milk protein
Not to exceed 3% by weight of ice cream containing ≥20% TMS. Any hydrolyzed whey proteins must be included in calculations of the maximal 25% whey solids replacement for NMS.

Mineral salts
Sodium salts of citric acid
Phosphates—disodium, tetrasodium, or hexameta forms (≤0.1% of product)
Ca or Mg oxides or $Ca(OH)_2$ (≤0.4% of any combination of these).

Water
May be added to or evaported from the mix.

Coloring
Natural or artificial colors are permitted with label restrictions on yellow #5 and yellow #6.

Sweetening ingredients
Safe and suitable that include:
Sugar
Dextrose
Invert sugar
Corn syrup, liquid or dried
Maple syrup, maple sugar
Honey
Brown sugar
Malt syrup, liquid or dried
Maltose syrup, liquid or dried
Refiners syrup
Lactose
Fructose
Aspartame, acesulfame K, Sucralose

Caseinates
Caseinates precipitated with gums
Ammonium caseinate
Calcium caseinate
Potassium caseinate
Sodium caseinate
Caseinates may be added in the liquid or dry form, but must be free of excess alkali.

Egg products
(Add before pasteurization)
Whole egg: liquid, dried, frozen
Yolk: Liquid, dried, frozen

Stabilizers, thickeners
Agar
Sodium alginate
Propylene glycol alginate
Calcium sulfate
Gelatin
Gum acacia
Guar gum
Gum karaya
Locust bean gum
Oat gum
Gum tragacanth
Carrageenan and its salts
Furcellaran and its salts
Lecithin
Carboxymethyl cellulose

Emulsifiers
Mono/diglycerides (≥0.2% of finished product)
Polyoxyethlene sorbitan monostearate (60) or monoloeate (≤0.1% of finished product)
Microcrystalline cellulose (≤1.5% of finished product
Dioctyl sodium sulfosuccinate (≤0.5% of weight of stabilizers)

5 ICE CREAM INGREDIENTS

Table 5.2. Approximate Composition of Bulk Cow's Milk

Constituents	Mean (%)	Normal variation (%)
Water	87.4	86.7–88.1
Fat	3.8	3.5–4.1
Protein	3.2	3.0–3.4
Lactose (milk sugar)	4.9	4.7–5.1
Ash (minerals)	0.7	0.67–0.73
Nonfat solids	8.8	8.37–9.23
Total solids	12.6	11.9–13.3

supplied via the blood. Some constituents filter or dialyze into the milk, but most are synthesized in the mammary tissues and are secreted by the gland. Consequently, any milk contains milkfat, casein, and lactose.

Milk is a structurally complex physiochemical system. Its components are dispersed in true solution (lactose, whey proteins, some minerals and minor components), as colloids (casein and complexed minerals), and as a coarse dispersion (milkfat). Table 5.3 gives particle sizes and states of dispersion of constituents of milk and ice cream.

Milkfat is suspended in milk as tiny globules that are held in an emulsified state. Charges on the globule membranes cause them to be repulsed from each other. They are lighter in weight than the serum (skim milk) that surrounds them; so they rise slowly in milk to form a layer of concentrated fat called cream. Some cows produce proteins called agglutinins that are adsorbed on surfaces of the fat globules. Agglutinins cause fat globules to stick together and this increases the rate of creaming (cream formation). Normal milk contains about 2.4 billion fat globules per milliliter, and the globules vary in size from 0.5–20 µm in diameter (Figure 5.1), depending on the breed of cow and stage of lactation. The layer of phospholipids, proteins, and other molecules that surrounds fat globules protects the glycerides from being hydrolyzed (split apart with the addition of water) by lipases from the cow in the milk. Milkfat is composed of fatty acids and glycerol, $C_3H_8O_3$, connected through ester linkages. Glycerol molecules combine with fatty acid molecules according to the reaction

$$1 \text{ glycerol} + 3 \text{ fatty acids} \rightarrow 3H_2O + 1 \text{ triglyceride}$$

Milkfat contains about 140 different fatty acids with carbon numbers ranging from 4 to 24. About one-third of these fatty acids contain one or more double bonds between carbons in the carbon chain. These are called unsaturated fatty acids, and they have lower melting points than do fatty acids of the same carbon chain length that are fully saturated with hydrogens. Milkfat is unique in its content of fatty acids that have short chains of carbon atoms. These fatty acids are butyric (C4), caproic (C6), and caprylic (C8). By combining long-chain saturated fatty acids with short-chain and unsaturated fatty acids, the cow's mammary tissues produce milkfat that is melted at body temperature (37°C) but is a semi-solid at room temperature (as in butter at 22°C).

Substances associated with milkfat include the phospholipids lecithin,

Table 5.3. Particle Size and State of Dispersion of the Constituents of Milk and Ice Cream[a]

1 nm (1/100 μm)	10 nm	100 nm	1 μm	10 μm	100 μm	1 mm (1000 μm)
Electron microscope	Ultramicroscope			Microscope		Visible
Passes filters and membranes	Pass filters but not membranes		Passes neither filters nor membranes			
Molecular movement	Brownian movement			Slow Brownian and gravitational movement		
Sedimentation and oil globules rise extremely slowly			Sedimentation and oil globules rise			
High osmotic pressure	Low osmotic pressure		No osmotic pressure			
True solution	Colloidal suspension			Coarse suspension		
			Milk constituents			
	Whey protein, albumin, globulin	Calcium caseinate		Fat globules		
	Colloidal phosphates					
			Ice cream constituents			
Lactose and soluble salts	Whey protein, albumin, globulin Stabilizers, colloidal phosphates	Casein		Fat globules		
Lactose, sucrose, corn sugar and soluble salts				Ice crystals Flavoring particles		

[a] 1 μm = 1/25,000 in.

5 ICE CREAM INGREDIENTS

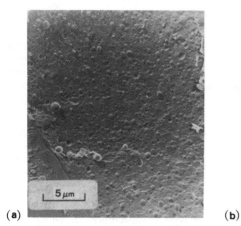
(a)

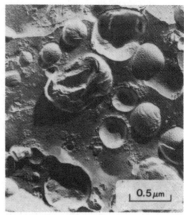
(b)

Figure 5.1. Electron micrograph of (a) fat globules in ice cream mix (inset, 5 μm), (b) fat globules showing casein micelles and subunits attached (inset, 5μm). [From K. G. Berger, A. W. White, G. Lyons & Co. Ltd., personal communication (1977).]

cephalin, and sphingomyelin; the sterols cholesterol and ergosterol; the carotenoids carotene and xanthophyll; and the vitamins, A, D, E, and K. Lecithin is formed by replacing one fatty acid of a triglyceride with phosphoric acid and choline, the latter being a nitrogenous base that is a part of the B vitamin complex. Milk contains about 0.075% lecithin and cephalin, and milkfat about 0.6% lecithin. Cholesterol is the principle sterol in milk, comprising about 0.015%, and 75–85% of it is associated with the lipid fraction. Ergosterol is the precursor of vitamin D, and carotene is the precursor of vitamin A. β-carotene imparts the yellow color to milkfat.

The proteins of milk are subdivided into casein and whey proteins. The major components of casein are α-, β- and γ-casein. Whey proteins are comprised of α-lactalbumin, β-lactoglobulin, serum albumin, immune globulins, and some minor proteins.

Casein comprises 75–80% of the total protein of milk and occurs only in milk. In the pure state casein is white, odorless and flavorless. It occurs as colloidal micelles in milk and can be removed by ultracentrifugation. Casein micelles can be observed by electron microscopy. Their diameters range from 1–100 nm with an average size of 40–50 nm. Micellar casein is combined with calcium, magnesium, phosphate, and citrate. It is precipitable with proteolytic enzymes (e.g. chymosin or rennin), alcohols that dehydrate the micelles, heat that denatures it, salts that combine with it and acids that neutralize its charge at approximately pH 4.6. A temperature of approximately 130°C is needed to coagulate the casein of high-quality milk.

Whey protein molecules are relatively large, ranging in size from 1–20 nm. These proteins are not precipitated with acid at pH 4.6 nor by proteolytic enzymes, but they begin to be precipitated by heat at about 77°C (170°F). Although they resemble casein in chemical composition, whey proteins contain no phosphate, whereas some fractions contain sulfur.

Lactose or milk sugar is dissolved in milk and as such is responsible for the major part of the reduction in freezing point of milk below that of water. Lactose, which is found only in milk, is a disaccharide made up of glucose and galactose. This carbohydrate is fermentable by many lactic acid-producing bacteria that are common in the environments where milk is produced and processed. Because these bacteria grow well at temperatures between 15–40°C, it is important to prevent their growth and acid production by keeping milk at 0–4°C.

The lactose content of milk normally ranges from 4.8–5.1%. It is a reducing sugar that is optically active, having a specific rotation of 52.53°. In solution lactose exists in two forms with the β-form comprising about 60% and the remainder in the α-form, depending on the temperature. Crystalline lactose exists in three forms, α-lactose hydrate, $C_{12}H_{22}O_{11} \cdot H_{20}$; α-lactose anhydride, $C_{12}H_{22}O_{11}$; and β-lactose anhydride, $C_{12}H_{22}O_{11}$. The crystals of lactose occur in many forms and are of high importance to ice cream manufacturers because of the possibility of their formation in the frozen product, making it have a "sandy" texture. The nonvolatile salts of milk are the minerals that are found in the ash that remains after milk is heated at a high temperature in a muffle furnace to completely oxidize the organic constituents. The mineral content of milk ranges from 0.65–0.75%, while the average content of minerals in the salt form is 0.9% (Table 5.4). These mineral salts, in the forms of citrates, phosphates, or oxides, affect the functional and nutritional properties of milk. Calcium and phosphorus are milk's most important minerals nutritionally and functionally. Milk contains many trace elements, and their concentrations are dependent on the type and composition of rations-fed cows.

Numerous other inorganic and organic substances occur in milk, some of which produce effects far out of proportion to their concentrations. These minor constituents include gases, enzymes, nonprotein nitrogenous substances, flavorful substances, nonlactose carbohydrates, vitamins, and pigments. The gases, carbon dioxide, nitrogen, and oxygen, are dissolved in milk in the approximate volume percentages of 4.5, 1.3, and 0.5%, respectively. The gas content decreases on heating or vacuumizing milk.

The enzymes of milk may be produced in the mammary glands during the secretory process or by bacteria growing in the milk. The enzymes that occur

Table 5.4. Average Amount of Minerals in Milk

Constituent	Amount (mg/100ml)
Calcium	123
Magnesium	12
Phosphorus	95
Sodium	58
Potassium	141
Chlorine	119
Sulfur	30
Citric acid	160

Source: National Research Council, Publication 254 (1953).

5 ICE CREAM INGREDIENTS

naturally in milk include acid and alkaline phosphatases, lipase, esterases A, B, and C, xanthine oxidase, protease, amylase, catalase, and carbonic anhydrase. Enzymes, being proteins, may be inactivated by heat, and they tend to be most active at the body temperature of the cow.

Of the 19 vitamins normally present in fresh raw milk, vitamins A, B_{12}, riboflavin, thiamine, and niacin are present in significant concentrations. Riboflavin is responsible for the green color of whey.

The nonprotein nitrogenous substances of milk include ammonia, urea, creatine, and free amino acids. The flavorful substances other than the main constituents include carbonyl compounds, lactones, and methyl sulfides.

MILK PRODUCTS USED IN ICE CREAM[1]

Sources of Fat

Both milkfat and NMS contribute significantly to the flavor of ice cream. Milkfat is more important than NMS because it provides the rich, full, and creamy flavor that ice cream should have. NMS, on the other hand, have an indirect effect on flavor. The proteins help give body and a smooth texture to the ice cream. Lactose displaces water and adds to the sweetness produced largely by the added sugars. The mineral salts carry a slightly salty flavor that rounds out the finished flavor of ice cream.

Whole milk is a desirable source of milk solids, but only fresh milk with a clean flavor and odor plus a low titratable acidity should be used. A combination of concentrated milk and sweet cream is usually the best source of fat for ice cream. The titratable acidity of cream containing 40% fat should not exceed 0.10%. A fat test should be made on each delivery so that buyers are assured of receiving the full amount of fat for their money. Sweet cream may be relatively expensive and difficult to obtain in good quality at certain seasons and in some markets.

Frozen cream is sometimes stored during the months of surplus and low price. Only the best fresh cream should be stored frozen. It should be pasteurized at 75°C for 15 min to minimize development of off-flavors. Milk and cream should be protected from contact with copper or iron. These metals dissolve in milk and cream catalyzing oxidation that imparts tallowy and metallic flavor during storage of frozen cream. This is the reason why dairy equipment is constructed of stainless steel, plastic, rubber, or glass. Frozen cream should be stored at –25°C for not more than 6 months. Although storing cream frozen may be economical, flavor is never quite as good as when the cream was fresh. Rancid, fishy, oily, and tallowy flavors are likely to result from oxidation of frozen cream. However, federal regulations prohibit addition of antioxidants to cream. Added sugar (10%) helps frozen cream retain freshness and melt fast with limited separation of fat. However, the added sugar represents capital, and

---

[1]Table 5.5 fives the approximate composition and weight of selected ingredients used in ice cream.

Table 5.5. Ice Cream Ingredients—Approximate Composition, Weights per Gallon, and Density

Ingredient	Fat (%)	NMS (%)	Sugar (%)	TS (%)	Moisture (%)	Weight per gallon (lb)	Density
Water	0.00	0.00	0.00	0.00	—	8.33	1.000
Skim Milk	0.00	8.60	0.00	8.60	—	8.63	1.036
Milk	3.00	8.33	0.00	11.33	—	8.60	1.032
	4.00	8.79	0.00	12.79	—	8.60	1.032
	5.00	9.10	0.00	14.10	—	8.60	1.032
Cream	18.00	7.31	0.00	25.31	—	8.45	1.014
	20.00	7.13	0.00	27.13	—	8.43	1.020
	25.00	6.68	0.00	31.68	—	8.39	1.007
	30.00	6.24	0.00	36.24	—	8.35	1.002
	35.00	5.69	0.00	40.69	—	8.31	0.997
	40.00	5.35	0.00	45.35	—	8.28	0.994
	50.00	4.45	0.00	54.95	—	8.20	0.984
frozen[a]	80.00	1.80	0.00	81.80	—	7.95	0.954
plastic	65.00	34.34	0.00	99.34	0.65	—	—
Dried cream	43.00	7.50	30.00	80.50	19.50	9.60	0.951
Concentrated sweet cream	82.50	0.50	0.00	83.00	18.05	7.92	0.900
Butter, unsalted	99.00	0.00	0.00	99.00	1.00	7.50	
Butter oil							
Evaporated milk							
canned	8.00	20.00	0.00	28.00	—	8.90	1.068
bulk	10.00	23.00	0.00	33.00	—	9.20	1.044
Sweetened condensed whole milk	8.00	23.00	42.00	73.00	—	10.87	1.305
Condensed skim, unsweetened	0.00	32.00	0.00	32.00	—	9.40	1.128
Condensed skim, sweetened	0.50	30.00	42.00	72.00	—	11.00	1.321
Condensed skim milk	0.00	20.00	0.00	20.00	—	8.98	1.078
	0.00	27.00	0.00	27.00	—	9.25	1.110
	0.00	30.00	0.00	30.00	—	9.35	1.122

Ingredient							
Condensed whole milk	8.00	22.00	0.00	30.00	—	8.99	1.079
Condensed whole milk	10.00	26.00	0.00	36.00	—	9.00	1.080
Condensed whole milk (blend)	19.00	21.00	0.00	40.00	—	8.90	1.068
Nonfat dry milk (skim milk powder)	0.00	97.00	0.00	97.00	—	—	—
Whole milk powder	26.00	72.00	0.00	98.00	—	—	—
Plain condensed whole milk	8.00	20.00	0.00	28.00	—	8.90	1.068
Anhydrous milkfat	99.90	0.00	0.00	99.90	—	—	—
Butter-powder mix	44.00	46.00	0.00	90.00	0.10	—	—
Butter-powder-sucrose mix	44.00	26.00	0.00	80.00	10.00	—	—
Dry buttermilk solids	5.00	91.00	0.00	96.00	20.00	—	—
Dry whey solids	0.00	93.00	0.00	93.00	—	—	—
Granulated sugar	0.00	0.00	100.00	100.00	—	7.50	0.900
Corn sugar	0.00	0.00	92.00	92.00	—	7.50	0.900
Invert sugar syrup	0.00	0.00	71.50	71.50	—	11.25	1.350
Corn syrup	0.00	0.00	82.00	82.00	—	11.98	1.438
30DE	0.00	—	—	79.45	—	11.81	1.417
36DE	0.00	—	—	79.86	—	11.81	1.417
42DE	0.00	—	—	80.30	—	11.81	1.417
high-fructose, 42	0.00	—	—	71.00	—	11.23	1.348
high-fructose, 55	0.00	—	—	77.00	—	11.55	1.386
high-fructose, 90	0.00	—	—	80.00	—	11.74	1.409
Honey	—	—	—	82.30	—	11.98	1.438
Maple syrup	—	—	—	68.00	—	11.14	1.337
Maple syrup	—	—	—	92.00	—	7.50	0.900
Dried egg yolk	58.00	0.00	0.00	98.00	—	—	—
Whole eggs	10.50	0.00	0.00	26.30	—	8.58	1.030
Cocoa powder	10–22	0.00	0.00	95.50	—	—	—
Chocolate liquor	49.00	0.00	0.00	94.00	—	—	—

[a] May contain 10% sugar.

appropriate accounting must include interest on the investment as well as the cost of space required for the added volume. Moreover, if such cream spoils, both the cream and the sugar are lost, and cost for their disposal must be paid. A good rule of thumb in making mixes is to use a maximum of 60% frozen cream with 40% fresh sweet cream.

Plastic cream is a source of concentrated fat for frozen desserts. It contains about 80% milkfat and has a consistency similar to that of butter. It is prepared by separating and then reseparating cream of 30 to 40% fat. Although plastic cream is stored and handled as butter, it has the advantage over butter of essentially retaining the fat-in-serum emulsion of cream. Because some of the emulsion may be broken during separation and cooling of the product in a swept-surface heat exchanger, mixes made with plastic cream as the sole source of fat may oil off or whip slowly, as do mixes made with butter, butter oil, or frozen cream.

Unsalted butter (sweet butter) can be an important source of milkfat for frozen desserts. It is comparatively inexpensive, can be transported at low cost, can be stored for several months with little loss in quality and is nearly always available. Butter of good quality can comprise from 50–75% of the fat of a mix. Such butter should score at least 92 on the USDA butter grading scale. The off-flavors common to defective butter—whey, storage, oxidized/metallic, old cream, rancid, and neutralizer (encountered infrequently because most butter is made from sweet cream)—will impart very undesirable flavors to ice cream. Finally, the use of butter in ice cream mixes often results in undesirable freezing properties. This is because both the composition and physical structure of butter differ from those of cream. Churning cream into butter releases much of the natural emulsifier, lecithin, from the ruptured fat globules into the buttermilk. Furthermore, the emulsion is broken and inverted during churning, and it must be restored to a fat-in-serum form in the ice cream mix. This can be done by proper re-emulsification.

Milkfat-sugar blends are made by mixing butter or anhydrous milk fat with sucrose alone or a blend of sucrose and NMS. In one study a milkfat-sugar blend kept well at $-18°C$ for up to one year. Generally, no more than 50% of milkfat-sugar blends should be used to replace the fat from fresh cream. The compositions of four types of milkfat-sugar blends are as follows: fat-sucrose blends with butter or with butter oil and fat-sucrose-NMS blends with butter or butter oil.

A small quantity of dry cream is manufactured in the United States, and this product is used in ice cream only under special circumstances.

The strong association between ingestion of the omega-3 fatty acids, eicosapentaenoic ($C20:5n-3$) and docosahexaenoic ($C22:6n-3$), and reduction of cardiovascular disease has raised the probability that ice creams will be marketed containing significant amounts of these fatty acids. To accomplish this requires two achievements, viz. (1) that cows be fed these fatty acids in a form protected from hydrogenation in the rumen of the cow so that the unsaturated form can be assimilated through the intestinal wall and transported to the mammary gland for synthesis into milkfat, and (2) that means be found to prevent oxidation of the unsaturated bonds in the milk and cream.

The opportunity to feed omega-3 fatty acids to cows became a reality in 1992 when a firm called OmegaTech was granted a U.S. patent for use, in animal feeds, of selected cold-water algae that are cultivated in fermenters. The Kelco Division of Merck & Company is licensed to develop the technology.

The author and his graduate students determined that the content of unsaturated fatty acids in frozen desserts could be increased considerably with minimal risk of oxidation of the stored ice cream. Im et al. (1994) found no effects on the sensory quality or functional properties when they used a mixture of canola oil and soybean oil with milkfat in making a frozen dessert containing 10% fat. The mixture of 7.8 parts milkfat:3parts canola oil:1.2 parts soybean oil provided a ratio of 3:2:1 of saturated to monounsaturated to polyunsaturated fatty acids. Frozen desserts with the fatty acids in a ratio of 1:1:1 of saturated to monounsaturated to polyunsaturated were deemed as oxidized by a sensory panel. The control had the same composition except that it contained milkfat only. Canola oil was chosen as the major oil additive because it contains mostly C18:1, oleic acid, a fatty acid that has favorable effects on the ratio of high-density lipoprotein to low-density lipoprotein. Furthermore, oleic acid resists oxidation much better than the more highly unsaturated fatty acids of soybean and corn oils.

Recently, Breeding (1993) made vanilla ice creams using liquid and solid fractions of milkfat produced by melt crystallization of butter oil at 18.5°C. She compared the flavor, body, and texture of these ice creams with control product made according to the same formula. Cream from which the butter oil was made provided the fat for the control ice cream. A trained sensory panel failed to find differences in flavor or texture among these products. Ice cream made with the liquid fraction contained higher amounts of unsaturated and lower amounts of saturated fatty acids than did ice creams made with cream or the solid (at room temperature) fraction. The liquid fraction was fully melted at 20°C, whereas 48°C was required to melt the solid fraction. Compared with the solid fraction, the liquid fraction contained one third more C18:1, C18:2, and C18:3 fatty acids combined than did the solid fraction. The difference was made up in higher amounts of C12:0 through C18:0 fatty acids in the solid fraction. The purpose of this experiment was to study the feasibility of enhancing the unsaturated fat content of ice cream with the liquid fraction of milkfat separated at 18.5°C. There were no significant effects of the type of milkfat on the intensity of either the vanilla flavor or the oxidized flavor of the ice creams.

Skim Milk and Buttermilk, Liquid and Dry

Fresh skim milk should be used in the mix whenever available at reasonable prices, because it is usually an economical and high-quality source of NMS. It must have a low titratable acidity, a low bacteria count, and a clean flavor. Skim milk should be purchased on a solids basis. Whereas fluid milk is usually traded on a price per hundredweight basis, ice cream is formulated to have a known NMS content. Therefore, it is imperative that ice cream processors purchase milk solids on a weight basis. One-hundred pounds of skim milk will normally contain 8.7 lb of NMS.

When skim milk is not available, a larger portion of nonfat dry milk (NDM) or concentrated skim milk is used to supply NMS. This practice increases risks that off-flavors or the "serum solids" flavor will be detectable in the finished product.

Sweet cream buttermilk is obtained by churning cream that has not developed detectable acidity, i.e., cream that is of a quality suitable for use in any retail product. Sweet cream buttermilk has beneficial effects on the whippability of mixes and contributes richness of flavor. Concentrated or dry sweet cream buttermilk is especially desirable in ice cream made low in fat content or with any form of de-emulsified milkfat (e.g., butter). Generally, sweet cream buttermilk can supply up to 20% of the NMS of ice cream with minimal risk of off-flavors. All buttermilk products should be examined for flavor on receipt and again before use if the time of storage is significant. Concentrated buttermilk should be treated as any other fresh pasteurized milk product. Dry buttermilk can be kept for several weeks to a few months depending on storage conditions, type of package, moisture content, and initial quality. The substantial fat content of the buttermilk needs to be considered in calculating a mix. The lecithin content of buttermilk approximates 0.1–0.2%.

As with dry buttermilk, nonfat dry milk should be purchased in only such quantities as will be used in several weeks and, preferably, should be kept in cold storage. Otherwise, staleness is likely to develop.

Manufacturers should specify extra grade dry products according to specifications of the American Dairy Products Institute (1990). Low, medium, or high heat powders can be used successfully in ice creams. However, the denaturation of whey proteins caused by high preheating temperatures tends to produce desirable body and texture characteristics in ice cream, especially when time and/or temperature of mix pasteurization is/are not held above the regulatory minimums. The storability, availability, and low moisture content of NDM are important advantages, but NDM, being hygroscopic and prone to oxidize, must be protected well from moisture, heat, and oxygen during storage. Although available in the instantized form, regular spray-dried NDM is used almost exclusively in frozen dessert mixes. High-quality NDM is bland, slightly cooked and mildly sweet in flavor, lightly cream colored, free of caking, and easily soluble.

Dry Whole Milk

Although great advances have been made in the technology of production, packaging, and storage of dry whole milk, it is used infrequently in frozen desserts. The major reason is the high risk that oxidation will render whole milk powders stale.

Concentrated Milks

Plain concentrated (formerly condensed) skim milk is used more frequently than any other source of concentrated milk products. It contains 25–35% NMS and is prepared by evaporating water from skim milk using vacuum and heat. The heat treatment is insufficient to sterilize, so the product must be refrigerated. Superheating concentrated milk to a high temperature increases the viscosity at ordinary temperatures. Its use in ice cream improves the whippability of the mix and increases firmness and resistance to melting of the ice cream but also increases cooked flavor (Tracy and Hahn 1938). Because superheated concentrated products cost more than plain concentrated products and carry a higher risk of being defective, they are mostly used when manufacturers wish to omit stabilizers. Another method of achieving the same result is to pasteurize the mix at an elevated temperature for an extended time.

Sweetened condensed milk is sometimes used as a source of concentrated NMS. The added sugar (40–44%) improves the keeping quality over that of plain concentrated products, because sugar raises the osmotic pressure and lowers the water activity (a_w). Sweetened condensed milk is highly viscous; hence it is not as easily handled as plain concentrated milk. A common defect in this milk is the formation of large lactose crystals. As the product is made, minute crystals of lactose are induced in it by seeding the concentrate with lactose powder while cooling. If these crystals grow excessively large, the texture will be sandy.

The titratable acidity test can be applied to concentrated and condensed milk products. When the products are diluted to the same NMS concentration as skim milk, the acidity should be approximately the same as fresh skim milk, about 0.15%.

Special Commercial Products

Special commercial products are sometimes used as constituents of ice cream mix. These products include sodium caseinate, delactosed milk products, modified NMS, certain mineral salts, concentrates and isolates of whey proteins, and combinations of some of these products. The products function to improve whippability, resistance to heat shock, and body and texture.

The amount of sodium caseinate used ranges from 0.5–1.0% of the mix. It serves as an aid to whipping and tends to improve texture. However, the risk of introducing a stale flavor via sodium caseinate can be significant.

The main purpose of using low-lactose products in ice cream is to reduce the risk that sandiness will develop during storage of the product. Additionally, because lactose, being dissolved, lowers the freezing point of the mix, removal of a portion of it permits drawing of ice cream from the freezer at a higher temperature than if all of the lactose is left in the product. The use of low-lactose milk solids may increase apparent mix acidity, because the increased protein concentration increases the concentration of buffering substances in the product. However, increasing the protein concentration of a mix can improve texture and resistance to heat shock.

Low-lactose milk solids are derived from skim milk that has been ultrafiltered or diafiltered. For example, concentrating skim milk to one-half its initial volume reduces the total lactose by one-half while concentrating the proteins to double their initial concentration. If the retentate from the process is brought back to volume with water, and the process of filtration under pressure is repeated to reduce the volume to one-half the initial volume, only one-fourth of the initial lactose will remain in the retentate. The remainder passes through the membrane in the permeate. Use of retentate in place of NMS has resulted in smoother texture (Hofi 1989), higher freezing point (Tong et al. 1989), and harder body (Kosikowski and Masters 1983). Higher protein in frozen desserts made with retentate increases the amount of bound water and can reduce the amount of stabilizer needed. By adding an alternative carbohydrate, such as glucose (corn sugar), to the mix, the freezing point can be lowered and the body softened (Geilman and Schmidt 1992). Use of ultrafiltered and, particularly, diafiltered retentate to replace NMS, and especially to replace whey solids, in frozen desserts greatly increases the amount of the dessert that can be eaten without the chance of discomfort by lactose malabsorbers.

Dry whey solids are widely used in frozen desserts, because they are relatively inexpensive sources of milk solids. Federal standards permit substitution of up to 25% of the NMS as whey solids. Typically, whey solids contain 72% lactose, 12% whey proteins, 11% minerals, 4% moisture, and 1% fat. The typical analysis of NDM is 52% lactose, 30% casein, 7.5% minerals, 6% whey proteins, 4% moisture, and 0.5% fat.

Mineral Salts

Certain mineral salts (Table 5.1) may increase the dryness, stiffness, and creaminess of ice cream. For example, calcium sulfate, added at the rate of 0.1% before pasteurization, produces a dry, stiff ice cream from the freezer and reduces the rate of melting. Titratable acidity of mixes is raised when mineral salts are added.

SWEETENERS

Many kinds of nutritive sweeteners are used in ice cream (Table 5.6). They include cane and beet sugars, many types of corn sweeteners, maple sugar, honey, invert sugar, fructose, molasses, malt syrup, brown sugar, lactose, and refiners syrup.

From 25–50% or more of sugar can be replaced by corn sugar (glucose) or corn syrup products. The major considerations are relative sweetness and some undesirable flavor notes carried by some corn syrup products.

The desired sweetness of ice cream, based on equivalency to sucrose, ranges from 13–16% (Leighton 1942). Sweetness depends on the concentration of sweetener in the water of the mix; thus, decreasing the water of the mix is equivalent to increasing the sweetness. Sweeteners, being dissolved, lower the freezing point of the mix and reduce its whippability. Furthermore, they tend to increase smoothness of texture and rate of melting. Table 5.6 describes sweeteners and bulking agents useful in frozen desserts.

Relative Sweetness

Dahlberg and Penczek (1941) published important results of comparisons of relative sweetness of sweeteners as affected by interactions, concentrations, and temperatures. Because there is no chemical test of sweetness, not all persons agree on the relative sweetness of various sweeteners. The common practice is to compare the sweetness of other sugars to sucrose, which is given

5 ICE CREAM INGREDIENTS

Table 5.6. Sweeteners and Bulking Agents of Frozen Desserts

Ingredient	Average molecular weight[a]	Relative sweetness[b]	Total solids (%)	Relative F.P. depression[c]	Maximum total sugar supplied[d] (%)
Dextrose	180	74	92	1.90	40
Fructose	180	173	100	1.90	40
Sucrose	342	100	100	1.00	100
Lactose	342	16	100	1.00	g
Maltose	342	32	100	1.00	
Honey		75	74	1.46	45
Invert sugar	261	95	77	1.12	30
HFCS[e]					
90%	180	125	77	1.88	50
55%	185	98	77	1.85	50
42%	190	86	71	1.80	50
HMCS[f]					
55DE	411	55	81	0.83	40
Corn syrups					
68DE	265	72	81	1.28	25–50
62DE	298	68	82	1.15	25–50
52DE	345	58	81	0.99	25–50
42DE	428	48	80–81	0.80	25–50
36DE	472	42	80	0.72	25–50
32DE	565	40	80	0.61	25–50
25DE	720	28	80	0.48	h
20DE	900	23	80	0.38	h
Maltodextrins					
18DE	1,000	21	95	0.34	h
15DE	1,200	17	95	0.29	h
10DE	1,800	11	95	0.19	h
5DE	3,600	6	95	0.10	h

[a]Average molecular weights of CSS and maltodextrins are estimated by dividing the average molecular weight of starch, 18,000, by the dextrose equivalent (DE) factor.
[b]Sweetness relative to sucrose (approximate) on an as is or product basis.
[c]Factor to estimate freezing point depression relative to solids equal in weight to sucrose.
[d]Percent of sugar on a sweetness basis generally acceptable from a quality viewpoint.
[e]HFCS—High-fructose corn syrup.
[f]HMCS—High-maltose corn syrup. 65% maltose.
[g]Lactose provides low sweetness but amount is limited by tendency to crystallize.
[h]Lower DE cornstarch products build body and provide bulk rather than sweetness.

the value of 100. Approximate relative sweetness values of commonly used sweeteners are given in Table 5.6.

Effect of Sweeteners on Freezing Point

Because sugars do not dissociate in solution, the freezing points of their solutions can be computed from the concentration and molecular weight. With given weights and volumes of solvent, the effect on freezing point will be inversely proportional to the molecular weight. For example, corn sugar (the monosaccharide known both as dextrose and glucose) has 6 carbons; whereas sucrose, a disaccharide containing glucose and fructose, has 12 carbons. Thus, sucrose weighs nearly twice as much as glucose and lowers the freezing point only about one-half as much (Figure 5.2). The molecular weights of representative sweeteners are given in Table 5.6.

Tharp (1982) gave formulas for calculating freezing points based on differ-

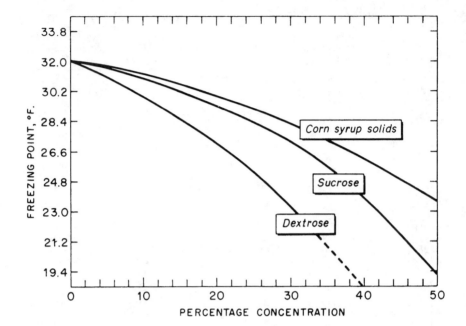

Figure 5.2. Freezing point depressions produced by selected sweeteners at increasing concentrations. (Courtesy American Maize Products Co.)

ences in molecular weights of carbohydrates and other dissolved substances. He noted that calculation of freezing points of multicomponent systems, e.g., frozen dessert mixes, is more complex than the simplicity of the calculations may suggest.

Based on mix compositions typical at the times, Tharp calculated freezing points of mixes for the years 1950, 1960, 1972, 1975, 1980, and 1981. They were 27.70, 27.00, 27.17, 27.07, 26.47, and 25.77°F, respectively. Differences in amounts of sweetener solids used in the mixes were suggested as primarily responsible for the differences in freezing points.

Sucrose

Sucrose, commonly known as granulated sugar, is made from sugar cane or sugar beets. Being crystalline, it is approximately 99.9% solids. It is highly soluble and has a density of 1.595 g/cc. (Sucrose is available also as a syrup containing approximately 67% solids.) Although sucrose depresses the freezing point, sweetening effect is the factor that limits its concentration in ice cream. Each 1% increase in sucrose in an ice cream mix lowers the freezing point about 0.2°F. Sucrose may be used as the sole sweetener in ice cream with excellent results. However, use of sucrose as the sole sweetener in ices or sherbets may result in formation of crystals on the surfaces. This defect in ices and sherbets can be avoided by using 1 part of dextrose to 3.5 parts of sucrose.

Corn Sweeteners and Related Ingredients

Three major forms of corn sweeteners are available to the food industry, viz., the crystalline forms of refined corn sugar (dextrose and fructose), dried corn syrup (corn syrup solids), and liquid corn syrup. Maltodextrins are also available in dried form. Each of the ingredients is made from starch, most of it from corn. Purified starch is hydrolyzed using acid or enzymes. Low degrees of hydrolysis produce maltodextrins. Because starch is composed of molecules organized into polymers, hydrolysis (conversion) at the 1,4 glucosidic linkages produces multiple polymers, the length of which depends on the number and location in the polymer of bonds hydrolyzed. Each bond hydrolyzed produces a free aldehyde group that has the same reducing ability as does glucose (dextrose). This makes it possible to monitor the process of hydrolysis, the extent of which is termed the dextrose equivalent or DE. Maltodextrins are only slightly hydrolyzed; consequently they range in DE from 4–20 and are only slightly sweet. They are made from food grade starches that can come from corn, potato, tapioca, rice, oat, or wheat. The process for hydrolyzing starch is random so that maltodextrins contain mixtures of saccharide polymers of varying chain lengths, from very short sugar chains to long complex carbohydrate chains, $(C_6H_{12}O_5)_n \cdot H_2O$. In general, the lower the DE, the longer the average saccharide polymer chain and the higher the average molecular weight (Table 5.6). Since starch, the material from which maltodextrins are made, is generally recognized as safe (GRAS), maltodextrins do not require FDA approval. Some maltodextrins are available in the agglomerated form to enhance dispensability. They are most advantageously used for making dry mixes in which they improve flowability and generate less dust. They decrease the bulk density, for example a maltodextrin from corn starch with a DE of 10 has a bulk density of 0.54 g/cc in the unagglomerated form and 0.34 g/cc in the agglomerated form.

Paselli SA2, developed by Avebe America, Inc., is a potato starch maltodextrin that dissolves in cold water. Sta-Slim is a modified potato starch product of A. E. Staley Manufacturing Co. Tapioca starch is the raw material used by National Starch and Chemical Corp. to make two dextrins, N-Oil and Instant N-Oil, as well as a maltodextrin, Instant N-Oil II. Maltrin is the maltodextrin made by Grain Processing Corp. from corn starch. Other firms that offer corn starch maltodextrins include American Maize Products Co., Archer Daniels Midland, Cargill, and Corn Products Division of CPC International.

Products having more than 20 to about 70% of the glucosidic linkages broken are known as corn syrups. They are classified based on degree of conversion as low conversion, 28–38 DE; regular conversion, 38–48 DE; intermediate conversion, 48–58 DE; and high conversion, 58–68 DE. The high-conversion syrups are further classified as acid conversion or acid-enzyme syrups, depending on the method of manufacture. More extensive enzymatic hydrolysis produces high-maltose sweeteners. Maltose is a disaccharide consisting of two glucose molecules.

Corn syrup solids (CSS) impart firmness to finished ice cream, provide an economical source of solids, and improve the shelf life of frozen desserts. Their sweetness comes primarily from dextrose. Dried corn syrup solids are produced by dehydration of corn syrup. Their chemical compositions are identical with those of the syrups from which they are made (Table 5.6). They contain dextrose, maltose, and dextrins (prosugars) but usually no starch. The higher dextrins or saccharides provide their cohesive and adhesive properties. CSS are white granular solids that have a tendency to become lumpy when exposed to moist air. The higher the DE the greater the tendency to cake. Because the dextrins contained in maltodextrins and corn syrup products are relatively larger molecules than are contained in the substances they sometimes replace, these ingredients have a tendency to raise the freezing point of mixes (Table 5.6). Usually, not more than 25–35% of the total sweetener is supplied by corn syrup solids. One reason for the limitation is that processing produces some substances that impart off-flavors and cause changes in color of corn syrups. Refining by passing the syrup over activated carbon removes much of these substances. The newer method of refining by passing the syrup through ion-exchange columns is a more powerful way of removing these impurities. As the industry moves to this type of processing, the limitation on amounts of corn syrup solids can be raised.

Complete hydrolysis of starch produces the monosaccharide glucose that is called corn sugar. Corn sugar or dextrose is a white granular material that contains approximately 99.8% sugar solids. Because it is only about 80% as sweet as sucrose, 1.25 parts of dextrose are required to replace 1 part of sucrose. Dextrose lowers the freezing point nearly twice as much as does sucrose on a weight-for-weight basis, because it has about one-half the molecular weight of sucrose. This effect on the freezing point limits to about 25% of the total sugar the amount of dextrose that can be used in ice cream. Usually dextrose is a more economical source of sweetener than is sucrose. However, corn sugar tends to lump more than does cane or beet sugar when exposed to moist air.

High-fructose corn syrups (HFCS) are made by enzymatically converting glucose of corn syrup to fructose. The most commonly used type is HFCS 42. It contains 42% fructose, 52% dextrose, and 6% higher saccharides. HFCS 90 is a super sweet mixture of 90% fructose, 7% dextrose, and 3% higher saccharides. Compared with sucrose, high-fructose corn syrups (42, 55, and 90%) are from 0.86 to 1.25 times as sweet and lower the freezing point nearly twice as much (Table 5.6).

Malt syrup contains up to 70% maltose. Dried maltose syrup, malt syrup, or malt extract, with properties similar to those imparted by maltose, carry a typical malt flavor and should be used only in products for which such a flavor is desired.

Blends of various sweeteners are offered by the several firms that process corn starch.

Maple and Brown Sugars

Maple sugar and brown sugar contain characteristic flavoring components that limit their use in ice cream. For example, only 6% of maple sugar in the mix will produce a distinct maple flavor. An additional impediment to their use is their comparative high cost.

Both maple and brown sugar are very high in sucrose. Maple sugar contains about 86% sucrose, 10% moisture, and 4% invert sugar; whereas maple syrup contains about 52% sucrose, 45% moisture, and 3% invert sugar.

Honey

Honey is composed of about 74.5% invert sugar, 17.5% moisture, 2% sucrose, 2% dextrin, and 4% miscellaneous matter. It is used in ice cream principally to provide honey flavor. Usually both the desired sweetness and honey flavor will be provided by 9 lb honey and 8 lb sucrose per 100 lb of mix. Honey flavor may blend poorly with other flavors, so the addition of other flavors to honey-flavored ice cream is seldom advisable.

Nonnutritive Sweeteners

Saccharin, discovered in 1879 at Johns Hopkins University, is an organic compound approximately 300 times as sweet as sucrose. It can withstand long periods of storage as well as heat and is the least expensive of the nonnutritive sweeteners. Although in one study large doses of saccharin were shown to cause bladder tumors in second generation rats, several studies and long use have failed to demonstrate a relationship between consumption of saccharin and cancer in humans. The American Medical Association, the American Dietetic Association, and the American Diabetes Association all have issued statements affirming the safe use of saccharin (Council on Scientific Affairs 1985a,b; American Dietetic Association 1987; American Diabetes Association 1987).

Aspartame was discovered in 1965 by a chemist who was attempting to develop an anti-ulcer medicine. It is the methyl ester of two amino acids, L-aspartic acid (aspartate) and L-phenylalanine, both of which occur naturally in foods. Furthermore, it is digested in the same manner as other amino acids. However, it contributes few calories to a serving of ice cream, because only about 0.5% as much of it is needed to sweeten as compared with sucrose (200 times as sweet as sucrose). Aspartame's taste is quite similar to that of sucrose. It enhances some flavors and, in combination with saccharin, results in a sweeter taste than either alone. At high temperature aspartame undergoes hydrolysis and loss of sweetness, so it should be added to ice cream mixes after pasteurization. More than 100 scientific studies were conducted to support the food additive petition made to the U.S. Food and Drug Administration, making it one of the most thoroughly studied food additives in history. Persons with the rare hereditary disease phenylketouria (PKU) must control their intake of phenylalanine from all sources, including aspartame (Newsome 1986). These persons, about 1 in 15,000, are diagnosed at birth by a blood test performed on all newborns. Because aspartame use in foods may introduce a new source of phenylalanine in the diet, products containing this sweetener must carry a warning statement on the label. Aspartame has been declared safe by the American Dietetic Association, the American Diabetes Association, the American Medical Association, and the Epilepsy Institute (American Dietetic Association 1987; American Diabetes Association 1987; Council on Scientific Affairs 1985b; Annonymous 1987).

Approval for use of acesulfame K in dairy product analogs was given by the U.S. FDA in 1988. More than 50 studies conducted internationally were submitted in support of the food additive petition. Acesulfame K is an organic salt, containing sulfur and nitrogen, that is 150 to 200 times as sweet as sucrose (Schlichter and Regan 1990). It is excreted through the human digestive system unchanged and, therefore, is nonnutritive and noncaloric (Anonymous 1988). Acesulfame K does not decompose on heating and has a synergistic sweetening effect with aspartame, cyclamate (a promising but yet unapproved nonnutritive sweetener), and several nutritive sweeteners. A slight aftertaste is sometimes noticed at high concentrations.

Sucralose is the generic name of a new high-intensity, noncaloric sweetener derived from ordinary sugar. It looks and tastes like sugar but is, on the average, 600 times sweeter. It is useful in a wide range of foods and beverages including frozen desserts. Although about one-hundred studies with animals and humans have demonstrated no adverse effects of sucralose, it has yet to be approved by U.S. FDA for use as a food additive in the United States. Canada has approved sucralose for use in frozen desserts and certain other nonstandardized foods at concentrations up to 0.025%. Sucralose must be declared on the principal display panel, along with any other sweetener with which it may be used, and food labels must also bear a statement of calories, protein, fat, and carbohydrate content. In general, the sweetness of sucralose decreases with increased sweetener concentration or pH and with decreased temperature. It remains stable during pasteurization treatment and at pH ranges common to frozen desserts, i.e. from 3–7. It was found on a 7-day study to not interact with common bases, oxidizing or reducing agents, aldehydes, ketones, or metal salts when stored in aqueous solutions at 40°C with pH ranging from 3–7. The request for regulatory approval of sucralose was submitted to U.S. FDA on February 9, 1987.

Fat Replacers

The recently recognized high demand for frozen desserts low in fat, cholesterol, and calories prompted food product researchers to work hard to develop ingredients that can function in many ways as does fat. Decreasing the fat in any food formulation will cause the loss of several properties of the fat, viz., structure, opacity, slip, fatty flavor, mouthcoating, and mouth fullness. However, ingredients have been developed that spare fat in various ways.

Several terms have been used to describe the extent and efficacy of fat replacement. "Fat replacer" is the generic term used to encompass all the related terms. The terms "extender" or "sparer" are used to define partial removal of fat from a particular food and substitution with a replacer ingredient. "Substitute" or "analog" have the characteristics of fats but contain fewer calories or no calories. "Mimetics" replace part of the fat of a product; they imitate a particular function(s) but not all functions of fat in a food.

Many experts classify fat replacers in three categories depending on the materials from which they are made—carbohydrate, protein, or fat. The hydrocolloids are water-soluble carbohydrate polymers. Among them, starches, cellu-

5 ICE CREAM INGREDIENTS

losic products, dextrins, and maltodextrins hold up to 3-plus times their weight in water, whereas some gums hold 100 times their weight.

Protein-based fat replacers are typically derived from whey or egg white. These proteins are usually processed to produce minute particles (0.1- to 3-µm dia., Figure 5.3) that can be perceived by the tongue similarly to fat globules. This process is called microparticulation. The size of the particulates is an important determinant of mouthfeel. The particles, being hydrophobic on their surfaces because of their content of nonpolar amino acids, resist interactions with each other. They can be spray dried and reconstituted without affecting particle integrity. They hold water and disperse well in aqueous systems. Particles of Simplesse®, a commercially microparticulated whey protein, have surfaces that repel fat globules so the spherical particles roll over each other with minimal force. Singer (1992) suggested that the creamy perception imparted by fat globules requires that a substantial amount of substance be sensed in the sample without detection of the individual particles. Bringe and Clark (1993) reported that nonaggregated particles of whey protein with mean diameters in the range of 0.1–2 µm cause a creamy sensation. Protein particles with diameters <0.1 µm leave a suspension watery and greasy, depending on the protein concentration. When the particles are larger than 3 µm in diameter, they impart a gritty or powdery mouthfeel.

Whey protein concentrates are used by at least two manufacturers of fat

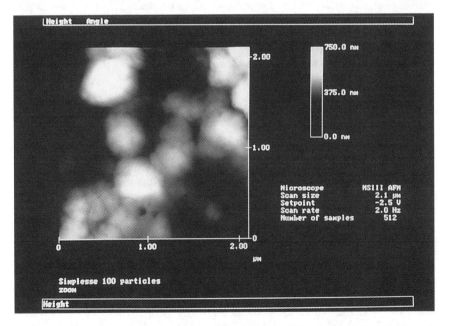

Figure 5.3. Atomic force micrograph of Simplesse® 100 microparticles. Simplesse® was diluted with water and air dried on a glass slide for analysis. (Courtesy Dr. Neal Bringe, St. Louis, MO.)

replacers who do not microparticulate the protein. The processes that cause these products to spare fat depend on the unfolding of the globular protein molecules and their binding to other substances.

The fat-based replacers currently available are food emulsifiers, mostly mono- and diglycerides. These surface active molecules stabilize emulsions by bringing the fat and water into close proximity to each other.

Both carbohydrate- and protein-based fat replacers, as dry ingredients, yield 4 kcal/g. Recall that carbohydrate-based fat replacers bind about 3 times their weight of water. Thus, it is theoretically possible to replace 4 parts of fat with 1 part of fat replacer and the water it binds. The net result: 9 kcal/g of fat are replaced with 1 kcal/g of hydrated fat replacer.

No fat replacer is satisfactory for all applications. Used in combinations, they can compensate for the loss of several functions of fat. Chief among their deficiencies are lack of creaminess and ability to carry fat-soluble flavors.

The maltodextrins, described in a preceding section, have been widely used as bulking agents in reduced fat ice creams.

Gums and cellulosic substances can mimic some of the functions of fat. They contribute to viscosity, to foam stability, and to control of crystal growth and of syneresis. For example, when cellulose gel is dispersed in water, a colloidal dispersion is formed that has an average particle size of 0.21 µm. By combining cellulose gel with guar gum, pliable spheres of 1–10 µm diameter will form in aqueous media (Calorie Control Council, undated). These spheres impart rheological properties of fat. Typical use levels for fat replacement are 1–2%.

Methylcellulose and hydroxymethylcellulose are surface active polymers that form films in solution, then gel upon heating. These molecules have both polar and nonpolar substituents that permit them to reduce interfacial tension between films that are formed between water and nonpolar constituents. Films that are formed assist in imparting creaminess to frozen desserts. Typical use levels are 0.2–0.8%.

Blends of gums have the ability to form microscopic spherical particles that mimic the rheology and mouthfeel characteristics of emulsified fat. These blends my contain guar, locust bean, and xanthan gums, carrageenan, sodium carboxymethyl cellulose, and microcrystalline cellulose. Xanthan gum is a thickener; guar and locust bean gums control crystallization and promote creaminess; carrageenan, a high molecular weight linear polysaccharide, provides structure and mouthfeel. These four components interact synergistically.

Polydextrose, sold as Litesse® by Pfizer (Pfizer Technical Information, undated), is a randomly bonded melt condensation polymer of dextrose and lesser amounts of sorbitol and citric acid. It resists breakdown by the enzymes of the human digestive tract so that it yields just 1 kcal/g. The small degree of digestion that produces this caloric output resides in the bacteria of the intestines. The primary function of polydextrose is as a bulking agent. It replaces the bulk of sugar when sugar is replaced by a high-potency sweetener. Polydextrose itself is not sweet, but it has some fat-sparing properties when used in frozen desserts. It has been approved for such use (21 CFR 172.841). Solutions of polydextrose have higher viscosities than those of sucrose or sorbitol at equivalent concentrations. This higher viscosity contributes to desirable mouthfeel. At

5 ICE CREAM INGREDIENTS

equivalent solids substitution levels, polydextrose combined 60:40 with sorbitol provides the same freezing point depression as does sucrose.

Sugar Alcohols

The group of mono- and disaccharide sugar alcohols, the polyols, include sorbitol, mannitol, isomalt, lactitol, xylitol, maltitol, and some related hydrogenated starch hydrolysates. These substances vary in their relative sweetness, laxation potential, solubility, heat of solution (cooling effect), stability, and cost. Although polyols are considered fully caloric in Canada (4 kcal/g), the European Community recognizes a caloric value of 2.4 kcal/g, as an acceptable average for labeling purposes. In the United States self-determination of the caloric content is permitted, and a major supplier lists the values as 2.6 cal/g for sorbitol, 1.6 cal/g for mannitol, and 2 cal/g for isomalt. Sorbitol and mannitol are stereoisomers of each other, the planar orientation of the OH groups on the second carbon being different.

```
        H                   H
        |                   |
     H-C-OH              H-C-OH
        |                   |
     H-C-OH              HO-C-H
        |                   |
     HO-C-H              HO-C-H
        |                   |
     H-C-OH              H-C-OH
        |                   |
     H-C-OH              H-C-OH
        |                   |
     H-C-OH              H-C-OH
        |                   |
        H                   H

     Sorbitol            Mannitol
```

These sugar alcohols are found naturally in fruit juices of apples, pears, cherries, and plums. Mannitol is the major constituent of manna, an exudate of the ash tree, or as a component of mushrooms and marine algae. Sorbitol and mannitol are produced by hydrogenation of glucose and fructose, respectively. Both weigh the same, but sorbitol is 0.6 times as sweet while mannitol is 0.5 times as sweet as sucrose. Both are generally recognized as safe (GRAS) in the United States and are approved for use in Canada. The laxation thresholds, in g/day, are considered to be 50 for sorbitol and 20 for mannitol.

During digestion, small amounts of sorbitol and mannitol are slowly absorbed through the wall of the small intestine and metabolized by the liver. However, most of the utilization occurs in the colon, where the polyols are fermented by bacteria that convert them to volatile fatty acids. Because they enter the glycogenolytic pathways without dependence on insulin, polyols do not cause appreciable increases in blood glucose levels when eaten.

Lactitol is the product of hydrogenation of lactose. It has 0.3 to 0.4 times the sweetness of sucrose. A GRAS affirmation petition has been filed.

Isomalt, known by the trade name Palatinit®, forms when sucrose is enzymatically rearranged to isomaltulose and the latter is chemically hydrogenated (PALATINIT® Süßungsmittel GmbH, 1990). Sorbitol and mannitol are equimolar building blocks of isomalt. It is approximately one-half as sweet as sucrose and is noncariogenic but tends to crystallize because of its low solubility in water (25% at 20°C). Isomalt is approved for use in more than 15 countries of Europe, and a GRAS affirmation petition was filed in the United States in 1990. The caloric content has been determined as 2 cal/g by industry, but Canadian regulations specify a value of 4 cal/g.

Maltitol is the product of hydrogenation of maltose. It has been used in foods in Japan, where it was developed in 1960. Among the sugar alcohols, it is comparatively sweet; 0.8 to 0.9 times the sweetness of sucrose. Its other major physical constants mimic those of sucrose, but the laxation threshold is 20 g/day. Polyols function in frozen desserts as bodying/bulking agents, sweeteners, and crystallization inhibitors.

Syrups

Syrups or liquid sweeteners provide the convenience of handling in closed systems that can be controlled with computers and in-line metering devices. The total solids content of sucrose syrups is usually measured by "degree Brix," which assumes all the solids to be sucrose. This is a safe assumption for practical purposes. Most other syrups are labeled with a Baumé number. This number is based on the specific gravity of the syrup and, therefore, is also a measure of the TS concentration. However it does not reflect the kinds or amounts of the sugars and dextrins present. Baumé and Brix are related for sucrose syrups as shown in Table 5.7. The Baumé number must be converted

Table 5.7. Relation of Baumé Reading to Brix Reading (Sucrose Syrup)

Degrees Baumé	Degrees Brix	Weight (lb/gal)	Sugar (lb/gal)	Water[a] (lb/gal)
33.0	61.0	10.78	6.58	4.20
33.5	62.0	10.83	6.71	4.12
34.0	63.0	10.88	6.85	4.03
34.5	63.9	10.92	6.98	3.94
35.0	64.9	10.97	7.12	3.85
35.5	65.9	11.02	7.26	3.76
36.0	66.9	11.07	7.41	3.66
36.5	67.9	11.13	7.56	3.57
37.0	68.9	11.18	7.70	3.48
37.5	68.9	11.23	7.85	3.38
38.0	70.9	11.28	8.00	3.28
38.5	71.9	11.33	8.15	3.18
39.0	72.9	11.39	8.30	3.09
39.5	73.9	11.44	8.45	2.99
40.0	74.9	11.49	8.61	2.88

[a]One U.S. gallon of water at 68°F weighs 8.322 lb.

5 ICE CREAM INGREDIENTS

Table 5.8. Approximate Physical Constants of Corn Syrups

Baumé	Specific gravity	Solids (%)	Weight (lb/gal)	Solids (lb/gal)
Low-conversion corn syrup—32 DE				
42	1.4049	77.64	11.700	9.084
43	1.4184	79.59	11.813	9.402
44	1.4322	81.53	11.928	9.725
45	1.4463	83.51	12.045	10.059
46	1.4605	85.49	12.163	10.389
Regular-conversion corn syrup—42 DE				
42	1.4049	78.30	11.700	9.161
43	1.4184	80.27	11.813	9.482
44	1.4322	82.25	11.928	9.811
45	1.4463	84.25	12.045	10.148
46	1.4605	86.26	12.163	10.492
Intermediate-conversion corn syrup—52 DE				
42	1.4049	78.94	11.700	9.236
43	1.4184	80.94	11.813	9.561
44	1.4322	82.96	11.928	9.895
45	1.4463	84.98	12.045	10.236
46	1.4605	87.03	12.163	10.586
High-conversion corn syrup—62 DE				
42	1.4049	79.59	11.700	9.312
43	1.4184	81.62	11.813	9.642
44	1.4322	83.67	11.928	9.980
45	1.4463	85.72	12.045	10.325
46	1.4605	87.80	12.163	10.679

to specific gravity before TS can be calculated (Table 5.8). Specific gravity is the ratio of the density of a liquid to the density of water at a given temperature, which for syrups is 39°C. It is given by $145/(145-d)$ where d is the Baumé reading at 68°F.

REFERENCES

American Dairy Products Institute. 1990. Standards for Grades Including Methods of Analysis. Bulletin 916 Revised. American Dairy Products Institute. Chicago, IL.
American Diabetes Association. 1987. Council on nutrition sciences and metabolism: Use of noncaloric sweeteners. *Diabetes Care.* 10:526
American Dietetic Association. 1987. Appropriate use of nutritive and non-nutritive sweeteners. *J. Am. Dietetic Assoc.* 87:1689–1694.
Anonymous. 1987. Seizures and aspartame: No connection. *NY J. Epilepsy* 5(1):1,9,13.
Anonymous. 1988. FDA clears Hoecht's non-caloric sweetener for use in dry foods. *Food Technol.* 42:108.
Breeding, C. J. 1993. Ice cream attributes affected by milk fat fractionation and dairy cow diets. Ph. D. dissertation. University of Missouri, Columbia. December.
Bringe, N. A., and D. R. Clarke. 1993. Simplesse®:Formation and properties of microparticulated whey protein. Chapter 5 In *Science for the Food Industry for the 21st Century.* M. Yalpani, ed. ATL Press.
Calorie Control Council. Undated. Fat Reduction in Foods. 110 pages. Calorie Control Council, Atlanta, GA.
Council on Scientific Affairs. 1985a. Saccharin, review of safety issues. *JAMA.* 254:2622–2624.

Council on Scientific Affairs. 1985b. Aspartame, review of safety issues. *JAMA.* 254:400–402.

Dahlberg, A. C., and E. S. Penczek. 1941. The relative sweetness of sugars as affected by concentration. New York State AES Bull. 258.

Geilman, W. G., and D. E. Schmidt. 1992. Physical characteristics of frozen desserts made from ultrafiltered milk and various carbohydrates. *J. Dairy Sci.* 75:2670–2675.

Hofi, M. A. 1989. The use of ultrafiltration in ice cream making. *Egypt. J. Dairy Sci.* 17:27.

Im, J. S., R. T. Marshall, and H. Heymann. 1994. Frozen dessert attribute changes with increased amounts of unsaturated fatty acids. *J. Food Sci.* 59(6):1222–1226.

Kosikowski, F. V., and A. R. Masters. 1983. Preparation of ice cream, skim milk and cream made from whole milk retentates. *J. Dairy Sci.* 66 (Suppl. 1):99. Abstr.

Leighton, A. 1942. A method of saving sugar in the manufacture of ice cream. *Ice Cream Trade J.* 38(9):12, 32–34.

Newsome, R. L. (ed) 1986. Sweeteners: Nutritive and non-nutritive. *Food Technol.* 39(8).

PALATINIT® Süßungsmittel. 1990. Isomalt. 3rd ed. PALATINIT® Süßungsmittel GmbH, Pfalz, Germany.

Schlichter, S. A., and C. Regan. 1990. Innovations in reduced-calorie foods: A review of fat and sugar replacement technology. *Topics in Clin. Nutr.* 6(1):50–60.

Singer, N. S. 1992. Advanced Food Ingredients Symposium. Rutgers University, New Brunswick, NJ. March 11.

Tharp, B. W. 1982. Use of freezing point calculations in evaluating dry mix compositions. *J. Dairy Sci.* 65 (Suppl. 1):19.

Tong, P. S., L. A. Jensen, and L. Harris. 1989. Characteristics of frozen desserts containing retentate from ultrafiltration of skim milk. II. Some physical properties. *J. Dairy Sci.* 72 (Suppl. 1):129. Abstr.

Tracy, P. H., and A. J. Hahn. 1938. A comparison of concentrated and superheated skim milk in the manufacture of ice cream. Dairy Mfg. Conf. Manual, Univ. of Illinois, Chicago.

6
Stabilizers and Emulsifiers

In 1915 Frandsen used the word *stabilizer* to designate a group of substances that at that time were known as holders, colloids, binders, and fillers (Frandsen and Markham 1915). The primary purposes of using stabilizers in ice cream are to produce smoothness in body and texture, retard or reduce ice crystal growth during storage, provide uniformity of product, and increase resistance to melting. Stabilizers function through their abilities to either form gel structures in water or to combine with water as water of hydration, thus increasing viscosity. Gels trap other particles within the three-dimensional network. Stabilizers increase viscosity also because they interact with other macromolecules in the frozen product. As ice freezes out of solution in frozen desserts, stabilizer concentration, which may be only 0.25–1.0% in the unfrozen mix, increases many-fold in the unfrozen milieu.

An emulsifier is a substance that produces a stable suspension of two liquids that do not mix naturally. The main function of an emulsifier in the manufacture of ice cream is to produce a dry, stiff and smooth product.

Excellent ice cream can be made, and considerable amounts are made, without the use of added stabilizer or emulsifier. Because milk and milk products contain natural stabilizing and emulsifying materials (proteins, phospholipids, phosphates, and citrates), mixes of certain composition that are given selected treatment produce ice cream of excellent quality.

USES OF STABILIZERS

The water of ice cream is never completely frozen. Furthermore, as temperatures of frozen product rise and fall, ice crystals melt then refreeze. This fluctuation in temperature causes undesirable textural changes. Stabilizers bind some of the water, thereby reducing the amount available to participate in the phase changes from ice to water and water to ice. However, Buyong and Fennema (1988) concluded that hydrocolloids (CMC, carrageenan, gelatin, guar gum, locust bean gum, and sodium alginate), at a concentration normally used in frozen desserts, alter only slightly (less than 2% as compared to pure water) the amount of ice that forms in hydrocolloid-water solutions during freezing to –43°C. Their experiment with ice cream involved only gelatin as a stabilizer.

They suggested that the desirable effects of stabilizers on the sensory properties of ice cream result from their abilities to alter the surface properties of ice crystals or to alter the perception of ice crystals in the mouth.

The amount and kind of stabilizer needed in ice cream varies with the mix composition; ingredients used; processing times, temperatures, and pressures; storage temperature and time; and many other factors. Usually 0.1–0.5% stabilizer is used in the ice cream mix.

Mixes high in fat or total solids (40%), chocolate mixes, or ultrahigh-temperature (104°C or above) pasteurized mixes require less stabilizer than do mixes that are low in total solids (37%), are HTST-pasteurized, or are to be stored for extended periods of time. Nonfat products require up to about 1.0%.

KINDS OF STABILIZERS

The hydrocolloids used as stabilizers in frozen desserts fall into the following categories (frequently used ones in **bold** type):

Proteins	gelatin
Plant exudates	arabic, ghatti, karaya, and tragacanth gums
Seed gums	**locust (carob) bean, guar**, psyllium
Microbial gums	xanthan
Seaweed extracts	agar, **alginates, carrageenan**
Pectins	**low and high methoxyl**
Cellulosics	**sodium carboxymethylcellulose, microcrystalline cellulose**, methyl and methylethyl celluloses, hydroxypropyl and hydroxypropylmethyl celluloses

Functionality of these hydrocolloids, all of which are polysaccharides except gelatin, varies and can be modified by changing the chemical structure of the natural forms. Individual hydrocolloids, regardless of type, seldom perform all of the desired functions. Each has a peculiar effect on body, texture, meltdown, and stability in storage. Therefore, to gain synergism in function, individual substances are usually combined as mixtures of stabilizers and emulsifiers. Components selected vary with the composition of the mixes and the outcomes expected from their use, including cost. Much experimentation is involved in determining the right combination and concentrations of the several available hydrocolloids to perform the functions desired for a given formula and market niche. Many professionals are involved in this type of work.

CHARACTERISTICS OF INDIVIDUAL STABILIZER INGREDIENTS

Agar: Must be dissolved in boiling water; helps provide gel structure; forms a firm and brittle gel that is heat reversable; not widely used in ice cream; use rate of 0.1–0.5%. Expensive.

Alginates: Dissolve in cold water; calcium of milk causes precipitation unless the product is hot or unless calcium in the mix is chelated with phosphates or citrates; thickens and gels in the presence of calcium and acid; reactive with

6 STABILIZERS AND EMULSIFIERS

certain carrageenans; use rate 0.1–0.5%; not widely used in ice creams. Expensive.

Carboxymethyl cellulose (CMC): known also as **cellulose gum**; easily dissolved in the mix; has a high water-holding capacity; forms weak gels by itself but gels well in combination with carrageenan, locust bean gum, or guar gum; acts also as an emulsifier; use rate of 0.1–0.2%; especially functional in sherbets and ices. Moderate cost.

Carrageenans: Known also as **Irish moss**; three forms are classified by the amount of sulfate in the molecules—lambda has the most sulfate, is soluble in cold milk, and forms weak gels; iota is soluble in hot (55°C) milk and forms elastic gels that resist syneresis; kappa dissolves in hot milk (<70°C) forming brittle gels that do not resist syneresis (Figure 6.1). They create low viscosities in hot mixes because they are individual molecules, but as mixes cool structures form that increase the viscosity; gel strength is increased by addition of potassium ions. Kappa functions well with locust bean gum to provide an elastic and cohesive gel without syneresis. These extracts of red seaweed are not found in pure form. Instead they are marketed as mixtures in which the dominant fraction determines the classification. Kappa and iota carrageenans react electrostatically with casein to form a three-dimensional network (Figure 6.2) that improves resistance to separation of the suspended phase in ice cream mixes stored for several days. Usage rate 0.02–0.15%. Expensive.

Cellulose gel: Known also as **microcrystalline cellulose**; tiny particles of purified cellulose, a polymer of glucose; binds water; partial resemblence of fat in the mouth as particles coat tongue but do not lubricate. Use in nonfat products at 0.2–0.8%. Moderate cost.

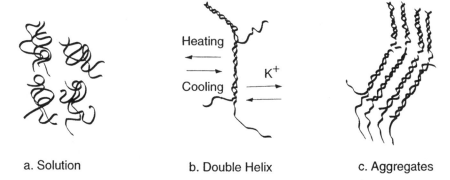

a. Solution   b. Double Helix   c. Aggregates

Figure 6.1. Theory of the mechanism of gel formation by kappa-carrageenan as affected by temperature and potassium ions. (a) Kappa-carrageenan is dissolved in an aqueous medium by heating. (b) It produces a rigid and brittle gel on cooling. (c) Cations, especially potassium, neutralize the sulfated units on the outsides of the coils allowing aggregates to form and the gel to shrink. (After Rizzotti *et al.* 1984.)

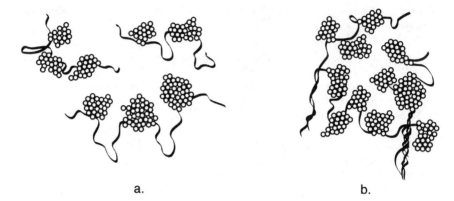

Figure 6.2. The interaction of kappa-carrageenan with casein. (Model adapted from Snoeren *et al.* 1975.)

Guar gum: Readily soluble in cold mixes, therefore favored for HTST-pasteurized mixes; may help produce excessive viscosity even at low concentrations; types available that develop maximal viscosities in from 2 to 20 hours; viscosity varies inversely with temperature; nonionic nature and straight chain structure enhance hydrogen bonding to other hydrophobic molecules. The polysaccharide polymer is composed of repeating mannose units with single galactose attached to alternate mannose units (Figure 6.3). Inexpensive.

Locust bean gum: Known also as **carob bean gum**; needs high temperature to dissolve; a nonionic galactomannan that is virtually unaffected by pH in frozen desserts; high capacity to hydrate provides high viscosity without gelling;

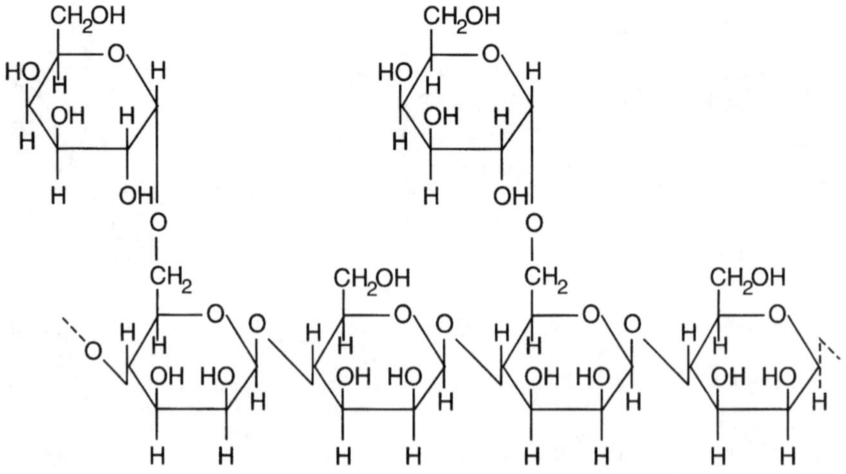

Figure 6.3. Basic repeating structure of guar gum, a galactomannan.

tends to cause milk proteins to precipitate unless used in conjunction with carrageenan; useful for restricting whipping in sherbets. Use rate 0.05–0.3% in ice creams; 0.25–0.5% in sherbets and ices. Expensive.

ICE CREAM IMPROVERS

The early ice cream textbooks and research papers report that products containing gums and rennet extract were offered to ice cream manufacturers under the general name of ice cream improvers. The usual conclusion of these reports was that the improvers failed to consistently impart the desired body and texture characteristics to ice cream products. In particular, stored ice creams tended to shrink when the process had included treatment with rennet extract. Recently, Chang et al. (1995) reported on experiments in which casein in ice cream mix was partially hydrolyzed with dilute chymosin (genetically engineered rennin) to affect partial aggregation of casein micelles. In the batch process chymosin was added to light ice cream (2% fat) mix as the temperature reached 60°C on the way to pasteurization temperature of 70°C. The 20 min that were allowed for the chymosin to act before it was completely inactivated at 65°C caused enough hydrolysis of the kappa-casein to increase the viscosity of the mix from 60 to 1,300 cP. Texture was significantly smoother and heat shock stability improved by the treatment. Research continues on adaptation of the process to HTST-pasteurized mixes. No shrinkage of the product was observed during storage at –20°C for three months. It is believed that shrinkage observed previously resulted from continued casein hydrolysis in the frozen products.

EMULSIFIERS

Emulsifiers are substances that concentrate and orient themselves in the interface between the fat and the plasma and reduce the interfacial tension of an immiscible system. They are amphiphilic substances, which, according to their chemical structures, possess both hydrophilic and lipophilic properties (Krog 1977). This interfacial adsorption is energetically more favorable than complete solution of the emulsifier in either the aqueous or lipid phase.

The natural emulsifiers of milk, cream, and most ice cream mixes are sufficient to enrobe and stabilize the fat globules. The principal function of added emulsifiers in ice cream is to destabilize the globular fat, resulting in a dry and stiff appearance of ice cream as it is extruded from the freezer. As interfacial tension produced by emulsifiers decreases, the destabilizing activity on fat globules increases.

Two major types of emulsifiers are used in ice cream: (1) mono- and diglycerides and (2) polyoxyethylene derivatives of hexahydric alcohols (usually sorbitol), glycol and glycol esters. The roles of the intrinsic and added emulsifiers have been the subjects of many studies, which have led us to understand the functions as follows.

Ice cream mixes are almost always homogenized. The purpose is to reduce fat globule sizes, the object of which is to impart smoothness to the ice cream. This process produces naked fat globules at the homogenizing valve. Immediately after new globules are produced, new membranes begin to form. The number of globules increases markedly as does the surface area. Therefore, a large amount of material is needed to adsorb to and stabilize the newly formed globules. The molecules that immediately adsorb are those in the microenvironment of the globules. It is probable that the subunits of casein micelles are spread by surface forces over many of the globules. Berger and White (1971) observed a layer of casein subunits, one unit thick (10 nm), surrounding some globules in freeze-fractured specimens of ice cream containing 10.5% NMS and 0.4% emulsifier.

As a homogenized mix is cooled then aged at about 4°C, the components of the membranes rearrange. Emulsifiers are adsorbed more strongly than are proteins, and they tend to replace proteins in the final membrane (Figure 6.4).

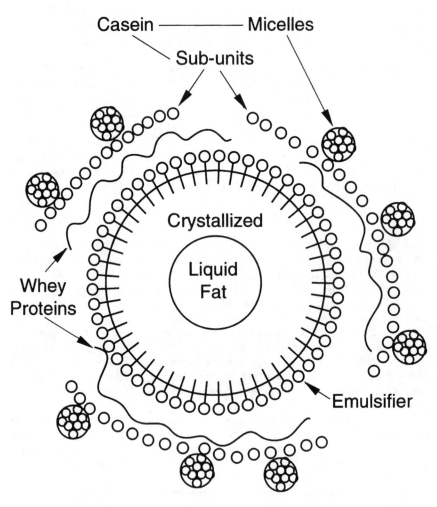

Figure 6.4. Illustration of the theory of how fat globules are emulsified after homogenization of an ice cream mix. (Adapted from S. Groven. 1987. Used by permission. Danisco, New Century, KS.)

This reformation lowers the net free energy of the emulsion and the amounts of casein adsorbed per unit of globular surface area (Barford and Krog 1987; Goff et al. 1987; Goff and Jordan 1989; Oortwijn and Walstra 1979). Thus, fat globules with adsorbed emulsifiers have an increased susceptibility to destabilization forces that exist in the ice cream freezer. However, addition of too much emulsifier reverses the situation and tends to stabilize the emulsion (Tong et al. 1989). About 60% of the fat is destabilized in obtaining optimal structure and texture in regular ice cream.

Polyoxyethylene sorbitan forms of emulsifiers (the Tweens) are stronger destabilizers than are the sorbitan forms (the Spans), and the latter are stronger than the mono- and diglyceride forms (Table 6.1). Among the monoglycerides, those containing short chain fatty acids are more active destabilizers than are those with long chains. Furthermore, Goff and Jordan (1989) demonstrated that both ice crystallization and agitation, as occurs in commercial ice cream freezers, are necessary for destabilization to occur. Strong churning conditions without ice crystallization destabilized little of the fat at $-4°C$. Furthermore, freezing quiescently failed to destabilize the fat in representative ice cream mixes containing 0.08% of (1) glycerol monostearate (GMS), (2) 40% mono-/60% diglyceride, or (3) polyoxyethylene sorbitan monooleate (Tween 80). They concluded that shear forces exerted by the ice crystals are a major factor in fat globule destabilization. Other contributing factors may be increases in viscosity and salt concentration that occur as water is removed from the serum by freezing.

Aging of ice cream mixes at $2-5°C$ is necessary to permit adequate hydration of certain added colloids and partial crystallization of the fat (Berger et al. 1972a). Crystals in the fat globules facilitate rupture and clumping of the globules (Darling 1982). On cooling, high melting glycerides form concentric crystalline layers at the innermost side of the fat globule membrane, and liquid fat occupies the core of the globule. The solid:liquid ratio increases as temperature decreases, but crystallization requires time. On rupture of globule membranes the liquid fat released becomes "cement" for crystals, and chains

Table 6.1. Pairs of Emulsifiers Chosen to Illustrate Effects of Structure on Hydrophilic-Lipophilic Balance (HLB) and on Interfacial Tension (IF) in a Model Two-Phase Emulsion[a]

Common name	Chemical name	HLB (dynes/cm)	IF[b]
GMS	Glycerol monostearate	3.8	5.52
bGMO	Glycerol monooleate	2.8	5.09
Span 60	orbitan monostearate	4.7	5.64
Span 80	Sorbitan monooleate	4.3	5.02
Tween 60	Polyoxyethylene sorbitan monostearate	14.9	2.42
Tween 80	Polyoxyethylene sorbitan monooleate	15.0	2.24

Adapted from Goff and Jordan, 1989.
[a]The two phases were 11% NMS and anhydrous butter oil. Without emulsifier the interfacial tension was 6.16.
[b]Interfacial tensions determined by the duNouy ring method at 70 C when emulsifiers were dispersed in the most soluble phase.

of them coalesce. Unbroken globules also stick together in clumps or clusters (Berger et al. 1972b). Trapped within this network are serum and air cells. As air is incorporated into an ice cream emulsion, the large differences in interfacial tension that exist between the air-serum interface and the fat-serum interface probably cause proteins to move from fat-serum interfaces to surfaces of air bubbles. In this dynamic system air bubbles continually form, collapse, and reform because of shear forces. Thus, desorption of protein and spreading of fat at the air-serum interfaces can lead eventually to complete churning of the fat. If allowed to proceed too far, churning can result in a product that imparts a greasy mouthfeel and has visible butter granules.

When polymers of amino acids (peptides and proteins) contact an interface with fat, their hydrophobic (lipophilic) regions adsorb and their hydrophilic regions extend into the serum phase. Protein molecules that adsorb find alternative ways of folding to minimize the hydrophobic free energy of the system (Phillips 1981). Emulsion stability is imparted, the extent of which is determined by the proportions of the adsorbed segments that protrude into the two phases (Friberg 1976). Halling (1981) referred to the hydrophilic regions of proteins adsorbed to fat globules as "hairy regions." When two so-emulsified fat globules approach each other, the hairy regions interact through interpenetration and compression if the distance is sufficiently small. Depending on the relative strengths of van der Waals attraction and steric repulsion, flocculation or stabilization will occur.

The amphiphilic proteins in ice cream mixes are the caseins and whey proteins. Both are adsorbed in the newly formed membrane following homogenization of ice cream mixes (Goff et al. 1989). The layer of milk protein on fat globules varies in thickness with the kind of protein in the aqueous phase at homogenization. When Oortwijn and Walstra (1979) emulsified milkfat in skim milk, the "surface excess" (amount of protein adsorbed per m²) was 20 mg/m²; whereas, surface excesses formed on the emulsification of milkfat in colloidal casein or in dissolved whey proteins were 40 mg/m² and 1 mg/m², respectively. Goff and Jordan (1989) extended these observations by withholding either skim milk proteins (as NDM) or whey proteins (as 75% whey protein concentrate, WPC) from ice cream mixes until after homogenization of the remainder of the mixes. Protein concentration was 2.8% in the finished mixes. After 15 min of freezing to –5°C in a batch freezer, amounts of fat destabilized were approximately 30% and 45% in products in which NDM and WPC, respectively, were withheld during homogenization. Additional experiments verified this principle with the conclusion that addition of whey proteins to mixes, especially in the form of whey protein isolates (95% protein), increases significantly the rate of destabilization of fat during freezing. Thus, emulsifier can be spared or even eliminated with the addition of about 3% WPI and 8% NMS. A potential problem with such a mix is excessive viscosity if the whey proteins are fully denatured by heat. In the same experiments addition of sodium caseinate to mixes improved whippability markedly while producing highly stable emulsions. Percentages of fat destabilized by freezing were 75–80% when 4.8% of the NMS (11% total) was WPI, but only about one-half this amount when a mixture of 3% WPI and 1.8% sodium caseinate was added. Thus, sodium caseinate was a powerful milkfat emulsifier. Goff et al. (1989) concluded that if mix viscosity can be

6 STABILIZERS AND EMULSIFIERS

controlled, WPI may replace all or a fraction of the ice cream stabilizer and emulsifier.

It has long been known that lecithin is a highly efficient emulsifier. In fact, milkfat is naturally emulsified with lecithin. Added egg yolk solids provide large amounts of lecithin. The unique smoothness of custard type ice creams can be attributed at least partially to the large amount of lecithin in the egg yolks. About 0.5% egg yolk solids is needed to produce noticeable effects, and 1.4% egg yolk solids is required in French vanilla or custard ice cream products.

Properly formulated and used at optimal concentrations, emulsifiers can be expected to perform the following functions:

- Improve dispersal of fat, especially in mixes made with butter or vegetable oil
- Control agglomeration of fat globules and coalesence of free fat
- Aid in whipping or incorporation of air
- Impart dryness to the extruded product
- Enhance smoothness of texture
- Increase resistance to shrinkage
- Increase resistance to melting.

INDUSTRIAL USAGE

Most commercial stabilizers are formulated by specialized firms with ingredients purchased from primary suppliers. They usually are combinations of stabilizers and emulsifiers but are referred to in short as stabilizers. The most frequently used ingredients in mixtures for regular ice cream are guar and locust bean gums, carrageenan, cellulose gum, mono/diglycerides, and polysorbate 80. The higher the fat content, the less stabilizer is needed. Blends of stabilizer for mixes containing 0–5% fat commonly contain both cellulose gel and cellulose gel combined with other ingredients, as shown in the two representative formulas that follow:

Blend 1	*Blend 2*
Mono/diglycerides	Cellulose gel
Cellulose gel	Whey protein concentrate
Cellulose gum	Mono/diglycerides
Locust bean gum	Modified food starch
	carrageenan
	Polysorbate 80

The use concentration recommended for these blends in nonfat and lowfat ice creams is 0.8–1.0%.

Industry has made many improvements in the stabilization and emulsification of ice cream and other frozen desserts since gelatin was the primary stabilizing ingredient. Blends are formulated to produce specific body and texture characteristics, to work best in mixes with high or low solids content, to partially substitute for fat, to disperse readily, to dissolve at selected temperatures, to minimize dust formation, and to be priced satisfactorily for the intended product. Food scientists working with the primary and secondary sources of these important ingredients will continue to contribute to their successful use and to the high quality of the frozen desserts that consumers demand.

REFERENCES

Barford, N. M., and N. Krog. 1987. Destabilization and fat crystallization of whippable emulsions (toppings) studied by pulsed NMR. *J. Am. Oil Chem. Soc.* 64(1):112.

Berger, K. G., B. K. Bullimore, G. W. White, and R. C. Wright. 1972a. The structure of ice cream by use of the electron microscope. Part I. *Dairy Indus.* 37:419–425.

Berger, K. G., B. K. Bullimore, G. W. White, and R. C. Wright. 1972b. The structure of ice cream by use of the electron microscope. Part II. *Dairy Indus.* 37:493–497.

Berger, K. G., and G. W. White. 1971. An electron microscopic investigation of fat destabilization in ice cream. *J. Food Technol.* 6:285–294.

Buyong, N., and O. Fennema. 1988. Amount and size of ice crystals in frozen samples as influenced by hydrocolloids. *J. Dairy Sci.* 71:2630–2639.

Chang, J-L., R. T. Marshall, and H. Heymann. 1995. Casein micelles partially hydrolyzed by chymosin to modify the texture of lowfat ice cream. *J. Dairy Sci.* 78:2617–2623.

Darling, D. F. 1982. Recent advances in the destabilization of fat emulsions. J. Dairy Res. 49:695.

Frandsen, J. H., and E. A. Markham. 1915. *The Manufacture of Ice Cream and Ices.* Orange Judd Publ. Co., New York.

Friberg, S. 1976. Emulsion stability. Chapter 1. In *Food Emulsions.* S. Friberg, ed. Marcel Dekker. New York.

Goff, H. D., M. Liboff, W. K. Jordan, and J. E. Kinsella. 1987. The effects of polysorbate 80 on the fat emulsion in ice cream mix: Evidence from transmission electron microscopy studies. *Food Microstruct.* 6:193.

Goff, H. D., J. E. Kinsella, and W. K. Jordan. 1989. Influence of various milk protein isolates on ice cream emulsion stability. *J. Dairy Sci.* 72:385–397.

Goff, H. D., and W. K. Jordan. 1989. Action of emulsifiers in promoting fat destabilization during the manufacture of ice cream. *J. Dairy Sci.* 72:18–29.

Halling, P. J. 1981. Protein stabilized foams and emulsions. *CRC Crit. Rev. Food Sci. Nutr.* 15:155.

Krog, N. 1977. Functions of emulsifiers in food systems. *J. Am. Oil Chem. Soc.* 54:124–131.

Oortwijn, H. and P. Walstra. 1979. The membranes of recombined fat globules. 2. Composition. *Neth. Milk Dairy J.* 33:134.

Phillips, M. C. 1981. Protein conformation at liquid interfaces and its role in stabilizing emulsions and foams. *Food Technol.* 35:50–57.

Rizzotti, R., G. Tilly, and R. A. Patterson. 1984. The use of hydrocolloids in the dairy industry. In G. O. Phillips. D. J. Wedlock, and P. A. Williams (eds.). *Gums and Stabilizers for the Food Industry #2.* Pergamon Press, Ltd. Heading Hill Hall, Oxford, England.

Snoeren, T. H. M., T. A. J. Sayens, J. Jenuink, and P. Both. 1975. Electrostatic interaction between kappa-carrageenan and kappa-casein. *Milchwissenschaft.* 30(7):393–396.

Tong, P. S., L. A. Jensen, and L. Harris. 1989. Characteristics of frozen desserts containing retentate from ultrafiltration of skim milk. II. Some physical properties. *J. Dairy Sci.* 72(Suppl. 1):129. Abstr.

7
Flavoring and Coloring Materials

Frozen desserts are valued mainly for their pleasing flavor and their cooling and refreshing effects. The many kinds of flavoring material and the many brands and market categories under which they are sold make it useful to understand their sources and to select and buy with great care. Among the important flavoring substances for frozen desserts are vanilla, chocolate and cocoa, fruits and fruit extracts, nuts, spices, and sweeteners.

American ice creams are known for their wide range of flavors and various flavor combinations that include a variety of confections, baked goods, liqueurs, and other materials. Some manufacturers have flavor lists consisting of as many as 500 or more flavor formulas. Flavor preferences change with time; some recent preferences are shown in Table 7.1. Vanilla and fruit flavors head the list.

Flavor is considered to have two important characteristics: type and intensity. Generally, the delicate flavors are easily blended and tend not to be objectionable at high concentrations. Harsh flavors tend to be objectionable, even at low concentrations. In any case flavors should be only intense enough to be easily recognized and to present a delicate, pleasing taste.

Table 7.1 Consumer Choices of Flavors of Selected Frozen Desserts, Percentage of Total U.S. Volume

FLAVOR GROUPS	Ice Cream[1]	Lowfat Ice Cream[2]	Frozen Yogurt[3]
Vanilla	28.1	25.3	28.4
Fruit	15.2	8.9	34.4
Chocolate	8.2	16.1	17.5
Nuts	13.5	15.4	-
Candy Mix-ins	12.5	15.7	10.6
Cake/Cookie	7.4	15.0	5.1
Neopolitan	6.9	0.7	2.6
Coffee/Mocha	2.7	2.9	-
All Other Flavors	5.5	-	1.4
	100.0	100.0	100.0

Source: International Ice Cream Association, from IRI data
[1]Regular, premium, and superpremium
[2]Premium
[3]Premium

Choosing a mix composition and the ingredients is often less of a problem for the manufacturer than is standardization of flavoring material for several reasons:

- The many flavoring materials available make it difficult to make a proper choice.
- The supply of flavors may vary from time to time in quality and availability.
- Serving conditions affect how pronounced a flavor will be perceived.
- No two consumers have exactly the same sense of taste, so choice of flavor varies widely.

Because flavor is so important in influencing consumer acceptance, it is easy to lose sales when a product is poorly flavored. Defects associated with flavor additions include too much or too little flavoring, unnatural or atypical flavoring, and lack or excess of sweetness. Flavorings also can affect appearance: lack or excess of particles, particles too large or too small, uneven distribution of particles or variegate, ribbon that is too thick or thin, and wrong ingredient or color.

Federal Standards of Identity divide flavoring materials and the labels of frozen desserts into three categories:

I   Pure extracts and flavors
II Pure extracts and flavors that dominate over synthetic component(s)
III Artificial flavor that predominates over natural flavor component(s)

Economy brands of ice cream commonly contain predominantly artificial flavoring as well as lower solids content than the average trade brand. The latter commonly contains pure and artificial flavor with pure flavoring predominating. Premium and superpremium products contain only pure extracts and flavors to complement the relatively high content of dairy solids and the very high qualities of all of the ingredients. The usage rate of flavorings is usually high in premium and superpremium products.

Establishing favorable product acceptance by consumers is vital to increasing product sales. The use of regular flavor evaluation clinics and consumer taste panels are good ways to help achieve this objective.

FLAVORS FOR FROZEN DESSERTS

An abundant variety of flavoring substances has been provided by nature, and flavor chemists and flavorists have attempted to duplicate and expand this variety. Natural flavoring substances are sometimes limited in supply, but the chemically produced types have been made available in almost unlimited quantities and at relatively low cost.

Flavorings of naturally and chemically produced origins are available mainly in mixtures for the proper flavoring of foods. Natural flavorings useful in frozen desserts derive from citrus and noncitrus fruits, tropical fruit, sugar-free fruit, natural flavorings from botanicals, spices, cocoa and chocolate, coffee, natural flavorings from vanilla beans, and nuts. The Code of Federal Regulations,

7 FLAVORING AND COLORING MATERIALS

paragraph 101.22 of Title 21, defines "natural flavor" or "natural flavoring" as the essential oil, oleoresin, essence or extractive, protein hydrolysate, distillate, or any product of roasting, heating, or enzymolysis that contains the flavoring constituents derived from a series of materials, which it then lists. The synthetic flavorings include aromatic chemicals and artificial flavors. Liqueur flavorings include alcohol, whiskey and distilled beverages, fruit brandy distillate, brandy flavor essence, and fruit liqueurs.

The delicate mild flavors imparted by natural flavoring materials usually produce a pleasing flavor even at high concentrations, whereas the overuse of imitation flavorings usually results in a harsh, objectionable flavor.

In addition to the great importance of flavoring in determining consumer preferences for ice creams, there are several other factors. Important among them are: mix composition, processing methods, serving temperature, and color hue and intensity.

A series of studies by Reid and Arbuckle (1938) conducted with sherbets of different flavors and ice creams of differing compositions showed that the majority of consumers enjoyed ice cream best when served at −12°C (8°F). As the serving temperature was increased from −14.4°C to −7.8°C (6°F to 18°F), vanilla flavor became more pronounced. The ice creams also became sweeter as serving temperature was raised. Sherbets and ice creams of high sugar concentrations or pronounced flavor were liked best when served at temperatures below −12°C (8°F).

Milk solids and sweeteners contribute substantially to flavor. In those frozen desserts containing fat, churning that takes place during freezing causes fat to concentrate around the air cells where it strengthens the air cells and imparts a sense of smoothness to the consumer's mouth (Cremers and Arbuckle 1954). Nonfat milk solids (NMS) contribute cooked and slightly salty flavors to ice cream because of the release of sulfhydryl groups during pasteurization and the natural milk salts contained in these solids, respectively.

A series of articles on the influence of sugar in vanilla ice cream (Pangborn et al. 1957) showed that numbers of panelists liking 15, 17, or 19% sugar did not differ, but each of these levels in the same ice cream formula was preferred over sugar concentrations of 11 or 13%. In strawberry ice creams evaluated by a household survey, preference was expressed for 19.2% sugar over 15.9, 17.6, and 20.8% sugar (Pangborn et al. 1957). A professional panel that studied the use of fruit concentrates and essences in ice cream found that a sugar concentration of 17% was optimum for the most desirable fruit flavor (Arbuckle et al. 1961). When the sugar content was over 17%, the fruit flavors tended to be submerged by the excess sweetness. Sweetening at the sucrose equivalencies of 15% for plain ice creams and 17–18% for bulky flavors of ice cream usually results in optimal acceptance.

VANILLA

The most popular of all ice cream flavors is vanilla. This delicate flavor makes up more than one-fourth of the total volume of ice creams sold in the United States. Of course, not all of those products are flavored with pure vanilla extract. Some contain a mixture of pure vanilla and synthetically produced vanillin while others contain artificial vanilla.

Vanilla is produced from the beans that are borne in pods of vines that belong to the orchid family. Requirements for soil, temperature, and humidity restrict the plant's growth to tropical regions within latitudes 10 and 20 degrees north and south of the equator. Vanilla was first grown in Mexico where the indigenous bee, the melipone, and hummingbirds pollinated it (Nielsen 1981). Self-pollination is not possible because a membrane exists between the stigma and stamens of the flower. Pollination by hand, developed in 1836 by Charles Morren of Leige, Belgium, made possible production of vanilla beans in other subtropical countries. Today 65–70% of the world's vanilla beans (the 10–25-cm-long pods) comes from the so-called Bourbon Islands (Madagascar, the major producer, Comoro, Reunion, and the Seychelles). Here the finest beans are grown from *Vanilla fragrans* (formerly *Vanilla planifolia*) vines. Indonesia also produces beans on *V. fragrans* plants and supplies 25–30% of the world's production. Mexico does not produce enough beans for export. Total world consumption of beans ready for extraction exceeds 1,800 metric tons, of which more than half is used in the United States, where average amounts of beans imported annually during 1980–84 and 1990–94 were 700 and 1,240 tons, respectively. About 5 lb of green uncured beans are required to make 1 lb of properly cured beans. Only about 2% of the vanilla bean is extractable as flavorful substance.

Production of vanilla is a long and labor-intensive process. The climbing vines are kept to heights of less than 6 ft or about 2 m. Otherwise, they can grow to heights of 10–15 m. The vines, which must be supported by a tree or post, require about three years to reach bearing stage and must be pruned regularly. After pollination by hand during the short blooming time of about 8 hours, pods develop in 4–6 weeks, but they must remain on the plant to full maturity, about 8 months. Each productive plant yields about 150 odorless beans that contain about 90% moisture. Fermentation and drying for two to three months produce a dark, supple pod with a slightly oily surface and a rich aroma of vanilla. Vanillin, p-hydroxybenzaldehyde, p-hydroxybenzoic acid, vanillic acid, and many more fragrant substances are formed as glycosidic complexes are broken down. Curing involves (1) heating to destroy chlorophyl, rupture cell walls, and trigger appropriate enzymatic reactions, (2) holding covered at a high temperature to enhance fermentation, (3) slowly drying as fragrances develop, and (4) final drying with sorting and selection. These steps are performed differently in the Bourbon and Mexican processes.

Aromas of vanilla differ markedly among the three commercial varieties, *V. fragrans, V. tahitensis,* and *V. pompona*. The major chemical substance in each of these varieties is vanillin, a molecule that can be synthesized by oxidative cleavage of the coniferin contained in the sap of conifers, thus from the extract of wood pulp (first done by W. Haarmann and F. Tiemann—Haarmann commercialized production of vanillin in 1874). More than 400 flavoring substances have been identified from extracts of vanilla beans. Emberger (1994) reported the olfactory profile of a series of fractions from an extract of Bourbon vanilla beans. Included in order of their occurrence in the fractions were weedy, terpeny, carrot, sweet, floral, medicinal, wintergreen, mildly phenolic, fatty, cinnamon, woody, cucumber, melon, bacon, smoky, bready, plum, sweet, honey, cocoa, lactone, animalic, beeswax, dusty, fecal, cheesy, isovalerianic acid, vanilla, and butyric.

Interesting comparisons of taste profiles were reported by Kindel (1994) of Bourbon, Indonesian, Mexican, and Tahitian vanillas. Bourbon was characterized by soft, buttery-creamy notes with only weak phenolic and floral notes. Although Indonesian vanilla is harvested from the same *V. fragrans* plant as is Bourbon vanilla, the former yields strong phenolic-smoky prune notes with weaker buttery-creamy notes. Vanilla grown on the islands of Tahiti and Moorea, *V. tahitensis*, is perceived quickly with anisic, floral, coumarinic, and sweet notes dominating. Although less than 10 tons of these beans are produced per year, some French and Italian markets prefer extract from them. A third variety, *V. pompona*, is grown almost exclusively on Guadaloupe and Martinique in very small amounts and is mainly employed for pharmaceuticals and perfumes. In recent years Indonesia has supplied about 46% and Madagascar 40% of the U.S. vanilla imports.

Vanilla flavorings are available in liquid and powder forms as pure vanilla extract, pure vanilla flavoring (21 CFR 169.175), vanilla reinforced with vanillin (21 CFR 169.177), and imitation vanilla (21 CFR 169.181). Single-fold vanilla extract is the solution in 1 gal of aqueous ethyl alcohol of the sapid and odorous principles extractable from 13.35 oz of properly cured vanilla beans (100 g/L). The beans may contain no more than 25% moisture. Twofold vanilla extract contains the extractive matter of 26.7 oz of vanilla bean per gallon of 35% ethanol. Vanilla flavoring and its concentrated forms conform to the definition and standard of identity of vanilla extract and have the same requirements for label statement except that the content of ethanol is less than 35%. Vanilla-vanillin blends are category II, but ethyl vanillin is not permitted in this category. Ethyl vanillin is made from guaiacol, a coal tar derivative.

Vanillin ($C_8H_8O_3$) may be added to vanilla extract at the rate of up to 1 oz per fold. For example, fourfold vanilla-vanillin would contain the extract of 26.7 oz of vanilla bean and 2 oz of vanillin in each gallon of 35% ethanol. Addition of vanillin at the rate of 1 oz per fold of vanilla reduces the usage of pure vanilla extract by about 50%. A 0.75% solution of vanillin is approximately equal in strength to single-fold vanilla.

Concentrated vanilla extract is made by distilling off a large part of the solvent, usually in a vacuum, until the strength reaches the desired concentration, which is then specified as fourfold, fivefold, etc. Each multiple must be derived from an original 13.35 oz of beans in the starting extract before concentration. The maximum strength of a direct extraction, without concentration, is 2.78 lb of vanilla beans in 1 gal of solvent. The amount of concentrated products used should be slightly higher than the amount of regular vanilla extract divided by the multiple of the concentrate.

Natural vanilla powders are made by mixing finely ground vanilla beans with sugar, or by incorporating the vanilla extractives with a dry carrier, evaporating the solvent, and drying. The amount used would correspond by weight to the number of ounces of a standard strength extract. For example, 1 gal of single-strength vanilla equals 8 lb of single-strength vanilla powder.

Vanilla paste is made by mixing the concentrated extractives with a dry carrier to form a paste. The amount used is the same as for powders.

Although vanillin is the primary flavorful substance of vanilla, the taste

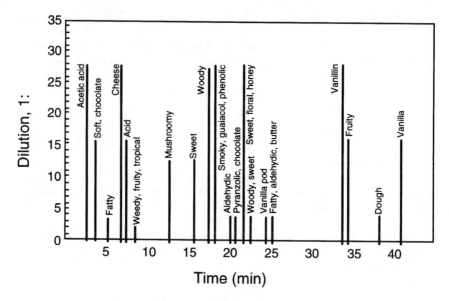

Figure 7.1. Important flavor notes of Bourbon vanilla diluted to 1:28. (From Emberger 1994.)

threshold for it is much higher than for several other components of vanilla. For example, guaiacol concentration is about two-thousandths that of vanillin, but the taste intensity is one thousand times higher than that of vanillin. Similarly, the taste threshold for anisaldehyde is three thousand times lower than that of vanillin. Anisaldehyde is particularly high in concentration in beans grown in Tahiti and is a major characterizing component of vanilla made from those beans (Emberger 1994).

Figure 7.1 shows the important flavor notes of Bourbon vanilla (Emberger 1994). In a series of steps the vanilla extract was diluted 1:2 with solvent, then subjected to capillary gas chromatography followed by olfactory evaluation. Thus, it was determined which of the peaks were detectable when dilution factors were 1, 2, 4, 8, 16, and 32. Those that remained detectable at the highest dilution, acetic acid, cheese, wood, smoke, honey, and vanillin, were deemed to contribute most to the vanilla flavor. Numbers of peaks in the chromatogram with detectable aromas were: 27 at no dilution, 8 at 1:4, 2 at 1:8, 3 at 1:16, and 6 at 1:32.

Imitation Vanilla Flavorings

Artificial vanilla flavorings have less than half of their flavoring derived from vanilla beans; some may contain no natural bean extractives. These preparations may contain water, ethanol, propylene glycol, vanillin, ethyl vanillin, propenyl guaethol, anisyl aldehyde, and heliotropine (piperonal).

Consistency in Vanilla Quality

Vanilla beans vary in their composition, and the methods of fermentation, curing, extraction, and blending introduce more variation. Therefore, it is quite

7 FLAVORING AND COLORING MATERIALS

important that the beans meet specifications for color, flavor, and moisture, and that extraction be done at the lowest practical temperature in a closed system for the minimal exposure time. Chill proofing and clarification by centrifugation or filtration are important for removing sediment and microbial spores. Aging for several weeks permits formation of esters from the acids and alcohols in the extract.

Vanilla Ice Cream

The plain ice cream mix is processed, cooled, aged at least 4 hr, then flavored by thoroughly blending the flavoring just prior to freezing. The amount of flavoring used depends on the composition of both the flavoring and the mix. For example, 5 fl oz of single-strength vanilla is used per 10 gal of mix (about 325 ml/100 lb or 710 ml/100 kg). For multi-fold vanillas the level of usage is slightly above that of single-fold vanilla divided by the number of folds. The quantity of vanilla must be increased as the concentration of milkfat is lowered and as the content of nonfat milk solids is increased. When the concentration of sweetener is low, vanilla must be increased in quantity, but at high intensities of sweetness variations in quantity of vanilla make comparatively small differences in consumer preference.

CHOCOLATE AND COCOA

Chocolate and cocoa are among the most popular flavors of frozen desserts. They are obtained from the cocao bean, the fruit of the *Theobroma cacao* tree, which grows in tropical regions about 10 degrees north and south of the equator. The major producing countries are the Ivory Coast, Brazil, Malaysia, Ghana, Indonesia, Nigeria, Cameroon, Ecuador, the Dominican Republic, and Mexico.

There are two main types of cocoa beans, criollo and forastero (Anonymous 1993). Criollos are lightly colored and mildly nutty in flavor. Forastero cocoas are dark brown, strongly flavored, slightly bitter, and comparatively high in fat content. Foresteros dominate the market, especially in their subtype, amelonado. The almond-sized cacao beans or seeds develop in a pulpy pod with 30–40 beans to the pod. The ripened pods, rich golden red in color, are cut from the trees, gathered in piles, and left to ripen further for about 48 hours, after which they are slashed open. The beans are removed and placed in bags, boxes, or vats to heat and ferment for about 10 days or until the characteristic flavor and cinnamon-red color develop. Pulp surrounding the beans is first degraded by yeasts and bacteria that convert sugars into alcohol and carbon dioxide. Subsequently, anaerobic bacteria convert the remaining sugars into alcohol and lactic acid. The *Acetobacter* convert ethanol to acetic acid. By this time the beans have lost their viability and are susceptible to several types of enzymatic degradation. Flavor precursors and pigments are formed and oxidation and condensation of polyphenolic compounds occurs. The beans are then washed clean of the dried pulp, dried slowly and sufficiently to prevent mold growth, and then sorted and graded prior to shipment to manufacturers of chocolate and cocoa. Bean quality is affected by genetics, environment, exposure to microorganisms, insects and insecticides, and rodents, and by size, integrity, and moisture content.

Processing Cocoa Beans

After being mechanically cleaned of foreign matter at the factory, the cocoa beans are blended to provide the characteristics and uniformity desired, and the shells are broken. This stream of beans and shells is sieved into fractions, and the beans are separated from the broken shell by streams of air in a winnowing machine. This process yields the kernals or "nibs," the seed part. These nibs are roasted, alkalinized (if this process is to be done), and ground into chocolate liquor (in Europe called chocolate mass), pure bitter chocolate containing 50–60% fat (called cocoa butter). This liquor, made hot by the friction of milling, may be cooled in molds that form it into large slabs or into smaller sizes that are sold in retail stores as bitter cooking chocolate.

Sweet milk chocolate is made from chocolate liquor by adding sugar and at least 3.39% milkfat and 12% total milk solids. The minimum content of chocolate flavoring is also specified.

Natural process cocoa is made from chocolate liquor by subjecting it to high pressure in hydraulic presses. This process removes a large amount of the cocoa butter, usually about 38–40% of the total, and leaves a hard, dry cake that contains 10–24% fat. Cocoa contains nearly all the flavoring material from the cocoa bean, the cocoa butter being practically flavorless. This cocoa cake is then milled into finely sifted cocoa powder. The pH of natural process cocoa is about 5.2.

Dutch process cocoa is made in the same manner as natural process cocoa except that the nibs are treated with alkali at the time of roasting. This treatment (with ammonium, potassium, or sodium bicarbonate, carbonate, or hydroxide, or with magnesium carbonate or oxide) makes the cocoa more soluble and darker in color. It also brings out a full fine chocolate flavor, reduces the potential for bitterness, and counteracts the acid flavor found in natural process cocoa. The pH of Dutch process cocoa may range from about 7–8.6, depending on the amount of alkali used.

A defect that sometimes appears in chocolate ice cream made with Dutch process cocoa is the formation of greenish-black discoloration where the ice cream comes in contact with exposed iron (modern packaging practices generally exclude containers that contain iron). In the "Dutching" process tannins of the cocoa are solubilized by the added alkalis, and they react with the iron to form ferric tannate, a colored compound.

Characteristics of Cocoa

Cocoa is a more concentrated source of chocolate flavor than is chocolate liquor, because it contains a higher proportion of chocolate flavoring material and less of the nearly flavorless fat. For example, 100 lb of regular cocoa contain 78–88 lb of "flavor" plus 10–22 lb of fat (most cocoa used in ice cream contains 10–12% fat); whereas 100 lb of chocolate liquor contain 40–50 lb of "flavor" and 50–60 lb of fat. Thus, 100 lb of cocoa contain about 40 lb more flavoring material than 100 lb of chocolate liquor. The usual high price of cocoa butter ordinarily makes it more economical to use cocoa than chocolate liquor. Furthermore, labeling regulations make necessary the declaration of the amounts of fat and saturated fat in foods. Many consumers select against products that are high in fat, especially saturated fat. This is why some manufacturers of chocolate ice cream are opting for lowfat cocoa which contains less than 10% fat (usually 7–9%). Another type of cocoa that is available, breakfast cocoa, contains at least 22% fat. Manufacturers of superpremium, high-fat ice creams may select breakfast cocoas. The lubricating effect that cocoa butter provides as it melts in the mouth at about 33°C (91°F) enhances the perception of richness in ice creams. If other fats are to be used in frozen desserts, they should melt within the range of 28–35°C (82–95°F).

Producers of nonfat ice creams can formulate with defatted cocoa that contains about 2.5% fat blended with cocoa containing 10–12% fat to produce lowfat cocoa containing about 5% fat. Use of such a blend at 2.5–3.0% of a mix facilitates production of chocolate ice cream with < 0.5% fat, and this meets the requirement for labeling the product as nonfat ice cream.

Defatted cocoa can be made by extracting the residual fat. This can be done without the risk of there being a residue of chemical solvents by a process called supercritical fluid extraction. In this process the cocoa is exposed to carbon dioxide (CO_2) at very high pressures (supercritical). At such pressures CO_2, a gas, acts as a liquid, becoming a highly efficient solvent for the cocoa butter. The CO_2 is exhausted from the cocoa into a chamber of lower pressure where it again functions as a gas instead of a liquid and loses its solvent power. Thus, the cocoa butter is deposited for collection. Defatted cocoa has the same flavor intensity on a nonfat solids basis as regular cocoa. However, because the fat absorbs light, defatted cocoa has a lighter color than cocoa. This is called its extrinsic color. When the defatted cocoa is dissolved in an ice cream mix, the intrinsic color becomes evident, and this color does not differ from that of a mix containing the same amount of cocoa on a nonfat solids basis. Defatted cocoa appears to be more fine in grind than the cocoa from which it is made. This is because it has virtually no fat to cause the particles to adhere to each other.

The color of cocoa is the result of several factors: (1) the source and blend of beans, (2) the fat content, (3) the process, natural or Dutch, (4) the fineness of the grind (coarser is darker), and (5) the rate of chilling. Rate of chilling affects color by determining the size of fat globules in the cocoa, but this property is lost when the cocoa is used. Flavinoids, especially anthocyanins and procyanidins, are the main precursors of chocolate color. During fermentation, hydrolysis and oxidation convert flavinoids to quinones. Quinones react with amino acids to form Maillard reaction products that are strongly colored. They also react with other flavinoids to form brown tannins.

Hue, saturation, and lightness of cocoa color are readily measured with a Hunterlab meter. A spatial representation of the color can be made using the system of the Commission Internationale de L'Eclairage (CIELab). Color is measured as reflectance of three successive flashes of a standardized light. The spectrophotometer translates the signals into L (psychometric hue), a (hues of green and red), and b (hues of blue and yellow) values.

Flavor of cocoa is also affected by formation of Maillard reaction products from sugars and amino acids produced during fermentation. To enhance the flavor of some cocoa products some manufacturers add aromatic substances such as cinnamon, oil of cloves, oil of bitter almond, or vanillin. Small quantities of these substances can impart desirable flavor notes to cocoa and chocolate. Among the desirable flavor notes of cocoa products are cocoa (the basic flavor note), bitter, rich, bouquet, sour, astringent, and acrid. Undesirable flavor notes include burnt, earthy/moldy, hammy, smoky, metallic, rancid, cardboard, and raw (Anonymous 1993).

Chocolate Ice Cream

Researchers of the 1920–1930 era provided the bases for many of today's manufacturing practices with chocolate ice creams (Dahle 1927; Dahlberg and Hening 1928; Fabricius 1930; Martin 1931; Reid and Painter 1931; and Tuckey et al. 1932).

Chocolate products used in flavoring ice cream are cocoa, chocolate liquor, blends of cocoa and chocolate liquor, chocolate syrups, and extracts from cocoa nibs. Chocolate extracts are made with alcohol and water so they contain little fat. However, it is permissible to substitute vegetable fat for milkfat in cocoa products provided the words vegetable fat appear in the names of the products. Federal Standards of Identity for cocoa products (21 CFR 163) do not permit sweetening them with nonnutritive sweeteners. However, nonnutritive sweeteners may be used in chocolate-flavored frozen desserts provided they are appropriately labeled.

The amount of chocolate flavoring to use in ice cream depends upon the desired strength of flavor and intensity of color. Consumer preference tests should be used to determine these parameters. Extra sweetener should be added to compensate for the bitter flavor of cocoa, the usual recommendation being the same weight of sucrose, or equivalent thereof, as of cocoa.

The Codex Alimentarius standard for ice cream and ices states that the flavor "must come from nonfat cocoa solids." The minimal quantity needed is 2–2.5% of the mix weight.

The simplest way of adding cocoa powder to a mix is to blend it together with other dry ingredients, especially granulated sugar, and incorporate them together with the liquid ingredients in the blending device or with a powder funnel (Chapter 9).

As with other mixes, chocolate-flavored ice creams that are low in fat require higher homogenization pressures that those containing high fat. In general, homogenization pressures should be 500 lb/in^2 (3.45 MPa) higher for chocolate mixes than for plain mixes.

Preparing Chocolate Syrup

The small manufacturer often prefers to flavor chocolate ice cream by adding syrup at the freezer. A desirable syrup can be made by adding 20 lb each of cocoa and sugar to enough water to make 10 gal. To compound the syrup, mix the sugar and cocoa together, then add enough water to make a heavy paste.

7 FLAVORING AND COLORING MATERIALS

Heat slowly and add water gradually. Heat to boiling, then cool to less than 4°C before addition to the mix to minimize heat removal in the freezer. This syrup should be used in a ratio of 1:5 to 1:7 in a plain ice cream mix before pasteurization. The ratio of 1:5 is the highest concentration of chocolate that can be used with a plain ice cream mix containing 10% milkfat and 20% total milk solids, because the minimums for these components are 8% and 16%, respectively, in bulky flavored ice cream with no modifying label (such as reduced fat, lowfat, or nonfat).

Whenever possible, a complete mix should be made by adding the cocoa or chocolate directly to the mix before pasteurization. The temperature of pasteurization is sufficient to incorporate cocoa properly. A chocolate mix made this way whips more rapidly than plain mix to which syrup is added at the freezer. Furthermore, the ice cream has better flavor, is more uniform, and contains fewer dark specks. A typical formula is milkfat, 10%; NMS, 10%; sugar, 18%; cocoa, 2.5%; chocolate liquor, 1.5%; stabilizer, 0.2%; and TS, 42.2%. Relatively low solids, 37.5%, is found in a mix composed of 9.5% milkfat, 9.5% NMS, 16.0% sugar, 2.0% cocoa powder, and 0.5% stabilizer.

Freezing Characteristics

Chocolate ice cream is one of the most difficult to freeze, because the mix whips comparatively slowly. The viscosity may be reduced and whipping time decreased by adding 0.1% of citrates or phosphates to the mix. Air in the ice cream dilutes the color so that in ice cream with high overrun it may be necessary to use dark cocoa to provide the intensity of color desired. However, very dark cocoa powders often do not impart the pleasing flavor to ice cream that less highly alkalinized powders provide.

Chocolate Confections

Chocolate is used in ice cream in many forms. Included among these are chocolate chips, swirl or variegate, chocolate-covered nuts, chocolate-coated baked goods and fruits, and miniature low-melt filled chocolates. An example of the latter is a 5/8-inch-long multi-colored chocolate base ice cream cone. There are about 400 pieces per pound, and they can be flavored according to the color of the chocolate. Because of their low melting point several of these products need to be refrigerated until the time they are placed in the ingredient feeder. Furthermore, the quantity added to the hopper of the feeder should be managed so that the pieces do not melt and stick together.

FRUITS IN FROZEN DESSERTS

The ice cream industry is a major market for fruits. Fruit ice creams rank second among flavors, representing about 15% of the total amount of ice cream (Table 7.1). About 34% of frozen yogurt is fruit flavored. Fruits are available in fresh, frozen, and heat-processed forms. Today's manufacturers of fruits and flavorings have taken advantage of new technologies to provide aseptically processed fruits that keep for months at room temperature with little change in quality. These fruits are processed in swept-surface heat exchangers that take the temperature of the fruit/sugar/acid/stabilizer mixture to 88–121°C

(190–250°F). After heating, the mixture is held in a holding tube for about 3 min, then is cooled in a series of swept-surface heat exchangers (Figure 7.2) to about 27°C (80°F). Movement from the cooling cylinders is direct to the filling machines where the product is metered into sterile containers inside a sterile filling chamber that is bathed in sterile air.

In a second system (Figure 7.3) fruit is pumped through heating, holding, and cooling coils that create secondary flow effects to evenly heat and cool the fruit. The integrity of fruit particles reportedly is better maintained than in swept-surface heat exchangers, because no scrapers are used to renew surface films. The containers are usually multi-layered bags made of polyethylene and foil. They come in sizes of 1, 2, 2.5, 3, 5, 55, and 300 gallons. Bags are placed in cardboard or rigid plastic containers. Large refillable totes are used where there is a high demand for fruit. In small operations containers of unused opened fruit must be refrigerated and can usually be held for a few weeks with no evidence of spoilage. Because the pH of fruits prepared aseptically must be 4.5 or below, the microorganisms most likely to grow are molds and yeasts.

Aseptic processing provides several advantages in the use of most fruits. Quality is usually much improved over that of kettle-type (Figure 7.4) heat processing. Kettles, being open to the atmosphere, permit heating to only about 100°C (212°F). This usually takes a minimum of 20 min, and about 20 more min are required at this temperature to destroy molds, yeasts, and acid-tolerant bacteria. Heat transfer is relatively inefficient in kettles, and volatiles are able to escape to the atmosphere causing a loss in natural flavor characteristics. Color usually darkens. Shelf life is often short, and refrigeration or preservatives may be needed to prevent spoilage. Scrape surface heat exchangers provide more uniform heating to fruits than do kettles. Furthermore, cooling is much faster than in kettle processing. Overall, quality, convenience, and economy are maximized by aseptic processing of fruits.

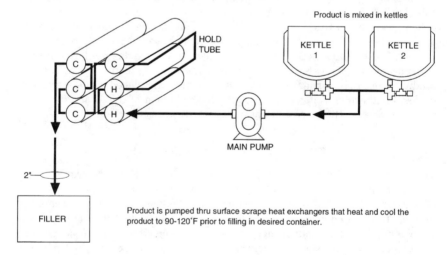

Figure 7.2. System for aseptically processing fruits by heating and cooling with a swept-surface heat exchanger. (Courtesy Lyons Magnus, Fresno, CA.)

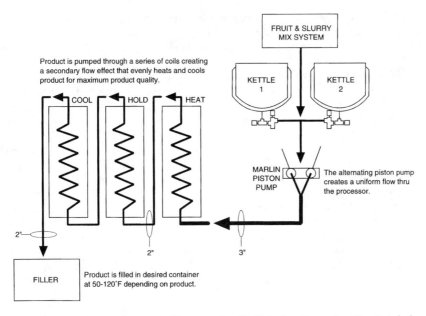

Figure 7.3. System for aseptically processing fruits by heating and cooling in tubular heat exchangers. (Courtesy Lyons Magnus, Fresno, CA.)

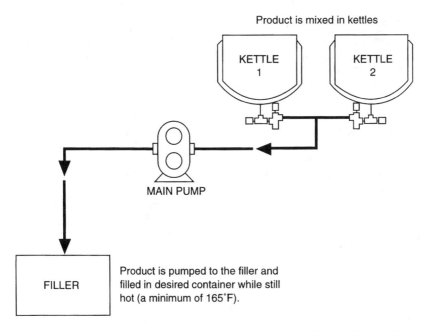

Figure 7.4. System for conventional kettle type processing of fruits. (Courtesy Lyons Magnus, Fresno, CA.)

Frozen fruits are quite suitable for use in frozen desserts. However, the cost of holding them frozen can be significant; furthermore, they must be thawed slowly in refrigerated space to maintain their quality. Once thawed, frozen fruits should be used within a few hours. Heating enhances the flavor of some fruits, whereas it degrades the quality and appeal of others. Those fruits whose flavors may be enhanced by heating include cherries and pineapple. Heating can lower the quality of strawberries, peaches, and, to a lesser extent, raspberries. Aseptically processed fruits suffer comparatively small changes in flavor and texture compared to kettle processed fruits.

Fruit flavors are available as (1) extracts from the prepared fruit, (2) artificial compounds, and (3) true extracts fortified artificially. These flavors supplement fruits in cases in which it is necessary to limit the amount of fruit, but they are often inferior in flavor to fruits and do not provide the desired fruit pulp. Defects in fruit-flavored frozen desserts may result from improperly handled fruit, use of insufficient fruit, poor incorporation of fruit, excessive fortified or artificial flavor, and/or poor quality base mix.

The sugar content of the fruit pack is the main determinant of how much sugar the mix should contain. Assuming a plain mix with 15% sugar is used with 20% of a 4:1 fruit pack, the overall sweetness would be calculated as follows:

$$
\begin{array}{rll}
80 \text{ lb of mix} \times 0.15 & = & 12 \text{ lb sugar} \\
\underline{20 \text{ lb}} \text{ fruit} \times 0.20 & = & \underline{4 \text{ lb sugar}} \\
100 \text{ lb} & & 16 \text{ lb sugar}
\end{array}
$$

and: $(100/16) \times 100 = 16\%$ sugar in the finished product. Use of a 2:1 fruit pack would increase the sugar concentration to 18.6%. Therefore, the sugar content of the mix would likely need to be lowered or the amount of fruit decreased.

In formulating a mix for fruit ice creams, total solids should be made higher than for vanilla ice creams of the same relative quality, e.g., super premium. This is necessary to offset the effect of dilution by the fruit pack. Federal Standards of Identity allow a reduction of fat and total milk solids in fruit ice creams to a minimum of 8% and 16%, respectively. However, they specify use of a factor of 1.4 in computing the reduction. A mix that works well with a 2:1 fruit pack has the following composition: 13.5% milkfat, 10.5% sucrose, 12.5% NMS, and 0.35% stabilizer/emulsifier.

Because fruit ice creams have a higher sugar content than plain ice creams they should be drawn from the freezer about 1°F colder. A drawing temperature of 23°F for the batch freezer and of 20°F for the continuous freezer is generally satisfactory.

Fresh Fruit

Fresh or fresh-frozen fruit is often the best source of flavor. Fresh fruit ice creams have a special sales appeal. The fruit should be washed and hulled or pealed and then mixed with sugar in the ratio of 2–7 lb fruit to 1 lb sugar (see

7 FLAVORING AND COLORING MATERIALS

Table 7.2) and held at about 4°C for 12–24 hr before using. During this time, a large part of the juice will combine with the sugar, because of osmotic action, to form syrup. This syrup will impart to the ice cream the full flavor of the fruit much better than would the fruit used immediately.

The fruit-sugar mixture used at the rate of 15–20% produces excellent results with many fruits. The fruit-sugar ratio may vary from 2:1 to as high as 9:1. Fruits may be used whole, sliced, crushed, diced, pureed (coarse, medium, or fine) or as juice. The use of whole or large-sliced fruit may result in a coarse-icy texture. Good fruit distribution, appearance, and desirable texture and flavor result from the use of diced or pureed fruit.

The amount of fruit required to impart the desired flavor depends on the characteristic of the flavor and varies from 10–25% of the weight of the finished product. In any case the minimum content is 3% by weight of clean, mature, sound fruits or their juice. It is also desirable to have pieces of fruit or pulp large enough for easy recognition in the finished product. Suggested use levels are given in Table 7.2.

When fruits are high in price, less fruit may be used and enough fruit extract added to impart approximately the same flavor intensity as would be provided

Table 7.2. Amount and Preparation of Fruits and Nuts for Ice Cream

Flavor	Fruit-Sugar ratio	Quantity of fruit in mix (%)	Kind of preparation	Added color
Apple	7:1	20–25	Sliced	Light yellow green
Apricot	3:1	20–25	Sliced, diced, or puree	Light orange yellow
Banana	—	18–20	Puree	[a]
Blackberry	3:1	20	Crushed or puree	Slight red
Blueberry	4:1	20	Crushed or puree	Light blue
Cherry	5:1	15–20	Whole or crushed	Light red
Fruit salad	3:1	15	Sliced or diced	[a]
Grape	—	25	Juice	Light purple
Peach	4:1	20–25	Sliced, diced, or puree	Light
Pineapple	4:1	12–15	Diced or crushed	[a]
Plum	4:1	25	Puree	Light red
Rasperry	2:1	10–12	Crushed or puree	Light purple
Strawberry	3:1–4:1	15–20	Sliced, crushed, or puree	Pink
Almond	[b]	3 lb:10 gal mix	Broken	[a]
Chocolate	[b]	2.7–3.5 lb cocoa	Powder	
		3.5–4.5 lb cocoa liquor blend	Syrup	[a]
		4.5–5.5 lb chocolate liquor per 10 gal mix		
Pecan	[b]	3 lb:10 gal mix	Broken	
Pistachio	[b]	4 lb:10 gal mix	Whole and broken	Light green
Walnut	[b]	4 lb:10 gal mix	Broken	[a]
Orange	5:1	14–18 oz:10 gal mix	Puree	Orange
Lemon	5:1	10–14 oz:10 gal mix	Puree	Yellow green
Lime	5:1	8–12 oz:10 gal mix	Puree	Green

[a] Natural, no color added.
[b] Sugar pack may range from 2:1 to 9:1.

by the natural fruit. The usual high price of raspberries often dictates the use of raspberry extract to enhance flavor. Another approach is to add fruit concentrates and essences. Popular ones are peach, blueberry, apple, grape, red raspberry, and strawberry. Supplemental use of 3.5–10% fruit equivalent of concentrates improves flavor. Adjustment of the acidity and sugar content of mixes may contribute to improved flavor when concentrates and essences are used (Arbuckle et al. 1961). For example, when the base mix contained 20% fruit pack, the most favorable pH, sugar concentration, and supplementation rate in experimental trials for selected fruits were found to be as follows, respectively:

> Blueberry ice cream—pH 5.7, 15% sugar and 5% blueberry juice concentrate
> Peach ice cream—pH 5.7, 16% sugar and 7.5% peach juice concentrate
> Cherry ice cream—pH 5.2, 15% sugar and 15% cherry juice concentrate
> Apple ice cream—pH 6.2. 15% sugar and 20% apple juice concentrate

Candied and Glaced Fruits

Candied or glaced fruits such as cherries, pineapple, and citron and such candied fruit peels as orange, lemon, and grapefruit are used chiefly in rich ice creams, puddings, aufaits, and mousses. They make excellent decorative materials on fancy molded ice creams, sherbets, and ices.

Dried Fruits

Dried fruits, especially apricots, figs, raisins, and prunes make tasty ice creams. They are continuously available, are shelf stable, and can be obtained in places where other types of fruit are expensive or unavailable. Dates, figs, and raisins have long been used in frozen puddings.

PROCEDURES AND RECIPES

Strawberry Ice Cream

Variety is the primary determinant of flavor imparted by strawberries. Maturity and promptness of processing are secondary factors (Fabricius 1931, Guadagni 1956). Mack and Fellers (1932) found that the amount of fruit used affects the body, texture, flavor, and appearance of strawberry ice cream. In making the ice cream using 6–20% fruit, they noted progressively better flavor up to 15% fruit. Higher amounts increased the flavor only slightly while decreasing the desirability of body and texture. A current recommendation of a major fruit supplier is 7.7% strawberry puree plus 15% solid pack strawberries.

In adding strawberries for continuous freezing, the juice should be drained from the berries and added to the mix prior to freezing. The berries, which should be chilled to −1 to 0°C, should be added with an ingredient feeder. If iciness of the berries in ice cream is a problem, the percentage sugar in the pack can be increased, thus lowering the freezing point. The consequences of adding excess amounts of sugar to fruit are increases in the amount of juice exuding from the berries and the necessity to reduce the sugar content of the mix to avoid excess sweetness. Arbuckle (1952) found a 3:1 berry:sugar ratio most favorable. For the fruit to have the same consistency as the ice cream at the serving temperature, the fruit should contain at least 21% sugar, and such is provided by a 3:1 to 4:1 ratio.

Raspberry Ice Cream

As with strawberries, variety is a major determinant of suitability of raspberries for flavoring frozen desserts. Purveyors of fruits select fruit from desirable cultivars and demand harvest at the peak of flavor. Mack and Fellers (1932) satisfactorily used 12–15% of raspberry puree from which 75% of the seeds had been removed. This gave a superior product in flavor, but the texture was coarse due to seeds. After reducing the puree content to 10% and adding raspberry extract, an equally fine flavor was observed and the texture was improved significantly. A current recommendation of a major fruit supply house is to add 10% of their aseptically processed raspberry puree. Additionally, they recommend adding 8.6 fl oz of raspberry ribbonette (a syrup) per gallon of finished ice cream if a swirl type product is desired.

Peach Ice Cream

Mack and Fellers (1932) reported that a good flavor of peach ice cream was obtained with the cultivars Hiley, J. H. Hale, Elberta, Champion, and Crawford. They suggested using 15–20% peaches in a 3:1 pack. Furthermore, yellow-fleshed peaches gave a greater concentration of flavor than white, and larger shreds of the peach were evident in the ice cream when puree was made from yellow peaches. No flavor was sufficient without the addition of flavoring extract along with the fruit. A current recommendation from industry calls for adding 7% nectarine puree and 13.6% peach cubes along with 0.095% peach flavor.

Cherry Ice Cream

Sour cultivars of cherries were found superior to sweet cherries for ice cream. Mack and Fellers (1932) listed in order of preference Montmorency, Early Richmond, and May Duke. A current recommendation of a major supply house is Bordeaux cherries. The addition of a small amount of cherry extract, oil of bitter almonds, or benzaldehyde can enhance cherry flavor. Additives should be well mixed with the fruit before adding to the ice cream. A current flavoring recommendation of a supplier of cherries is 12.5% of Bordeaux cherry halves. However, Hening (1935) recommended 19.5% Montmorency cherries plus 10.5% added sugar when the mix contained 14% milkfat, 11% NMS, 14% sugar, and 0.5% gelatin.

Ice Cream with Complex Flavors

As technologies have developed, it has become possible to produce many multi-flavored frozen desserts. One favorite known for many years is butter pecan. Such products are commonly made by adding the background flavoring to the mix then adding the nuts or fruits to the frozen product with an ingredient feeder when the system is continuous. When batch freezing is done, solid additives may be added to the nearly frozen mix before discharging from the freezer, or they may be stirred in gently as the soft-frozen product is discharged into the packages. Variegates (stripes, ribbons, swirls) are formed in continuous processing by pumping the variegate through a special tubular apparatus with a variable speed positive displacement pump. Depending on the speed of production such equipment may have from one to five outlet tubes which may be turned or left stationary within the stream of soft-frozen product. Distributing variegate syrups uniformly and in desired amounts into batch-frozen ice creams requires planning and practice.

Although the task is difficult, it is possible to produce complex flavors of ice cream from a batch freezer. An example of the process for making *Cherry Chocolate Ice Cream* follows (courtesy Fantasy BlankeBaer). To make 10 gal of finished ice cream; chill 128 oz of red maraschino cherries to below 4.4°C (40°F). Drain the juice from the cherries and add it to 9 gal of plain mix. Optionally add ½ to 1 oz red color to the mix. Freeze the mix to –5°C (23°F) or proper stiffness. As containers are being filled, add liquid chocolate chip mix [preheated to 38°C (100°F); 9 fl oz/3 gal container) to the product while stirring vigorously. Stirring will cause the chocolate to shatter as it hardens in the cold product. Add drained cherries (14.5 oz of drained weight) while adding the chocolate.

The following are selected examples of formulas for other complex flavors (courtesy Ramsey Laboratories):

Razzle Dazzle Krisp: Add 10% black raspberry puree to mix and freeze; add 5% pecan krisp candy to the soft-frozen product.

Peanut Butter Toffee Creme: Add 10% creme caramel penuche syrup and freeze; add 5% chilled English toffee krisp candy by ingredient feeder and inject 7–10% peanut butter sundae pack as a variegate to the soft-frozen product.

Cookie Dough Triple Delight: Add 1 fl oz vanilla flavor per 10 gal plain mix and freeze; blend in 5% frozen chocolate chip cookie dough and inject 7–10% of chilled peanut butter sundae pack.

Hawaiian Praline Cheesecake: Add 7.8% cheesecake base to ice cream mix and freeze; inject 5% praline almonds and 18% pineapple tidbits to the soft-frozen product.

Major suppliers of flavorings have suggested many combinations of flavors, fruit, candies, and nuts. Selected examples follow.

1. From Bunge Foods:

    Boysenberry Crunch: Vanilla ice cream with graham cracker crunch pieces and a flowing ribbon of boysenberry sauce.

    Honey Nut and Cream: Buttery caramel variegate and crunchy honey-roasted peanuts in vanilla ice cream.

    Mocha Macadamia Crunch: Coffee and chocolate blend to create mocha ice cream, topped with macadamia nut coffee pieces.

7 FLAVORING AND COLORING MATERIALS

2. From Consolidated Flavor Corporation:
    Amaretto Orange: Exotic amaretto flavor with the tangy taste of orange.
    Carrot Cake: Carrots, raisins, walnuts, and spices combined to provide the taste and aroma of carrot cake.
    Triple Chocolate Neopolitan: A combination of three chocolate flavors—chocolate, dark chocolate, and white chocolate—to fulfill the fantasies of the ardent chocolate buff.
3. From Creative Flavors:
    Canadian Maple: Real Canadian maple syrup as background flavor with maple-coated walnuts and maple swirl in plain ice cream base.
    Cherry Cola: A cherry cola swirl and cherries in rich vanilla ice cream.
    Praline Cashew Trifle: Rich chocolate ice cream with banana ripple, pralines, cashews, and chocolate-coated bananas.
4. From David Michael and Co.:
    Mandarin Orange-Coconut: Mandarin orange ice cream strewn through with rich coconut.
    Sassyberry Swirl: A tart berry flavor blended throughout vanilla ice cream.
    Irish Cream Sundae: Creamy luscious chocolate and coffee flavor with a definite zing.
5. From Fantasy BlankeBaer:
    Cake and Ice Cream: Thick, chocolaty, fudge frosting and chocolate cake pieces blended into creamy vanilla ice cream.
    Chunky Cherry Almond: Black cherry ice cream full of chopped roasted almonds and chunky black cherry sauce swirled throughout.
    Raspberry Chocolate Macaroon: Swirls of red raspberry sauce in coconut macaroon-flavored ice cream, sprinkled with chocolate candy coconut bits.
6. From Fruitcrown Products Corp.:
    Briar Patch Berry: A blend of blueberries, strawberries, raspberries, blackberries, and cranberries "streaked" through wildberry-flavored ice cream.
    Dixieland Peach Melba: Red raspberry-flavored ice cream injected with solid-pack peaches.
    Tangy Almond: A combination of almond bits in a tangy lemon sauce.
7. From Guernsey Dell:
    Crunchy Munchy Chocolate: Choco-graham pieces and marshmallow ribbon in chocolate ice cream.
    Granola Berry: Granola clusters and raspberry revel in vanilla ice cream.
    Purely Praline: Praline pecans in light praline ice cream.
8. From Pecan Deluxe Candy Co.:
    Brownies, Cookies and Cream: Brownie fudge, cookie pieces and English walnuts in vanilla ice cream.

English Caramels and Pecans: Vanilla ice cream with caramel cube pieces and chocolate pecans.
   Raspberry Moussecake: Plain ice cream mix flavored with raspberry and mousse-mix bases to which a mixture of brownie fudge pieces and chocolate-coated chocolate cookies is added.
9. From Star Kay White:
   Banana Split: Contains hot-packed banana fruit base with almond crunch candy pieces and chocolate fudge variegate.
   New York Style Strawberry Cheesecake: Cheesecake background complemented with strawberry variegate.
   Triple Play Chocolate: A combination of chocolate background flavor with chocolate fudge chips and chocolate variegate.
10. From Virginia Dare:
    Apple Appeal: A fresh, juicy, natural apple flavor that combines ideally with cinnamon, rum raisin, or cream.
    Blueberry Bonanza: A true-to-nature blueberry flavor that goes great with cheesecake, banana, and cream notes.
    Citrus Cream: A natural tangerine emulsion and three-fold vanilla extract.

Sugar Free

With the marketing of sugar-free soft drinks and the emphasis among nutritionists on reduction of caloric intake has come increasing demand for sugar-free frozen desserts. Suppliers continue to make available increasing numbers and varieties of sugar-free fruits and chocolate products. A sample of ingredients sweetened with aspartame follows: blueberry chunks, raspberry and strawberry revels and purees, chocolate flakes and chunks, chocolate revel, and chocolate-coated peanuts and almonds.

NUTS

Nutmeats and nut extracts are used extensively in frozen desserts. Among the most popular are pecans, walnuts (English and black), almonds, pistachios, filberts, and peanuts. Nutmeats should be sound, clean, free from rancid flavor, low in count of microorganisms, and free of pathogenic bacteria. Methods of eliminating microorganisms other than careful hygienic control during processing include application of dry heat, dipping in a boiling, slightly salty sugar solution for a few seconds or treating with ethylene oxide. The latter is highly effective, but the gas is toxic and must be used under carefully controlled conditions. To prevent sogginess, nuts treated in boiling water should be dried for 3–4 min at 121–149°C (250–300°F).

Nuts should be stored in a cool dry place until used. Almonds, filberts, and pistachios should be blanched to remove their skin prior to use. Specifications of permitted pieces of shell should be checked carefully by manufacturers in purchasing prepared nutmeats.

Use concentrations of nuts range from 1–5% depending on the nut and the

7 FLAVORING AND COLORING MATERIALS

accompanying flavor(s). The following are flavors and recommended amounts of nuts (calculated as percentage of the unflavored mix): banana nut, 2.2%; caramel praline, 5%; chocolate caramel nut, 1.7%; maple nut, 2.2%; mud nut, 1.7%; butter pecan, 3.3%; pecan pie, 2.2%; and black walnut, 2.8%.

SPICES AND SALT

Spices such as cinnamon, cloves, nutmeg, allspice, and ginger are used sparingly as flavors in frozen desserts. Ginger ice cream is a favorite in some localities. Cinnamon, nutmeg, and cloves are often used to enhance or modify the flavor of chocolate products, and they complement puddings, eggnog, and certain flavors of punch.

Spices may be purchased in either the finely ground dry form or as extracts. Their flavors are strong so that only small amounts are needed to produce the desired effect.

Salt, although not a spice, is often used in small quantities to enhance certain flavors of ice cream, especially those containing eggs—custards and rich puddings—and in nut ice creams. The recent tendencies of Americans to reduce their intake of sodium coupled with the requirement to indicate sodium on the nutrition label has caused many manufacturers to minimize the amount of salt added to frozen desserts.

COLOR IN FROZEN DESSERTS

Ice cream should have a delicate, attractive color that readily suggests to the consumer what the flavor is. Both "certified" (FD&C) and "exempt from certfication" (natural) colors are approved by the U.S. Food and Drug Administration for use in ice cream. Labeling regulations for the certified colors have changed in conjunction with the NLEA. All FD&C colors must be shown separately on the ingredient legend for most foods. However, ice cream, butter, and cheese have been exempt. With the discovery that Yellow No. 5 and Yellow No. 6 can cause allergic-type reactions in a limited number of sensitive persons, the U.S. FDA has required that ice cream ingredient labels shall declare the presence of these colors when they are used. Exempt colors can be shown together as "color added" or "artificial color" or they may be listed by name. Generally the use of the term "natural color" is not permitted.

Most flavors of ice cream require the addition of at least a small amount of color. Enough yellow color is usually added to vanilla ice cream to give it the golden shade of cream produced by the "colored" breeds of cattle (the Jersey and Guernsey breeds in the United States). Fruit ice creams need to be colored, because the usual amounts of fruit added are insufficient to impart adequate color. Chocolate ice cream, on the contrary, seldom needs added color. Dutch process cocoa, in particular, imparts high color at the use concentration.

Certified colors are most often used in coloring ice cream. Annatto color is about the only exempt color used in ice cream. However, annatto tends toward the pink rather than the desired "egg-shade" yellow, especially in mixes that have a high free calcium level.

If colors are purchased in the powder form and made up by the ice cream processor, they should be dissolved in boiling water and stored refrigerated for short times. Longer times of holding are possible when 0.1% sodium benzoate is added to limit the growth of microorganisms. Although solutions of colors contain little nutrient material, it is possible for some microorganisms to grow in them. Solutions of colors of the strength normally used may be prepared by dissolving 4 oz of primary color per gallon of water to make a solution of about 3%. A normal use concentration may range in dilution in the mix from 1:500 to 1:1,000.

FLAVORING LOWFAT AND NONFAT ICE CREAM

Milkfat alters the maximum intensity of flavors and modifies the timing and rates of onset and diminution of a flavor experience (Mela and Marshall 1992). Flavors soluble in fat, such as vanilla, are carried with the fat to the tongue of the consumer. There, as fat melts, such flavors are released to the olfactory senses. If there is insufficient fat to carry these flavors, they are perceived quickly and tend to disappear relatively quickly from the flavor profile. In addition, fat provides richness of flavor and lubricity of mouthfeel.

Fat replacers used in nonfat and lowfat ice creams usually consist of modified whey proteins or starch hydrolysates. Both tend to bind and to mask delicate flavors. Whey proteins, for example, even in concentrations as low as 0.5%, are prone to bind aldehydes. Damodaran and Kinsella (1980) explained that as the chain length of the aldehyde was increased, the tendency to bind to whey proteins also increased. Thus, they believed the binding was the result of hydrophobic interactions. As little as 1% milkfat can reduce the vapor pressure of flavorful substances. Schirle-Keller and Reineccius (1992) compared relative vapor pressures (RVP) when ketones of various chain lengths were placed in Simplesse®, microparticulated whey protein, or in emulsified milkfat. Simplesse® had little effect on the RVP (0.8 with nonanone and 0.9 with decanone) whereas 5% milkfat reduced the RVP to about 0.1. Because of these types of differences in effects on volatility of flavorful substances in ice cream, concentrations and, sometimes, balance of volatile components need to be varied considerably in frozen desserts depending on the ingredients.

Ohmes and Marshall (1995) showed that vanillin intensities did not vary among three whey-based fat replacers added at the rate of 4.8% to nonfat ice cream, the control lowfat ice cream that contained 4.8% milkfat, and a second control that contained 4.8% additional nonfat milk solid. However, the control that contained milkfat had lower flavor scores for "syrup, whey, and cooked milk" than the other four samples. Samples containing Simplesse® did not differ from either control in texture, while those containing Prolo 11® were smoother and more gummy than the controls.

Some flavors are quite compatible with fat replacers. Among them are butterscotch, butter pecan, and cheesecake. These "heavier" flavors tend to cover

7 FLAVORING AND COLORING MATERIALS

flavors contributed by the fat replacers while providing flavor notes that blend well with those of the typical nonfat product.

REFERENCES

Anonymous. 1993. The Cocoa Manual. Cocoa De Zaan B. V., Holland.

Arbuckle, W. S. 1952. Stabilized fruits for ice cream. *Ice Cream Trade J.* 48(5):34, 36, 86.

Arbuckle, W. S., W. Venter, Jr., J. F. Mattick, and N. C. Aceto. 1961. The technology of utilizing concentrated fruit juices and essences in ice cream and related products. AES Bulletin A-118, University of Maryland.

Cremers, L. F. M., and W. S. Arbuckle. 1954. The identification of fat globules in the internal structure of ice cream. *J. Dairy Sci.* 37:642.

Dahlberg, A. C., and J. C. Hening. 1928. Chocolate ice cream. *Ice Cream Trade J.* 24(5):42.

Dahle, C. 1927. The manufacture of chocolate ice cream. *Ice Cream Rev.* 11:August, p. 90.

Damadoran, S., and J. E. Kinsella. 1980. Flavor protein interactions. Binding of carbonyls to bovine serum albumin: Thermodynamic and conformational effects. *J. Agr. Food Chem.* 28:567.

Emberger, R. 1994. Research into the flavor of vanilla. H & R Contact. Haarmann and Reimer, Inc. Springfield, NJ. pp. 4–8.

Fabricius, N. E. 1930. Improving chocolate ice cream. *Ice Cream Rev.* 14:August, p. 72.

Fabricius, N. E. 1931. Strawberries for ice cream manufacture. Iowa AES Circ. 132.

Guadagni, D. G. 1956. Some quality factors in strawberries for ice cream. Quick Frozen Foods 18(7):211–212.

Hening, J. C. 1935. Using frozen cherries in cherry ice cream. *Ice Cream Trade J.* 30(11):16.

Kindel, G. 1994. Vanilla the queen of spices. H & R Contact. Haarmann and Reimer, Inc. Springfield, NJ. pp. 15–18.

Mack, M. J., and C. R. Fellers. 1932. Frozen fruits and their utilization in frozen products. Massachusetts AES Bull. 287.

Martin, W. H. 1931. The selection and use of flavoring in making chocolate ice cream. *Ice Cream Trade J.* 27(4):39.

Mela, D. J., and R. J. Marshall. 1992. Sensory properties and perceptions of fat. In *Dietary Fats: Determinants of Preference, Selection and Consumption.* D. J. Mela (ed.). Elsevier Applied Science, New York.

Nielsen, C., Jr. 1981. The story of vanilla. Nielsen-Massey Vanillas, Inc. Chicago, IL, 9 pages.

Ohmes, R. R., and R. T. Marshall. 1995. Response of sensory panel to vanillin in nonfat ice creams containing whey-based fat replacers. *J. Dairy Sci.* 78 (Suppl. 1):Abstr. 129.

Pangborn, R. M., M. Simone, and T. A. Nickerson. 1957. Influence of sugar in ice cream. I. Consumer preference for vanilla ice cream. *Food Technol.* 11:679–682.

Reid, W. H. E., and W. S. Arbuckle. 1938. The effects of serving temperature upon consumer acceptance of ice cream and sherbets. Missouri AES Bull. 272.

Reid, W. H. E., and W. E. Painter. 1931. Freezing properties, stability, and physical qualities of chocolate ice cream. *Ice Cream Rev.* 15(12):40.

Schirle-Keller, J. P., and G. A. Reineccius. 1992. Flavor interactions with protein and carbohydrate-based fat replacers. Inst. Food Technol. Ann. Mtg. New Orleans, Jun 20–24. Abstr. 183.

Tuckey, S. L., P. H. Tracy, and R. A. Ruehe. 1932. Studies in the manufacture of chocolate ice cream. *Ice Cream Trade J.* 28(8):39.

8
Calculation of Ice Cream Mixes

THE IMPORTANCE OF CALCULATIONS

Ice cream makers need to know how much it costs to make each gallon of ice cream; how much ice cream can be made from a gallon of mix; how much cream, sugar, etc., it takes to make the desired amount and quality of product. Answers to such questions can be obtained only if it is known how to make at least simple mix calculations. A knowledge of calculations is also helpful in properly balancing a mix to provide uniform quality and to meet legal standards.

The method and procedures of making calculations are demonstrated by a few typical problems along with some detail to aid in understanding. The presentation focuses on the method and procedure with minimum effort in performing the arithmetic. The reader is assumed to be most interested in learning the quick and easy way to arrive at the correct answer. Therefore, little explanation is given of the mathematics involved in deriving the formulas or the logic in setting up the equations.

Much practice is usually necessary to develop speed and accuracy in making calculations. This practice can be obtained by using the demonstrated problems as patterns for setting up and solving similar problems.

Ice cream mixes can be divided into simple and complex groups. Simple mixes are composed of ingredients that supply one constituent each, whereas complex mixes include at least one constituent that is obtained from two or more ingredients. For example, fat could be supplied from cream only for a simple mix but be supplied by milk, cream, and whole concentrated milk for a complex mix. Therefore, complex mixes are more difficult to calculate than simple mixes. Simple mixes can be figured by multiplication, addition, subtraction, or division, whereas complex mixes require use of the Pearson square method, the serum point method, or algebra.

MATHEMATICAL PROCESSES MOST FREQUENTLY USED

Even though the discussion on calculations assumes a working knowledge of arithmetic, it may be helpful to review the following mathematical facts, which will be used frequently:

8 CALCULATION OF ICE CREAM MIXES

1. When a percentage figure is used in division or multiplication, the percent sign is dropped and the decimal point is moved two places to the left. Thus, 94% = 0.94 and 2.25/94% indicates that 2.25 is divided by 0.94, or

$$\frac{2.25}{0.94} = 2.25 \div 0.94 = 2.39$$

2. When the amount of milk and the percentage of a constituent are given, the amount of that constituent is obtained by multiplication:

$$50 \text{ lb milk} \times 4\% \text{ milkfat} = 50 \times 0.04 = 2.00 \text{ lb fat}$$

3. When the amount and the percentage of fat in milk or cream are known, the amount of milk or cream is obtained by division:

$$\frac{2.00 \text{ lb fat}}{4\%} = \frac{2.00}{0.04} = 50 \text{ lb milk}$$

4. When the amount of fat and milk are given, the percentage fat is obtained by division:

$$\frac{2.00 \text{ lb fat}}{50 \text{ lb milk}} = 0.04 = 4\%$$

METHODS OF CALCULATING MIXES

The Pearson square method is useful when the calculation is limited to the proportion of milk and cream needed, but it will not readily calculate the amount of nonfat milk solids (NMS) needed. The algebraic method is equally accurate and applicable to the most complex problems, but it involves rather lengthy calculations and a thorough knowledge of setting up and solving simultaneous equations. The serum point method is basically identical to the formulas presented in these chapters. However, the use of formulas simplifies the procedure, making it easier to learn.

Another advantage of the formulas presented in this book is that they make possible the calculation of tables needed by a manufacturer who, because of particular repetitive conditions, can justify the effort of their preparation. Tables that indicate the amount of each ingredient to be used have the advantage of eliminating errors and saving time. However, a new table must be prepared for every change in composition of the mix or of its ingredients. This means that a large number of lengthy tables would likely be required if only tables were to be used in mix calculation.

STANDARDIZING MILK AND CREAM

For convenience in calculating mixes it may be desirable to use an easy method for standardizing milk and cream so that stocks are always of the same fat content. Either of the following two methods is satisfactory.

Pearson Square Method

Draw a rectangle with two diagonals. At the upper left-hand corner write the test (fat percentage) of the cream to be standardized. At the lower left-hand corner write the test of the milk to be used in standardizing. In the center of the rectangle, place the desired test of the standardized product. "Cross-subtract" and at the right-hand corners write the differences between the numbers at the diagonally opposed left-hand corners and the number in the center. The number at the upper right-hand corner represents the proportion of cream of the test indicated by the number at the upper left-hand corner. The number at the lower right-hand corner indicates the proportion of milk of the test indicated by the number in the lower left-hand corner. By mixing the milk and cream in these proportions the desired fat content will be obtained.

For example, 35% cream is to be standardized to 20% using 4% milk, as shown in Figure 8.1. The difference 20 – 4 = 16 represents the amount of 35% cream that must be mixed with 15 lb of 4% milk to make 31 lb of 20% cream. When the proportions of milk and cream have been found, any amount of 35% cream may be standardized to 20% by mixing with 4% milk in the proportions of 16/31 of cream to 15/31 of milk. For example, if 310 lb of 20% cream is

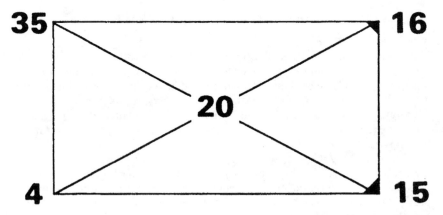

Figure 8.1. Pearson square method for standardizing 35% cream to 20% using 4% milk. The number in the lower right corner is the difference between the number in the upper left corner and the number in the center. The number in the upper right corner is the difference between the number in the lower left corner and the number in the center.

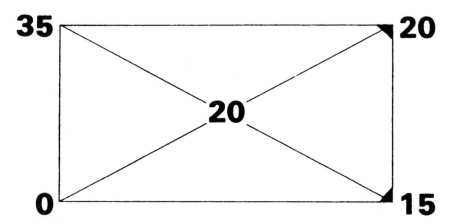

Figure 8.2. Pearson square method for standardizing 35% cream to 20% using skim milk (0% fat).

wanted, then 16/31 × 310 = 160 lb, the amount of 35% cream needed, and 15/31 × 310 = 150 lb, the amount of 4% milk needed.

If skim milk is used instead of milk, the figures would be as shown in Figure 8.2. In this case a mixture of 20/35 cream and 15/35 skim milk will test 20%.

Arithmetical Method

The arithmetical method of standardizing is quite simple and accurate. It involves multiplying the amount of cream by the difference between the test of the cream and the required test, then dividing the product by the test of the milk to be added.

Problem 1. Standardize 120 lb of 30% cream to 20% using 3.8% milk.

Solution. The difference between the test of the cream and the required test is 30 − 20 = 10. The difference between the required test and the test of the milk is 20 − 3.8 = 16.2.

Using these figures, we have 120 × 10 ÷ 16.2 = 74.07 lb. Therefore, 74.07 lb of 3.8% milk are required to reduce the test of 120 lb of 30% cream to 20%. The accuracy of this method is shown in the following calculations. The amount of fat in 74.07 lb of 3.8% milk is

$$74.07 \times 0.038 = 2.8148 \text{ lb}$$

The amount of fat in 120 lb of 30% cream is

$$120 \times 0.30 = 36 \text{ lb}$$

Thus, we have a total of 194.07 lb of milk and cream containing 31.8148 lb of fat. The same amount of 20% cream contains

$$194.07 \times 0.20 = 38.814 \text{ lb}$$

Problem 2. Calculate the quantities of ingredients needed for 100 lb of mix containing 12% fat, 15% sugar, and 0.5% stabilizer when 40% cream and 4% milk are used.

Solution

Step 1. First prepare a proof sheet (Table 8.1). It will be useful in calculating the mix and checking the accuracy of the calculations. Enter the amount of sugar and stabilizer required: 15 and 0.5 lb, respectively, for a total of 15.5 lb. This leaves 84.5 lb of milk and cream.

Step 2. Calculate to the nearest 0.1% the percentage of fat needed in the mixture of milk and cream to provide 12% fat in the 100 lb of mix:

$$(12/84.5) \times 100 = 14.2\%$$

Step 3. Use the Pearson square method to determine the proportions of milk and cream (see Figure 8.3). Then calculate the amounts of each to produce 84.5 lb:

$$25.8 + 10.2 = 36.0 \text{ parts}$$
$$(25.8/36) \times 84.5 = 60.56 \text{ lb } 4\% \text{ milk}$$
$$(10.2/36) \times 84.5 = 23.94 \text{ lb } 40\% \text{ cream}$$

The completed proof sheet is shown in Table 8.2. However, this mix has insufficient nonfat milk solids to bring the total milk solids (TMS) to 20%. If we simply add enough nonfat dry milk to raise the TMS to 20% (1.2% would be required), the mix is no longer balanced. It would then weigh 101.2 lb, and the percentages of fat, NMS, sugar, stabilizer, and total solids would then be reduced to 11.86, 9.88, 14.82, 0.49 and 34.88, respectively. This example establishes that a method is often needed that takes into consideration both fat and NMS to correctly balance an ice cream mix. Such a method is the one that follows.

CALCULATING MIXES WITH THE SERUM POINT METHOD

This method is based on the principle that the amounts of NMS and fat contributed by milk of any composition can be subtracted from the entire quantity of fat and NMS needed in a mix, leaving the remainder to be supplied by concentrated sources of fat or NMS. Rules for the serum point method are as follows:

Table 8.1. Sample Proof Sheet

Ingredients	Ingredient weight (lb)	Calculated constituents				
		Fat (lb)	NMS (lb)	Sugar (lb)	Stabilizer (lb)	TS (lb)	Cost ($)

Total

Calculated %

Check with desired wt. desired %

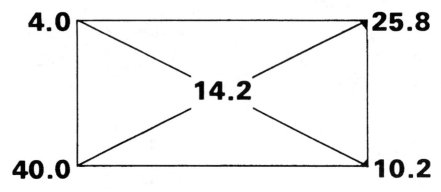

Figure 8.3. Application of Pearson square method to Problem 2.

1. List the amount and composition of the desired mix.
2. Determine the amounts of fat, serum solids (NMS), stabilizer/emulsifier, etc. by using the formula

amount of ingredient = (amount of mix) × (% of ingredient)

3. Calculate the amount of serum of the mix (recall that serum is water and serum solids or NMS) by subtracting the weights of *all* of the other ingredients from the weight of total mix:

serum = mix − (fat + sweeteners + stab/emul)

4. To calculate the amount of concentrated milk needed, it is necessary to know the amount of serum solids and the amount of serum in 1 lb of concentrated milk as well as amounts of the same components in the mix (see steps 2 and 3). The formula for the amount of concentrated milk is

$$\text{amount of conc. milk} = \frac{(\text{NMS needed}) - (\text{serum of mix} \times 0.09^1)}{(\text{NMS/lb conc. milk}) - (\text{serum/lb conc. milk} \times 0.09)}$$

5. If there is any fat or sugar in the concentrated milk[2], it/they must be calculated:

fat = (amount of conc. milk) × (% fat)
sugar = (amount of conc. milk) × (% sugar)

---

[1] The figure 0.09 represents the % NMS of skim milk, in this case 9%. However, if skim milk tests 8.6% NMS (or other %), use the appropriate figure.

[2] The general term *concentrated milk* as used here refers both to plain concentrated milk, which is often called plain condensed milk in the industry, and to sweetened condensed milk. Additionally, usage of the term in these illustrations is meant to apply to all concentrations of fat.

Table 8.2. Completed Proof Sheet For Problem 2

Ingredients	Ingredient weight (lb)	Calculated constituents					
		Fat (lb)	NMS (lb)	Sugar (lb)	Stabilizer (lb)	TS (lb)	Cost ($)
Sugar	15.0			15.0		15.0	
Stabilizer	0.5				0.5	0.5	
Cream, 40%	24.0	9.6	1.3			10.9	
Milk, 4%	60.5	2.4	5.3			7.7	
Total	100.00	12.0	6.8	15.0	0.5	34.1	—
Calculated %		12.0	6.8	15.0	0.5	34.1	
Check with desired wt. desired %		12.0		15.0	0.5		

6. Calculate the amount of fat needed from the milk and cream by subtracting the amount of fat in the concentrated milk from the total amount of fat needed in the mix:

fat (milk and cream) = fat (mix) − fat (concentrated milk)

7. Calculate the amount of sugar that must be added to the mix by subtracting the amount of sugar in the concentrated milk from the total amount needed in the mix:

sugar (needed) = sugar (total) − sugar (concentrated milk)

8. If there is no fat or sugar in the concentrated milk, omit steps 6 and 7.
9. Calculate the amount of milk and cream needed by subtracting the total of all other ingredients from the total amount of the mix:

milk and cream = (total amount of mix) − (total amounts of other ingredients)

10. The amount of cream needed is then calculated as follows:

$$\text{cream} = \frac{(\text{fat needed}) - [(\text{milk and cream needed}) \times (\% \text{ fat in milk})]}{(\text{amount of fat/lb cream}) - (\text{amount of fat/lb milk})}$$

11. Calculate the amount of milk needed by subtracting the amount of cream from the total amount of milk and cream.
12. The total weight of all ingredients will now equal the amount of mix required.
13. Verify the calculations.

For example, use the serum point method to calculate the ingredients needed for 100 lb of mix containing 12% fat, 11% NMS, 15% sugar, and 0.3% stabilizer. Consider the following ingredients to be on hand: 40% cream, 4% milk, concentrated milk (27% NMS), sugar, and stabilizer.

1. We have listed the composition and amounts of the mix.
2. Step 2 is easily determined from the desired mix proportions as 12 lb fat, 11 lb NMS, 15 lb sugar, and 0.3 lb stabilizer.
3. Total serum equals 100 − (15 + 12 + 0.3) = 72.7 lb.
4. The amount of concentrated milk is given by the serum point formula as

$$\frac{11 - (72.7 \times 0.09)}{0.27 - (1 \times 0.09)} = 24.76 \text{ lb}$$

5. There is no sugar or fat in the concentrated milk, so we go on to step 9.
9. Amounts of milk and cream are calculated as 100 − (24.76 + 15 + 0.3) = 59.94 lb.

8 CALCULATION OF ICE CREAM MIXES

10. The amount of cream needed is

$$\frac{12 - (59.94 \times 0.04)}{0.40 - 0.04} = 26.67 \text{ lb}$$

11. The amount of milk needed is then $59.94 - 26.67 = 33.27$ lb.
12. The sum of the weights of the ingredients is now 100 lb.
13. All of the calculations should then be verified by completing a proofsheet.

MIX DECISIONS

The following are key decisions an ice cream manufacturer must make before calculating an ice cream mix:

1. *Composition of the mix*: the proportionate amount of each constituent must be specified, e.g. the % fat.
2. *Size of batch*: batch size is usually constant within a factory. Calculation on a 100-lb (kg) basis permits easy conversion to larger batches.
3. *Choice of ingredients*: decisions should be based on qualities, functions, costs, and availabilities.
4. *Composition of ingredients*: although tables offer general information about composition, best results are obtained when calculations are based on analyses of the ingredients used.
5. *Classification of the mix*: the above information permits classification of the mix as simple or complex, and calculations can be made accordingly.

After these decisions have been made, the following steps are taken to calculate the mix:

1. Prepare a proof sheet and record each subsequent calculation.
2. Calculate the amount of each ingredient that supplies only one constituent of the mix (usually sugar and stabilizer).
3. Calculate the amount of each ingredient that supplies concentrated serum solids or fat. Record the amounts of constituents supplied.
4. Calculate the amount of each remaining dairy product. Record the amounts of constituents supplied.
5. Sum each column of the proof sheet and compare the "needed" and "total" sums.

SIMPLE MIXES

Simple mixes require minimal calculations and include such mixes as one made of stabilizer, sugar, cream, and concentrated skim milk or nonfat dry milk (NDM). Mixes having two sources of one constituent are more complex, but

114 ICE CREAM

sometimes the calculation can be reduced to the same simple procedure as is used for the simplest mixes.

Problem 3. How much cream, NDM, sugar, dried egg yolk, stabilizer, and water will be needed to make 450 lb of mix containing 10% fat, 11% NMS, 14% sugar, 0.5% egg yolk solids, and 0.5% stabilizer? Compositions of the ingredients are given in Table 8.3.

Solution

Step 1. List the available ingredients and the amounts of each constituent in the desired mix, as in Table 8.3.

Step 2. Make a proof sheet and enter the weight and percentage of each constituent in the desired weight and percent columns. Enter the information obtained in each succeeding step. (See completed proof sheet, Table 8.4.)

Step 3. Compute the amount of each ingredient that supplies only one constituent. In this problem these will be the stabilizer and sugar.

The weight of moisture-free stabilizer must equal 2.25 lb. Since the stabilizer is 90% solids, it follows that

$$2.25/0.90 = 2.50 \text{ lb}$$

Enter 2.50 in the weight column of the proof sheet, and enter 2.25 in the stabilizer and TS columns. Because sugar contains no water, 63 lb of either cane or beet sugar provides the full amount of sweetener required. Enter this figure in the weight, sugar, and TS columns of the proof sheet.

Step 4. Compute the amounts of those ingredients that supply more than one constituent of the mix but are the only remaining source of some one constituent—in this problem, the dried egg yolk, the only source of egg yolk solids.

$$2.25/0.94 = 2.39 \text{ lb}$$

Table 8.3. Available Ingredients/Desired Composition Table for Problem 3

Available ingredients		Desired mix composition		
Ingredient	Constituent (%)[a]	Constituent	Proportion (%)	Weight (lb)
Stabilizer	TS (90%)	Fat	10	45.0
Dried egg yolks[b]	Fat (62.5%)	NMS	11	49.5
	TS (94%)			
Sugar	Sugar (100%)	Sugar	14	49.5
Nonfat dry milk	TS (97%)	Egg yolk solids[b]	0.5	2.25
	NMS (97%)	Stabilizer[c]	0.5	2.25
Cream	Fat (30%)			
	NMS (6.24%)			
Water				

[a]The figures are approximate, and are taken from Table 5.5.
[b]Although egg products are not widely used in ice cream manufacture, they are included here for illustrative purposes.
[c]Stabilizer in these calculations means the moisture-free part only. It also includes the emulsifier.

Table 8.4. Completed Proof Sheet for Problem 3

Ingredients	Ingredient weight (lb)	Calculated constituents					
		Fat (lb)	NMS (lb)	Sugar (lb)	Egg solids (lb)	Stabilizer (lb)	TS (lb)	Cost ($)
Stabilizer	2.50					2.25	2.25	
Dried egg yolk	2.39	1.49			2.25		2.25	
Sugar	63.00			63.00			63.00	
Cream (30%)	145.03	43.51	9.05				52.56	
Nonfat dry milk solids	41.70		40.45				40.45	
Water	195.38							
Total	450.00	45.00	49.50	63.00	2.25	2.25	160.51	
Calculated %		10.00	11.00	14.00	0.50	0.50	35.67	
Check with desired wt desired %	450.00	45.00 10.00	49.50 11.00	63.00 14.00	2.25 0.50	2.25 0.50	162.00 36.00	

Enter this in the weight column and 2.25 in the egg solids and TS columns of the proof sheet.

Also calculate the amount of the other constituent supplied by the egg yolk solids, namely, fat:

$$2.39 \times 0.625 = 1.49 \text{ lb}$$

Enter in the proof sheet. This 1.49 lb of fat is part of the 2.25 lb of egg solids or TS. Because nutritional labeling regulations require that all fat be accounted for on the label, this 1.49 lb of fat should be entered as part of the fat. This leaves 43.51 lb of fat to be supplied by other ingredients, in this case the cream. Thus, the amount of 30% cream needed is

$$43.51/0.30 = 145.03$$

Enter in the proof sheet. This cream furnishes not only fat but also NMS, which is computed as follows:

$$145.03 \times 0.0624 = 9.05 \text{ lb}$$

Enter this in the proof sheet. The total 49.5 lb of NMS that is needed minus the 9.05 lb from the cream leaves 40.45 lb of NMS to be supplied by the remaining ingredients, in this case the NDM. It follows that the weight of NDM needed is

$$40.45/0.97 = 41.70 \text{ lb}$$

Enter this in the proof sheet. Enter 40.45 lb in the NMS and TS columns of the proof sheet.

Finally, total the calculated amounts of each ingredient, and subtract this sum from the total 450 lb of mix desired. The difference represents the amount of water needed:

$$450 - (2.50 + 2.39 + 63.00 + 145.03 + 41.70) = 450.00 - 254.62 = 195.38 \text{ lb}$$

Enter in the proof sheet.

Step 5. At this time it may be desirable to calculate the cost of the mix by entering in the proof sheet the costs of the ingredients used.

Step 6. To find the percentage (test) of each constituent in the mix, each column should now be totaled. The calculated test (percentage) of any constituent is obtained by dividing the total of that column by the total weight of all the ingredients. For example, the calculated test of NMS is obtained by dividing 49.50 lb of NMS by 450 lb, giving 11.00%. Similarly, 160.51 lb, the total weight of TS, is divided by 450 lb to give 35.67% TS. In a similar manner find percentages of each of the other constituents and enter them in their respective columns of the proof sheet.

8 CALCULATION OF ICE CREAM MIXES

Step 7. Determine the correctness of the calculations from the proof sheet by comparing the calculated percentage for each constituent with the corresponding desired test. Usually the mix is said to be correctly calculated when the two percentages agree within 0.1% (See Table 8.4.)

COMPLEX MIXES

We have identified complex mixes as those for which at least one constituent is obtained from more than one ingredient. These complex mixes are most rapidly calculated by using formulas in the following steps:

1. List the available ingredients and the number of pounds of each constituent in the desired mix (see Table 8.3).
2. Make a proof sheet and enter the weight and percentage of each constituent of the desired mix. Enter the information obtained in each succeeding step.
3. Compute the amount of each ingredient that supplies only one constituent. This usually includes the stabilizer, egg products, and sugar substitutes.
4. Calculate, using one of the following formulas, the amounts of concentrated and/or dry dairy products needed to supplement liquid milk or skim milk:

Amount of NDM:[3]
$$\frac{(\text{NMS needed}) - (\text{serum of mix} \times 0.09)}{\text{NMS per pound of NDM}} \quad (\text{I})$$

Amount of concentrated milk or concentrated skim milk:
$$\frac{(\text{NMS needed}) - (\text{serum of mix} \times 0.09)}{\text{NMS/lb conc. milk} - (\text{serum/lb conc. milk} \times 0.09)} \quad (\text{II})$$

5. Calculate the amount of sugar needed.
6. Use the following formula to calculate the amount of cream or butter needed to supplement milk:

$$\frac{(\text{fat needed}) - (\text{milk and cream needed} \times \text{fat/lb milk})}{(\text{fat/lb cream}) - (\text{fat/lb milk})} \quad (\text{III})$$

7. Calculate the amount of sweetened condensed milk or skim milk needed.
8. Add each column of the proof sheet and compute the percentages.
9. Examine the figures for accuracy by checking them against the desired composition.
10. Compute the cost of the mix (lb/ingredient × $/lb).

Problem 4. NMS from Three Sources. Calculate the amounts of ingredients needed to make 2,000 lb of mix testing 14% fat, 9% NMS, 13% sugar, and

---

[3]Although not absolutely accurate, this formula can be used for practical operations when butter is the source of fat.

0.5% stabilizer. Available ingredients are 40% cream, 4% milk, concentrated skim milk (27% NMS, 27% TS), sugar, and stabilizer.

Solution. This is a complex mix since the NMS is to be obtained from three sources (concentrated skim milk, cream, and milk).

Step 1. Make a list of available ingredients and the constituents desired (Table 8.5).
Step 2. Make a proof sheet and enter the information from step 1 (Table 8.6).
Step 3. In this problem there is only one source of stabilizer, and it contains 90% TS. Because 90% of the amount of stabilizer used must equal 10 lb, the amount of stabilizer needed is 10.0/0.90 = 11.11 lb. This figure is entered in the weight column of the proof sheet.

The 11.11 lb of stabilizer contains 10 lb of dry stabilizer, thus it provides 10 lb of dry stabilizer and of TS. Enter these figures in the stabilizer and TS columns, respectively.

In this problem no egg products and no sugar substitutes are used, so they require no calculation. Also, since sugar is the only source of sweetener, the

Table 8.5. Available Ingredients/Desired Composition Table for Problem 4

Available ingredients		Desired mix composition		
Ingredient	Constituent (%)	Constituent	Proportion (%)	Weight (lb)
Stabilizer	TS (90%)	Fat	14	280
Cane sugar	Sugar (100%)	NMS	9	180
Condensed skim milk	TS (27%)	Sugar	13	260
	NMS (27%)	Stabilizer	0.5	10
Cream	Fat (40%)			
	NMS (5.35%)			
Whole milk	Fat (4%)			
	NMS (8.79%)			

Table 8.6. Completed Proof Sheet for Problem 4

		Calculated constituents				
Ingredient	Ingredients weight (lb)	Fat (lb)	NMS (lb)	Sugar (lb)	Stab (lb)	TS (lb)
Stabilizer	11.11				10.00	10.00
Sugar	260.00			260.00		260.00
Conc. skim milk	275.56		74.40			74.40
Cream 40%	616.31	246.52	32.97			279.49
Milk, 4%	837.02	33.48	73.57			107.05
Total	2,000.00	280.00	180.94	260.00	10.00	730.94
Calculated %		14.00	9.05	13.00	0.50	36.55
Check with desired wt	2,000.00	280.00	180.00	260.00	10.00	730.00
desired %		14.00	9.00	13.00	0.50	36.50

8 CALCULATION OF ICE CREAM MIXES

calculation of 260 lb is easily completed, and the amount is added to the sugar and TS columns of the proof sheet.

Step 4. To find the amount of concentrated skim milk needed, we use formula (I). The figures we need for the formula are obtained as follows:

From step 1, the amount of NMS needed is 180 lb. To obtain the amount of serum of the mix, first find the sum of the fat (280 lb, step 1), the stabilizer (11.11 lb, step 3), and the sugar (260 lb, step 3); a total of 551.11 lb. Subtract this total from the desired total weight of the mix, 2,000.00 lb, leaving 1448.89 lb, the amount of serum of the mix. We obtain the amount of NMS in 1 lb of concentrated skim milk by multiplying the 1 lb by 27%, which gives 0.27.

We substitute these figures in formula (I):

$$\frac{180 \text{ lb} - (1448.89 \text{ lb} \times 0.09)}{0.27 - 0.09} = \frac{180 - 130.40}{0.27 - 0.09} = \frac{49.60}{0.18} = 275.56 \text{ lb}$$

This number, the amount of concentrated skim milk needed, is entered in the weight column of the proof sheet (Table 8.6). We also enter in their respective columns the amounts of NMS and TS supplied by the concentrated skim milk. These are obtained by multiplying 275.56 lb by the respective percentages of NMS and TS:

$$275.56 \times 27\% \text{ NMS} = 74.40 \text{ lb NMS}$$
$$275.56 \times 27\% \text{ TS} = 74.40 \text{ lb TS}$$

Step 5. The amount of sugar needed has been calculated in step 3.

Step 6. The amount of cream needed is calculated with the aid of formula (III):

From step 1, the amount of fat needed is 280 lb. To calculate the amount of milk and cream needed, we first find the sum of the stabilizer (11.11 lb, step 3), the concentrated skim milk (275.56 lb, step 4), and the sugar (260 lb, step 3). The total is 546.67 lb, and this is subtracted from the desired weight of the mix, 2,000 lb, leaving 1453.33 lb, the amount of milk and cream needed. The amounts of fat in 1 lb of 40% cream and 1 lb of 4% milk are obtained as follows:

$$\text{Fat in cream} = 1 \text{ lb} \times 0.40 = 0.40 \text{ lb}$$
$$\text{Fat in milk} = 1 \text{ lb} \times 0.04 = 0.04 \text{ lb}$$

These figures are substituted in formula (III):

$$\frac{280 \text{ lb} - (1453.33 \times 0.04)}{0.40 - 0.04} = \frac{280 \text{ lb} - 58.13 \text{ lb}}{0.36} = \frac{221.87}{0.36} = 616.31 \text{ lb}$$

This number, the amount of cream needed, is entered in the proof sheet, and the amounts of fat and NMS supplied by the cream are calculated as follows:

$$616.31 \text{ lb} \times 40\% = 246.52 \text{ lb fat}$$
$$616.31 \text{ lb} \times 5.35\% = 32.97 \text{ lb NMS}$$

and adding:

$$246.52 + 32.97 = 279.49 \text{ lb TS}$$

Step 7. The amount of milk needed is computed by subtracting the amount of cream (616.31 lb, step 6) from the amount of milk and cream needed (1453.33 lb, step 6), to obtain 837.02 lb. This is entered in the weight column. The amounts of fat, NMS, and TS contained in the milk are calculated by multiplying 837.02 by the respective percentages:

$$837.02 \times 0.04 = 33.48 \text{ lb fat}$$
$$837.02 \times 0.0879 = 73.57 \text{ lb NMS}$$

and adding:

$$33.48 \text{ lb} + 73.57 \text{ lb} = 107.05 \text{ lb TS}$$

and these are entered in the respective columns of the proof sheet.

Step 8. Each column of the proof sheet is now totaled. These totals should agree with the corresponding figures of the desired mix (see Table 8.6). (The calculations in this book are done using four decimal places, although only two are recorded in most places. One unit (of a pound) in the second decimal place is equivalent to about one-sixth of an ounce.)

The percentage of TS is calculated as follows:

$$\frac{\text{lb TS} \times 100}{\text{lb mix}} = \frac{730.94 \times 100}{2{,}000.00} = 36.55\% \text{ TS}$$

Similar computations would yield the calculated percentages for each of the constituents.

Step 9. The correctness of the calculations may be determined by comparing the calculated percentage for each constituent with the corresponding desired test of the mix. For example, in this problem the calculated percentage of NMS is 9.05% (180.94 lb NMS/2,000 lb mix), compared with 9.00% NMS of the desired mix. Since the difference between the two percentages is less than 0.10%, it is not deemed significant. Similarly, the other differences are not significant. Therefore, the calculations are considered correct.

Step 10. The cost of a mix depends on many factors, so the figures are not included in this example.

Problem 5. More Than One Source of Sugar, NMS, and Fat. Calculate the amount of ingredients needed to make 900 lb (about 100 gal) of mix testing 12% fat, 10% NMS, 14% sugar, and 0.4% stabilizer. The available ingredients are 40% cream, 4.0% milk, dehydrated corn syrup solids to supply 25% of the

8 CALCULATION OF ICE CREAM MIXES

sugar (sweetness in proportion to that of sucrose), sweetened condensed skim milk (42% sugar, 30% NMS), sugar, and stabilizer.

Solution. This complex mix has three sources each of sweetener and NMS and two sources of fat.

Step 1. List the available ingredients and desired constituents (Table 8.7).
Step 2. Make a proof sheet and enter the information from step 1.
Step 3. In this problem there is one source of the 3.6 lb of dry weight of stabilizer. Since the stabilizer contains 90% TS, the wet weight needed equals 3.6/0.90 = 4.00 lb. Enter this figure in the weight column of the proof sheet, and enter 3.6 into both the stabilizer and TS columns (Table 8.8).

Dehydrated corn syrup solids (CSS) are used to supply 25% of the desired sweetener. Therefore, 126 × 0.25 = 31.50 lb of "sugar" to be obtained from this source. Enter this 31.50 lb in the sugar column. Because these CSS have a sucrose equivalency of 47%, it follows that the amount of CSS needed is 31.50/0.47 = 67.02 lb. Enter this number in the weight column of the proof sheet.

To obtain the TS of the CSS multiply the weight, 67.02 lb, by the percentage of TS, 96.5%, to obtain 64.68 lb TS. Enter this number in the TS column.

Step 4. Use formula (II) to find the amount of sweetened condensed skim milk needed. Obtain the figures needed for the formula as follows:

From step 1, the amount of NMS needed is 90 lb. To obtain the amount of serum of the mix, first sum the fat (108 lb, step 1 or 2), stabilizer (4.0 lb, step 3), dehydrated CSS (67.02 lb, step 3), and the sugar not supplied by the CSS (126 − 31.50 = 94.50). This sum, 273.52 lb, is subtracted from the total weight of the mix: 900 − 273.52 = 626.48 lb serum.

Obtain the amount of NMS in 1 lb of sweetened condensed milk by multiplying the 1 lb by 30%, which gives 0.30. To obtain the amount of serum in 1 lb of sweetened condensed milk, multiply 1 lb by 42% sugar to get 0.42 lb sugar. Then subtract 0.42 lb sugar from the 1 lb sweetened condensed milk. leaving 0.58 lb serum (water + NMS).

Now, substitute these figures into formula (II):

$$\frac{90.00 \text{ lb} - (626.48 \text{ lb} \times 0.09)}{0.30 - (0.58 \times 0.09)} = \frac{90 - 56.38}{0.30 - 0.052} = \frac{33.62}{0.248} = 135.56 \text{ lb}$$

Table 8.7. Available Ingredients/Desired Composition Table for Problem 5

Available ingredients		Desired mix composition		
Ingredient	Constituent (%)	Constituent	Proportion (%)	Weight (lb)
Stabilizer	TS (90%)	Fat	12	108
Dehydrated corn		NMS	10	90
syrup solids	Sugar (47%)	Sugar	14	126
	TS (96.5%)	Stabilizer	0.4	3.6
Sweetened condensed				
skim milk	Sugar (42%)			
	NMS (30%)			
Cane sugar	Sugar (100%)			
Cream	Fat (40%)			
Milk	Fat (4%)			

Table 8.8. Completed Proof Sheet For Problem 5

Ingredients	Ingredient weight (lb)	Calculated constituents					
		Fat (lb)	NMS (lb)	Sugar (lb)	Stabilizer (lb)	TS (lb)	Cost ($)
Stabilizer	4.00				3.60	3.60	
Dehydrated corn syrup solids	67.02					64.68	
Sweetened condensed skim milk	135.56		40.67	31.50		97.61	
Sugar	37.56			56.94		37.56	
Cream, 40%	227.13	90.85	12.15	37.56		103.00	
Milk, 4%	428.73	17.15	37.68			54.83	
Total	900.00	108.00	90.50	126.00	3.60	361.28	
Calculated %		12.00	10.06	14.00	0.40	40.14	
Check with desired wt.	900.00	108.00	90.00	126.00	3.60	327.60	
desired %		12.00	10.00	14.00	0.40	36.40	

8 CALCULATION OF ICE CREAM MIXES 123

Enter this amount of sweetened condensed skim milk in the proof sheet under weight, and enter the amounts of NMS, sugar, and TS calculated as follows:

$$135.56 \text{ lb} \times 0.30 = 40.67 \text{ lb NMS}$$
$$135.56 \text{ lb} \times 0.42 = 56.94 \text{ lb sugar}$$
$$135.56 \text{ lb} \times 0.72 = 97.61 \text{ lb TS}$$

Step 5. Calculate the amount of sugar needed by adding the amounts provided by the CSS (31.5 lb) and the sweetened condensed skim milk (56.94 lb), and subtract from the total amount of sugar desired:

$$126 - (31.5 + 56.94) = 37.56 \text{ lb sugar}$$

Since sugar furnishes 100% sweetness (compared to other sweeteners sucrose has a relative sweetness of 100), the amount required is 37.56 lb. Enter this amount in the weight, sugar, and TS columns of the proof sheet.

Step 6. Calculate the amount of cream using formula (III) and the following figures:

From step 1, the amount of fat needed is 108 lb. To calculate the amount of milk and cream needed, first find the sum of the stabilizer (4 lb, step 3), the dehydrated CSS (67.02 lb, step 3), the sweetened condensed skim milk (135.56 lb, step 4), and the sugar (37.56 lb, step 5). Subtract this total from the weight of the mix: 900 − 244.14 = 655.86 lb of milk and cream.

Obtain the amount of fat in 1 lb of cream (1 lb × 40% = 0.40 lb) and the amount of fat in 1 lb of milk (1 lb × 4% = 0.04 lb).

Now, substitute these figures into formula (III):

$$\frac{108.00 \text{ lb} - (655.86 \times 0.04)}{0.40 - 0.04} = \frac{108.00 \text{ lb} - 26.23 \text{ lb}}{0.36} = \frac{81.77}{0.36} = 227.13 \text{ lb}$$

Enter this amount of cream needed in the weight column of the proof sheet. Also, enter in their respective columns the amounts of fat, NMS, and TS contained in the cream obtained as follows:

$$227.13 \text{ lb cream} \times 0.40 \text{ fat} = 90.85 \text{ lb fat}$$
$$227.13 \text{ lb cream} \times 0.0535 \text{ NMS} = 12.15 \text{ lb NMS}$$
$$90.85 \text{ lb fat} + 12.15 \text{ lb fat} = 103.00 \text{ lb TS}$$

Step 7. Find the amount of milk needed by subtracting the amount of cream (277.13 lb) from the amount of milk and cream (655.86 lb, step 6), which leaves 428.73 lb. Enter this in the weight column, and calculate the amounts of fat and NMS contained in the milk by multiplying 428.73 lb by the respective percentages:

$$428.73 \text{ lb} \times 0.04 = 17.15 \text{ lb ft}$$
$$428.73 \text{ lb} \times 0.0879 = 37.68 \text{ lb NMS}$$

Sum the fat and NMS to obtain the TS:

$$17.15 + 37.68 = 54.83 \text{ TS}$$

Step 8. Total each column of the proof sheet and check that they agree with the corresponding figures of the desired mix (see Table 8.8).

The percentage of TS is computed as follows;

$$\frac{361.28 \text{ lb TS}}{900 \text{ lb mix}} \times 100 = 40.14\% \text{ TS}$$

Similarly, compute the percentages of the remaining constituents.

Step 9. The correctness of the calculations is determined by comparing the calculated percentages for each constituent with the corresponding test of the desired mix.

Step 10. If desired, calculate the cost of the mix.

There are times when it is necessary to use odd lots of ice cream mixes, cream, and concentrated milk products, etc. These materials can be used to make mixes if they are unquestionably of high quality. Accurate information must be available to permit reliable calculations of a mix using such ingredients. Problem 6 illustrates an example of such a case.

Problem 6. Using Leftovers. This problem shows how to use leftover products or small amounts of several ingredients.

A 900-lb mix containing 12% fat, 10% NMS, 14% sugar, 0.5% egg yolk solids, and 0.3% stabilizer is desired. The following leftover materials are to be used in their entirety: 40 lb frozen 50% cream, 90 lb 30% cream, 25 lb sweetened condensed skim milk (42% sugar, 28% NMS), 100 lb 3.0% milk, and 30 lb corn syrup. The following ingredients are also available: stabilizer, sugar, frozen egg yolk (33% fat, 50% TS), concentrated skim milk (32% NMS), 40% cream, and 4% milk.

Solution

Step 1. List the available ingredients and desired constituents (Table 8.9).
Step 2. Make a proof sheet and enter the information from step 1.
Step 3. Since there are several ingredients to be used completely, enter the amount of each of these in the weight column of the proof sheet. Then compute the amount of each constituent contained in each ingredient and enter the amounts in their respective columns. For example, 40 lb of frozen cream multiplied by its fat test (50%) gives 20 lb fat, multiplied by its NMS test (4.45%) gives 1.78 lb NMS, and multiplied by its TS test (54.45%) gives 21.78 lb TS. Similar multiplications are made for 30% cream, sweetened condensed skim milk, 3% milk, and corn syrup.

8 CALCULATION OF ICE CREAM MIXES

Table 8.9. Available Ingredients/Desired Composition Table for Problem 6

Available ingredients		Desired mix composition		
Ingredient	Composition (%)	Constituent	Proportion (%)	Weight (lb)
Frozen cream	Fat (50%)	Fat	12	108
	NMS (4.45%)	NMS	10	90
Cream	Fat (30%)	Sugar	14	126
	NMS (6.24%)	Egg solids	0.5	4.5
Sweetened condensed skim milk		Stabilizer	0.3	2.7
	NMS (28%)			
	Sugar (42%)			
Milk	Fat (3%)			
	NMS (8.33%)			
Corn syrup	Sugar (67%)			
	TS (83%)			
Stabilizer	TS (90%)			
Frozen egg yolk	Fat (33%)			
	TS (50%)			
Cane sugar	Sugar (100%)			
Concentrated skim milk				
	NMS (32%)			
	TS (32%)			
Cream	Fat (40%)			
	NMS (5.35%)			
Milk	Fat (4%)			
	NMS (8.79%)			

When these data have been added in the proof sheet, it is apparent that no stabilizer has been added. In this problem there is only one source of stabilizer, and it contains 90% TS. Thus, 2.7/0.90 = 3.0 lb of stabilizer. Enter this figure in the weight column of the proof sheet, and enter the 2.7 lb of dry weight of the stabilizer in the stabilizer and TS columns of the proof sheet.

The partially completed proof sheet now reveals that all of the 4.5 lb of egg yolk solids must be obtained from the frozen egg yolk. It contains 50% egg yolk solids and 50% TS. Therefore, 4.50/0.50 = 9.0 lb of frozen egg yolk. Enter this figure in the weight column of the proof sheet. Then, obtain the amounts of each constituent furnished by the egg yolk by multiplying the 9 lb by the respective percentages of fat, egg solids, and TS.

$$9 \text{ lb} \times 0.33 = 2.97 \text{ lb fat}$$
$$9 \text{ lb} \times 0.50 = 4.50 \text{ lb egg solids}$$
$$9 \text{ lb} \times 0.50 = 4.50 \text{ lb TS}$$

Enter these amounts in their respective columns.

Examination of the proof sheet will reveal that provision has been made for all of the stabilizer and egg solids as well as 30.60 lb of sugar equivalent (from the corn syrup and sweetened condensed milk). Since the only remaining source of sweetener is sugar, subtraction of 30.60 lb from the total amount of sugar needed, 126 lb, leaves 95.40 lb to be furnished by this source. Enter this figure in the weight, sugar, and TS columns of the proof sheet.

Addition of each column reveals that the components considered above furnish 392.4 lb weight, 52.97 lb fat, 22.73 lb NMS, 126 lb sugar, 4.50 lb egg yolk solids, and 2.7 lb stabilizer. Subtraction of each of these from their corresponding figures in the desired mix reveals that the remaining 507.6 lb of mix ingredients must furnish 55.03 lb fat and 67.27 lb NMS. Use these figures in the following steps:

Step 4. Using formula (I), find the amount of concentrated skim milk needed. Obtain the figures to substitute into the formula as follows:

Step 3 has provided the amount of NMS needed, 67.27 lb. Obtain the amount of serum by subtracting the amount of fat (step 3) from the total weight (step 3): 507.60 − 55.03 = 452.57 lb of serum. The amount of NMS in 1 lb of concentrated skim milk is 1 lb × 32% = 0.32 lb. Substitute these figures in formula (I):

$$\frac{67.27 \text{ lb} - (452.57 \text{ lb} \times 0.09)}{0.32 - 0.09} = \frac{67.27 - 40.73}{0.23} = \frac{26.54}{0.23} = 115.39 \text{ lb}$$

Enter this number in the weight column. Multiply this number by the percentages of NMS and TS in the concentrated skim milk:

$$115.39 \times 0.32 = 36.92 \text{ lb NMS}$$
$$115.39 \times 0.32 = 36.92 \text{ lb TS}$$

and enter these numbers in the proof sheet.

Step 5. The amount of sugar was calculated in step 3.

Step 6. Calculate the amount of cream using formula (III), with the following figures:

The amount of fat needed is 55.03 lb (step 3). Calculate the amount of milk and cream needed by subtracting the amount of concentrated skim milk needed (115.39, step 4) from the total weight (507.60, step 3), yielding 392.21 lb. The amounts of fat in 1 lb of 40% cream and 1 lb of 4% milk are 0.40 lb and 0.04 lb, respectively. Substitute these figures in formula (III):

$$\frac{55.03 - (392.21 \times 0.04)}{0.40 - 0.04} = \frac{55.03 - 15.69}{0.40 - 0.04} = \frac{39.34}{0.36} = 109.28 \text{ lb}$$

Enter this amount of cream in the weight column of the proof sheet, and calculate the amounts of fat, NMS, and TS furnished by it using the respective percentages of the components in the cream:

$$109.28 \times 0.40 = 43.71 \text{ lb fat}$$
$$109.28 \times 0.0535 = 5.85 \text{ lb NMS}$$
$$109.28 \times 0.4535 = 49.56 \text{ lb TS}$$

Step 7. Compute the amount of milk needed by subtracting the amount of cream (109.28 lb, step 6) from the amount of milk and cream (392.21 lb, step 6), which is 282.93 lb. Enter this in the weight column, and calculate the

8 CALCULATION OF ICE CREAM MIXES

amounts of fat, NMS, and TS contained in the milk by multiplying by the respective percentages:

$$282.93 \times 0.04 = 11.32 \text{ lb fat}$$
$$282.93 \times 0.0879 = 24.87 \text{ lb NMS}$$
$$282.93 \times 0.1279 = 36.19 \text{ lb TS}$$

Enter these numbers in the proof sheet.

Step 8. Total each column of the proof sheet and check that they agree with the corresponding figures of the desired mix (see Table 8.10).

The percentage of TS is calculated as follows:

$$\frac{333.40 \text{ lb TS} \times 100}{900 \text{ lb mix}} = 37.04 \text{ \% TS}$$

Similarly, calculate the percentages of the remaining constituents.

Step 9. Determine the correctness of the calculations by comparing the calculated percentage with the corresponding test of the desired mix.

Step 10. If desired, calculate the cost of the mix.

USE OF COMPUTERS IN ICE CREAM PRODUCTION

Computers and appropriate programs provide a quick and reliable way to calculate mixes. It is possible to have the computer calculate the amounts of each available ingredient to prepare a mix at a minimal cost. Also the computer will specify the cost range within which it is possible to use each ingredient. This is accomplished through a technique known as linear programming. The quantities of ingredients available and their composition are entered into the computer along with the acceptable compositional range for each element of the finished product. These parameters are automatically inserted into linear programming equations. The computer will then calculate the amount of each ingredient that will minimize the cost of the mix within the specified compositional and quality restrictions.

Byars (1969) gave the following instructions for using the computer to minimize the cost of production:

(1) Establish the desired formula and amount of mix with minimum and maximum or exact limitations for each ingredient.
(2) Establish the ingredients available, costs per pound, and compositional analyses of the ingredients.
(3) Specify maximum amount of each product available.
(4) Establish abbreviations to be used for each ingredient for computer analysis.
(5) Develop linear programming equations with the objectives of minimizing costs and the constraint equations involved.

Table 8.10. Completed Proof Sheet for Problem 6

| Ingredients | Ingredient weight (lb) | Calculated constituents |||||||
		Fat (lb)	NMS (lb)	Sugar (lb)	Egg solids (lb)	Stabilizer (lb)	TS (lb)	Cost ($)
Cream, frozen, 50%	40.00	20.00	1.78				21.78	
30%	90.00	27.00	5.62				32.62	
Sweetened condensed skim milk	25.00		7.00	10.50			17.50	
Milk, 3%	100.00	3.00	8.33				11.33	
Corn syrup	30.00			20.10			24.90[a]	
Stabilizer	3.00					2.70	2.70	
Egg yolk, frozen	9.00	2.97			4.50		4.50[a]	
Sugar	95.40			95.40			95.40	
Condensed skim milk	115.39		36.92				36.92	
Cream, 40%	109.28	43.71	5.85				49.56	
Milk, 4%	282.93	11.32	24.87				36.19	
Total	900.00	108.00	90.37	126.00	4.50	2.70	333.40	
Calculated %		12.00	10.04	14.00	0.50	0.30	37.04	
Check with desired wt	900.00	108.00	90.00	126.00	4.50	2.70	331.20	
desired %		12.00	10.00	14.00	0.50	0.30	36.80	

8 CALCULATION OF ICE CREAM MIXES

With a wide variety of product lines, very high production capacities, many new ingredients such as fat replacers, and consumer demand for special products, the number of variables that must be considered is not limited to sources of milkfat and NMS.

Hand calculation of formulas becomes more and more cumbersome with each increase in complexity of product line. Limitations imposed by government regulations, availability and costs of raw materials, storage capacity, inventory, shipping schedules, and other such variables argue for computerization of each aspect of control of the processes.

Furthermore, materials still account for the majority of product cost. This fact makes essential the use of the most efficient methods of minimizing costs while enssuring highest quality within the range of quality attributes designed into the product.

Whereas computerized formulation is a great time saver, the computer merely serves as a tool for the process. Therefore, to produce valid formulas, information fed into the system must be current and accurate.

Each product formula should contain applicable limitations on the proportions of milkfat, fat replacers, NMS, sweeteners, stabilizers, emulsifiers, colorings, and flavorings, etc. These limitations are called constraints in the examples that follow. As in doing hand calculations, each ingredient should be described in terms of its content of fat, NMS, sweeteners (with sucrose considered to have a relative sweetness of 100), stabilizer/emulsifier, and TS. The linear program reviews each material in terms of its possible contribution to the formula. This contribution is its characteristic value.

Use of a linear program to solve formulations could yield several solutions for each formula. If there are several ingredients with similar characteristics, these could be combined in a variety of ways to meet the quality restrictions placed on each product. However, only one solution will result in least cost. Therefore, the program looks at all possible combinations of ingredients and gives the one combination that satisfies the cost limitation.

In the examples that follow (Table 8.11), three mixes have been set up for use in computer formulation runs.

Ingredients are listed with their characteristic compositions in the available ingredients file (Table 8.12) and some are common to each of the formulas. Other characteristics can be entered as well.

Each formula has been established with constraint values that must be met (see Table 8.13). With seven dairy ingredients in the available ingredients file, there are several combinations that will satisfy the constraints of each formula,

Table 8.11. Production Run. Copyright © Computer Concepts Corp., 1983. All rights reserved.

Sequence number	Formula name	Finished weight	Percent shrink	Batch number/ number of batches
1	5 10% white mix	9200.00	0.00000	1.0
2	14 3% white mix	4600.00	0.00000	1.0
3	17 10% choc mix	4600.00	0.00000	1.0

Table 8.12. Available Ingredients File Listing. Copyright © Computer Concepts Corp., 1983. All rights reserved.

Ingredient	Cost/unit	% Fat	% Moisture	% Milk solids N/F	% Total solids	% Dairy	Minimum	Maximum
1 Whole milk	0.13014*	3.25000	88.0500	8.70000	11.9500	100.000		
5 Cream	0.72087*	38.3500*	56.2500	5.40000*	43.7500	100.000		
10 Skim milk	0.07275*	0.00000*	91.0000*	9.00000*	9.00000*	100.000		
12 Condensed	0.30455	0.00000	66.1000	33.9000	33.9000	100.000		
13 Butter milk cond	0.28000	4.00000	0.00000	30.0000	34.0000	100.000		
14 Skim milk pwdr	0.96750*	0.00000*	3.00000	97.0000*	97.0000	100.000		
16 Whey powder	0.14500*	1.40000*	6.13500	92.4650*	93.8650	100.000		
40 Corn syrup	0.08867*	0.00000	19.3000*	0.00000	80.7000*	0.00000		
43 Liquid cane sugar	0.30000*	0.00000*	33.7000*	0.00000*	69.8600	0.00000		
71 Emul/stab	1.20000	0.00000	0.00000	0.00000	100.000	0.00000		
74 Frodex	0.25430	0.00000*	0.00000	0.00000	100.000	0.00000		
134 Cocoa	0.62000	0.00000*	0.00000	0.00000	100.000	0.00000		
300 Water	0.00001	0.00000	100.000	0.00000	0.00000	0.00000		

8 CALCULATION OF ICE CREAM MIXES

Table 8.13. Composition Constraints (%) on Seven Dairy Product Ingredients

	Type of mix (fat content)		
Ingredient	White	White	Chocolate
Fat	10.1	3.1	9.4
NMS	10.0	12.0	9.5
Solids	9.1	11.0	12.4
Corn syrup solids	9.5		4.8
Stabilizer/emulsifier	0.5	0.4	0.6
Cocoa			2.6
Frodex	3.0		
Whey solids		3.0	2.375
Finished weight (lb)	9200	4600	4600

but there is only one least cost solution. All formulas are submitted to the computer at the same time in what is called a multiproduct run. Taking ingredients from the available ingredients file, the program satisfies the constraints of each formula.

The examples show formulas after being solved by the linear program. In the first example (Tables 8.14–8.16) all ingredients were considered to have been available in unlimited quantities; therefore, the program could pick materials that would solve the formula constraints and do so at the lowest cost. This low cost does not mean that quality was sacrificed, because quality was considered in setting limitations in the program.

Even though not all ingredients are available at all times or some ingredients must be forced into use, the program can easily solve the same formulas based on additional restrictions. In the available ingredients listing that follows (Table 8.17), all of the milk needed to be used. Therefore, a minimum on that ingredient was forced in. The amount of cream that could be used was 5000 lb, and this was given as a maximum. When the same three formulas were calculated with the least cost program, the constraints were all met, but the costs as compared with the "unlimited supply" run had increased. However, the solutions were still least-cost based on the new limitations. (The same three formulas with the new minimums and maximums are shown in Tables 8.18–8.20.)

In plant situations use of a computer for formulation of ice cream mixes can provide a significant time savings, make optimal use of ingredients by considering those that must be forced into use and those that have limited availability, maintain product quality by use of limitations set by the firm, and provide these benefits on a least-cost basis.

REFERENCES

Byars, L. L. 1969. Computerized linear programming for minimum cost ice cream blending. Proc. 65th Annu. Conv. IAICM, New Orleans.

Table 8.14. Multiproduct Run (Plain Mix, 10%) with No Ingredient Minimums or Maximums. Copyright © Computer Concepts Corp., 1983.

Ingredient	Total Weight	Batch Weight	Gallons	Percent Batch	Cost/Unit Low	Cost/Unit Actual	Cost/Unit High	Penalty
1 Whole milk	2143.33	2143.33	258.86	23.30 FW	0.11844	0.13014		0.01170
5 Cream					0.40341	0.72087	0.93143	
10 Skim milk					0.06263	0.07275		0.01012
12 Condensed					0.23587	0.30455		0.06868
13 Butter milk cond	2680.87	2680.87		29.14 FW	0.07520	0.28000	0.31311	
14 Skim milk pwdr					0.67488	0.96750		0.29262
40 Corn syrup	1083.02	1083.02	97.22	11.77 FW		0.08867		
43 Liquid sugar	1196.00	1196.00	107.36	13.00 FW		0.30000		
71 Emul/stab	46.00	46.00	4.60	0.50 FW		1.20000		
300 Water	2050.78	2050.78	246.19	22.29 FW	−0.14994	0.00001	0.01395	
Finished Weight	9200.00	9200.00	714.23	100.00 FW		0.30497	Total Cost=2805.74	

Constraint	Solution Value	Minimum Value	Maximum Value	Penalty Cost
2 % fat (total)	10.10	10.10	FW	0.01782
54 % NMS (inc whey)	10.00	10.00	FW	0.00696
35 % Emul & % stab	0.50	0.50	0.50 FW	−0.01200
29 % Sugar solids	9.10	9.10	9.10 FW	−0.00429
30 % Corn syrup slds	9.50	9.50	9.50 FW	−0.00110
55 % Total solids	39.18		FW	
98 Finished weight	9200.00	9200.00	9200.00 FW	0.30497

Table 8.15. Multiproduct Run (Plain Mix, 3%) with No Ingredient Minimums or Maximums. Copyright © Computer Concepts Corp., 1983.

Ingredient	Total Weight	Batch Weight	Gallons	Percent Batch	Low	Cost/Unit Actual	High	Penalty
1 Whole milk					0.11844	0.13014		0.01170
5 Cream	223.44	223.44	26.99	4.86 FW	0.40341	0.72087	0.93143	0.01012
10 Skim milk					0.06263	0.07275		0.06868
12 Condensed					0.23587	0.30455		
13 Butter milk cond	1374.44	1374.44		29.88 FW	0.11300	0.28600	0.31311	
14 Skim milk pwdr					0.67488	0.96750		0.29262
16 Whey powder	138.00	138.00	15.13	3.00 FW		0.14500		
43 Liquid sugar	722.86	722.86	64.89	15.71 FW		0.30000	0.66827	
71 Emul/stab	18.40	18.40	1.84	0.40 FW		1.20000		
74 Frodex	138.00	138.00	15.13	3.00 FW		0.25430		
300 Water	1984.86	1984.86	238.28	43.15 FW	14994	0.00001	0.01395	
Finished Weight	4600.00	4600.00	362.26	100.00 FW		0.18260	Total Cost=839.98	

Constraint	Solution Value	Minimum Value	Maximum Value	Penalty Cost
2 % Fat (total)	3.10	3.10	FW	0.01782
54 % NMS (inc whey)	12.00	12.00	FW	0.00696
17 % Whey solids	3.00		3.00 FW	0.00523
29 % Sugar solids	11.00	11.00	11.00 FW	−0.00429
32 % Frodex	3.00	3.00	3.00 FW	−0.00254
35 % Emul/stab	0.40	0.40	0.40 FW	−0.01200
98 Finished weight	4600.00	4600.00	4600.00 FW	0.18260

Table 8.16. Multiproduct Run (Chocolate Mix, 10%) with No Ingredient Minimums or Maximums.
Copyright © Computer Concepts Corp., 1983.

Ingredient	Total Weight	Batch Weight	Gallons	Percent Batch Low	Cost/Unit Actual	Cost/Unit High	Penalty
1 Whole milk	1123.52	1123.52	135.69	24.42 FW	0.12772	0.13014	
5 Cream	1796.34	1796.34	196.97	39.05 FW	0.03924	0.72087	0.74939
10 Skim milk	335.22	335.22	35.85	7.29 FW	0.04294	0.07275	0.07539
12 Condensed					0.27400	0.30455	0.32592
14 Skim milk pwdr					0.89196	0.96750	
16 Whey powder	109.25	109.25	11.98	2.37 FW		0.14500	0.87463
40 Corn syrup	273.61	273.61	24.56	5.95 FW		0.08867	
43 Liquid sugar	814.86	814.86	73.15	17.71 FW		0.30000	
71 Emul/stab	27.60	27.60	2.76	0.60 FW		1.20000	
134 Cocoa	119.60	119.60	13.11	2.60 FW		0.62000	
300 Water					−0.01103	0.00001	

							0.00242
							0.07554
							0.01104

| Finished Weight | 4600.00 | 4600.00 | 494.07 | 100.00 FW | | 0.31185 | 0.31185 |

Total Cost=1434.51

Constraint	Solution Value	Minimum Value	Maximum Value	Penalty Cost
1 Cost/unit ($/lb.)	0.00			
2 % Fat (total)	9.40	9.40	FW	0.01777
54 % NMS (inc whey)	9.50	9.50	FW	0.00931
55 % Total solids	39.28		FW	
17 % Whey solids	2.37		2.37 FW	0.00730
27 % Cocoa flavoring	2.60	2.60	2.60 FW	−0.00631
29 % Sugar solids	12.40	12.40	12.40 FW	−0.00444
30 % Corn syrup slds	4.80	4.80	4.80 FW	−0.00124
35 % Emul & % stab	0.60	0.60	0.60 FW	−0.01211
98 Finished weight	4600.00	4600.00	4600.00 FW	0.31185

Table 8.17. Available Ingredients File Listing for Examples in Tables 8.18–8.20.

Ingredient	Cost/unit	% Fat	% Moisture	% Milk solids N/F	% Total solids	% Dairy	Minimum	Maximum
1 Whole milk	0.13014*	3.25000	88.0500	8.70000	11.9500	100.000	2500.00	5000.00
5 Cream	0.72087*	38.3500*	56.2500	5.40000*	43.7500	100.000		
10 Skim milk	0.07275*	0.00000*	91.0000*	9.00000*	9.00000*	100.000		
12 Condensed	0.30455	0.00000	66.1000	33.9000	33.9000	100.000		
13 Butter milk cond	0.28000	4.00000	0.00000	30.0000	34.0000	100.000		
14 Skim milk pwdr	0.96750*	0.00000*	3.00000	97.0000*	97.0000	100.000		
16 Whey powder	0.14500*	1.40000*	6.13500	92.4650*	93.8650	100.000		
40 Corn syrup	0.08867*	0.00000	19.3000*	0.00000	80.7000*	0.00000		
43 Liquid sugar	0.30000*	0.00000	33.7000*	0.00000*	69.8600	0.00000		
71 Emul/stab	1.20000	0.00000*	0.00000*	0.00000*	100.000*	0.00000		
74 Frodex	0.25430	0.00000	0.00000	0.00000	100.000	0.00000		
134 Cocoa	0.62000	0.00000*	0.00000	0.00000	100.000	0.00000		
300 Water	0.00001	0.00000	100.000	0.00000	0.00000	0.00000		

Table 8.18. Multiproduct Run (White Mix, 10%) with Restrictions on Whole Milk and Cream. Copyright © Computer Concepts Corp., 1983.

Ingredient	Total Weight	Batch Weight	Gallons	Percent Batch	Cost/Unit Low	Cost/Unit Actual	Cost/Unit High	Penalty
1 Whole milk	2143.33	2143.33	258.86	23.30 FW	0.12087	0.13014		0.00927
5 Cream					0.40341	0.72087	0.88779	
10 Skim milk					0.06263	0.07275		0.01012
12 Condensed					0.23587	0.30455		0.06868
13 Butter milk cond	2680.87	2680.87		29.14 FW	0.07520	0.28000		
14 Skim milk pwdr					0.67488	0.96750	0.31311	0.29262
40 Corn syrup	1083.02	1083.02	97.22	11.77 FW		0.08867		
43 Liquid sugar	1196.00	1196.00	107.36	13.00 FW		0.30000		
71 Emul/stab	46.00	46.00	4.60	0.50 FW		1.20000		
300 Water	2050.78	2050.78	246.19	22.29 FW	−0.14994	0.00001	0.01395	
Finished Weight	9200.00	9200.00	714.23	100.00 FW		0.30497	Total Cost=2805.74	

Constraint	Solution Value	Minimum Value	Maximum Value	Penalty Cost
2 % Fat (total)	10.10	10.10	FW	0.01782
54 % NMS (inc whey)	10.00	10.00	FW	0.00696
35 % Emul & % stab	0.50	0.50	0.50 FW	−0.01200
29 % Sugar solids	9.10	9.10	9.10 FW	−0.00429
30 % Corn syrup slds	9.50	9.50	9.50 FW	−0.00110
55 % Total solids	39.18		FW	
98 Finished weight	9200.00	9200.00	9200.00 FW	0.30497

Table 8.19. Multiproduct Run (White Mix, 3%) with Restrictions on Whole Milk and Cream.
Copyright © Computer Concepts Corp., 1983.

Ingredient	Total Weight	Batch Weight	Gallons	Percent Batch	Cost/Unit Low	Cost/Unit Actual	Cost/Unit High	Penalty
1 Whole milk		223.44	26.99	4.86 FW	0.12087	0.13014		0.00927
5 Cream	223.44				0.40341	0.72087	0.88779	
10 Skim milk					0.06263	0.07275		0.01012
12 Condensed					0.23587	0.30455		0.06868
13 Butter milk cond	1374.44	1374.44		29.88 FW	0.11306	0.28000	0.31311	
14 Skim milk pwcr					0.67488	0.96750		0.29262
16 Whey powder	138.00	138.00	15.13	3.00 FW		0.14500	0.66827	
43 Liquid sugar	722.86	722.86	64.89	15.71 FW		0.30000		
71 Emul/stab	18.40	18.40	1.84	0.40 FW		1.20000		
74 Frodex	138.00	138.00	15.13	3.00 FW		0.25430		
300 Water	1984.86	1984.86	238.28	43.15 FW	−0.14994	0.00001	0.01395	
Finished Weight	4600.00	4600.00	362.26	100.00 FW		0.18260	Total Cost=839.98	

Constraint	Solution Value	Minimum Value	Maximum Value	Penalty Cost
2 % Fat (total)	3.10	3.10	FW	0.01782
54 % NMS (inc whey)	12.00	12.00	FW	0.00696
17 % Whey solids	3.00		3.00 FW	0.00523
29 % Sugar solids	11.00	11.00	11.00 FW	−0.00429
32 % Frodex	3.00	3.00	3.00 FW	−0.00254
35 % Emul/Stab	0.40	0.40	0.40 FW	−0.01200
98 Finished weight	4600.00	4600.00	4600.00 FW	0.18260

Table 8.20. Multiproduct Run (Chocolate Mix, 10%) with Restrictions on Whole Milk and Cream. Copyright © Computer Concepts Corp., 1983.

Ingredient	Total Weight	Batch Weight	Gallons	Percent Batch	Cost/Unit Low	Cost/Unit Actual	Cost/Unit High	Penalty
1 Whole milk	1963.11	1963.11	238.53	42.68 FW	0.10287	0.13014	0.13015	
5 Cream	957.16	957.16	115.60	20.81 FW	0.72079	0.72087	1.04264	
10 Skim milk						0.07275		0.00001
12 Condensed	334.82	334.82	35.81	7.28 FW	0.27274	0.30455	0.32592	
14 Skim milk pwdr					0.89198	0.96750	0.87465	0.07552
16 Whey powder	109.25	109.25	11.98	2.37 FW		0.14500		
40 Corn syrup	273.61	273.61	24.56	5.95 FW		0.00867		
43 Liquid sugar	814.86	814.86	73.15	17.71 FW		0.30000		
71 Emul/stab	27.60	27.60	2.76	0.60 FW		1.20000		
134 Cocoa	119.60	119.60	13.11	2.60 FW		0.62000		
300 Water					−0.01104	0.00001		0.01105
Finished Weight	4600.00	4600.00	505.50	100.00 FW		0.31288	Total Cost=1439.26	

Constraint	Solution Value	Minimum Value	Maximum Value	Penalty Cost
1 Cost/unit ($/lb.)	0.00		FW	
2 % Fat (total)	9.40	9.40	FW	0.01777
54 % NMS (inc whey)	9.50	9.50	FW	0.00931
55 % Total solids	39.28		FW	
17 % Whey solids	2.37		2.37 FW	0.00730
27 % Cocoa flavoring	2.60	2.60	2.60 FW	−0.00631
29 % Sugar solids	12.40	12.40	12.40 FW	−0.00444
30 % Corn syrup slds	4.80	4.80	4.80 FW	−0.00124
35 % Emul & % stab	0.60	0.60	0.60 FW	−0.01211
98 Finished weight	4600.00	4600.00	4600.00 FW	0.31185

9
Mix Processing

In preceding chapters we have emphasized that mix composition, ingredient quality, and accurate calculations are each prerequisites for the manufacture of desirable ice cream. Once the compositional requirements related to quality and quantity are met, the mix is ready for processing.

Mix processing begins with combining the ingredients into a homogeneous suspension/solution that can be pasteurized, homogenized, cooled, aged, flavored, and frozen. The flowchart of a batch pasteurizing operation (Figure 9.1) provides a simplified scheme of mix processing and ice cream freezing, packaging, hardening, and storage. Figure 9.2 shows a continuous flow operation with options to produce packaged ice cream and molded or extruded novelties.

The first step in processing is composing the mix. The procedure may range in scope from a small batch operation, in which each ingredient is weighed or measured individually into a pasteurizing vat, to a large, automatic, continuous operation in which liquid ingredients are metered into a batching tank. Continuous mix-making operations vary considerably in their characteristics; some of them are adaptations of batch operations. Liquid ingredients, including stabilizers, and product-blending equipment have been developed to accommodate continuous operations. Pumping of ingredients and mix through a closed system cuts costs of handling, reduces some important risks of contamination, and makes possible automated cleaning in place of the equipment.

PREPARATION OF THE MIX

Preparing the mix involves moving the ingredients from the storage areas to the mix preparation area, weighing, measuring or metering them, and mixing or blending them. Undissolved components must be kept in suspension until they are fully hydrated or are dispersed in such small sizes that they remain suspended in the finished mix.

Combining the Ingredients

All liquid ingredients (milk, cream, concentrated milk, syrup, etc.) are placed in the vat, and the agitation and heating are started at once. Amounts of

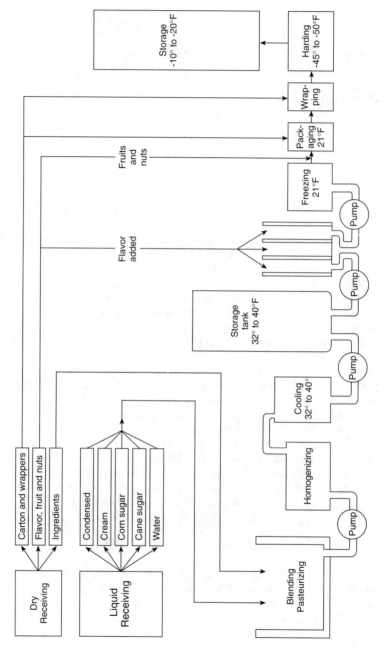

Figure 9.1. Flowchart for ice cream processing in a batch-type system. Batch pasteurization is at 69°C (155°F) for 30 min.

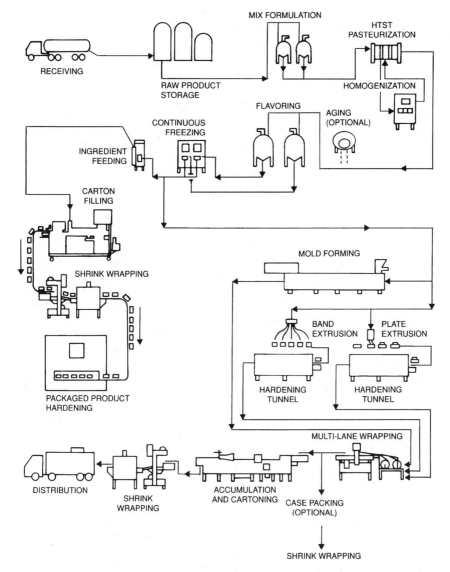

Figure 9.2. Schematic drawing of a continuous process system for making, packaging, hardening, and shipping ice cream and ice cream novelties. (Courtesy APV Crepaco, Austin, TX.)

liquid ingredients can be measured with a calibrated measuring stick, pumped through a volumetric or mass flow meter or directly added as predetermined volumes or weights. Systems that employ a meter on each inlet line provide the most rapid means of compounding a mix, because each liquid ingredient can be added simultaneously. If all ingredients are in the liquid form, the process is very time-efficient and accurate, provided the compositions and densities of the ingredients are consistent from batch to batch and the information is used to control the metering operation. Flow meters supply the information needed to permit electronic or manual operation of pumps or valves to control ingredient flow. Automated systems of mix manufacture commonly employ microprocessors to compute the amount of each ingredient for a specific formula, start and stop the flow when the desired quantity has been transferred, and record the data useful for future reference. In plants of relatively small capacity the number of meters may be limited so that one meter is used for more than one ingredient. If so, the densities of the ingredients must be nearly the same to provide sufficient accuracy.

To facilitate operation with a fully liquid system, it is often necessary to dissolve some dry ingredients. These are usually made in concentrated form and stored refrigerated until needed. It is of utmost importance that such ingredients be hydrated to the same concentration from batch to batch or that any change in composition be an input variable that is included in the formulation step. Liquefying dry ingredients well ahead of time for their use permits foam to dissipate and colloidal substances to hydrate fully. When small amounts are being made in a large vessel, the depth of the liquid on the agitator can determine the extent of incorporation of ingredients and the amount of foaming.

If they are not liquefied first, dry ingredients, including NDM, dry whey, dried eggs, cocoa, sugar, and stabilizer, are added while the liquid materials are being agitated and before the temperature reaches 50°C (122°F). Proper suspension to avoid lumpiness of the dry ingredients can be obtained by (1) mixing the dry ingredients with part of the crystalline sugar before adding slowly to the liquid, or (2) sifting the dry ingredients slowly into the liquid. The liquid should be cool (< 30°C, < 86°F) when NMS, cocoa, or similar ingredients are added. Instructions of the manufacturer should be followed for addition of stabilizer/emulsifier blends to the mix. Some blends are capable of dissolving at relatively low temperatures (Figure 9.3), whereas others should not be added to a mix until the temperature reaches about 65°C (149°F). Added frozen products, e.g., butter, frozen cream, plastic cream, should be cut into small pieces and allowed sufficient time to melt before pasteurization is started. With few exceptions coloring and flavoring materials are added at the time the mix is frozen.

Dry ingredients can be blended into the liquid materials with an emulsifying agitator that is mounted inside the mix tank. More efficient blending is done with high shear mixers or with "powder funnels." High shear blenders (Figure 9.4), which have a relatively small capacity compared with that of a mix tank, suspend dry ingredients in a small portion of liquid components. The suspension is then transferred to the mix tank. These dry ingredient incorporators are constructed with agitators (Figure 9.5) that form a deep vortex. Air incorporation is minimized by the flow pattern of mix away from the vortex.

9 MIX PROCESSING

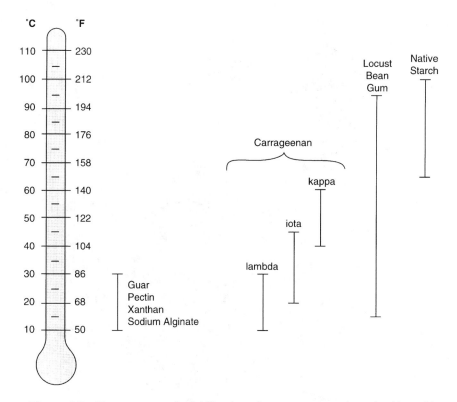

Figure 9.3. Temperatures of solubilization of some common polysaccharide stabilizers.

A simple and inexpensive device for incorporating dry ingredients employs a funnel attached to a tee in a pipe immediately upstream from a centrifugal pump which, in turn, is connected to the mix vat or blending tank (Figure 9.6). Liquid flows from the tank past the tee to the pump and is returned to the tank. The partial vacuum that is created by the operating pump draws the dry ingredient into the flowing liquid and disperses it into the liquid. Avoidance of excess foaming is important when the mix is pasteurized soon after blending.

Stabilizers are best dispersed into a mixture of low water activity (0.86 or lower) such as liquid sugar or corn syrup with about 70% solids. A suitable liquid sugar can be made by blending 16–17 lb of sugar per gallon of water to make a syrup containing 66–68% solids. Sorbitol and polydextrose will work as well. One gallon of the mixture is adequate to disperse 1 lb of stabilizer. In this environment stabilizer should be well dispersed by the blender in no more than one minute. Longer times allow too much hydration and the buildup of excessive viscosity making dispersion of suspension into the remaining liquid quite difficult. To minimize foaming, the high shear blender should be filled to three-fourths capacity before stabilizer is introduced, and the blender must be stopped as soon as dispersion is complete. Other dry ingredients should be

Figure 9.4. High shear blender for incorporating dry ingredients into ice cream mix. (Courtesy American Ingredients Company, Kansas City, MO.)

Figure 9.5. Agitator for a high shear dry ingredient mixer. (Courtesy American Ingredients Company, Kansas City, MO.)

incorporated before the stabilizer, and the blender should be operating as the stabilizer is introduced. Batch operation is far more satisfactory than continuous. After blending, the dispersion should be delivered beneath the surface of the mix in the batching tank or tangentially onto the sidewall of that tank to minimize foaming. In no case should cream be admitted to the high shear blender, because churning is almost certain to occur.

Incorporation of stabilizers into nonfat mixes is particularly difficult, espe-

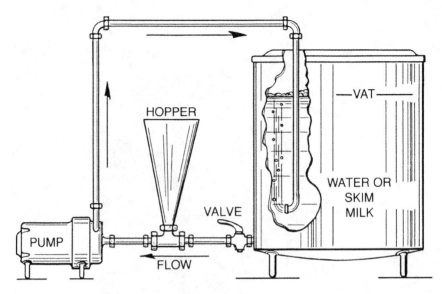

Figure 9.6. "Powder funnel" type device for incorporating dry ingredients with liquid components of frozen dessert mixes. (Courtesy American Dairy Products Institute, Chicago, IL.)

cially if the mix is to be pasteurized in a plate-type heat exchanger. The reason is that nonfat mixes foam liberally and become highly viscous. As a mix containing foam goes through the homogenizer, it creates a high amount of noise, and burn-on may occur in the heating section of the pasteurizer. Stabilizers for nonfat mixes usually contain microparticulated cellulose. These minute particles migrate quickly to and stabilize the lamellae of air cells. The more air cells formed in the blender, the more are the problems encountered in the homogenizer and plate pasteurizer.

When batch or vat pasteurization is employed, heating during blending is the appropriate practice. However, for continuous pasteurization, economic and quality considerations should prompt the decision to blend into cold or into heated mix. Regenerative pasteurization is most efficient when the mix that enters the heat exchanger is cold rather than warm. As an example, assume that mix enters the regenerator section of the heat exchanger at 5°C, where it receives heat from the outflowing mix that has been heated to 80°C in the heating section of the pasteurizer. If the efficiency of heat transfer in the regenerator is 80%, the inflowing mix will be heated to 60°C, thus requiring that the temperature be raised another 20°C in the pasteurizing heat exchanger. The outflowing mix will be cooled from 80°C to 25°C in the regenerator by the incoming cold mix, and it will need to lose another 21°C in the cooling section of the pasteurizer to bring the temperature to the desired 4°C. Contrast this situation to one in which the temperature of the inflowing mix is 40°C, each of the other parameters being the same as just described. The difference in temperatures of the outflowing and inflowing mixes is now only 40°C, and

9 MIX PROCESSING

80% regeneration will result in recovery of 32°C during both the heating and the cooling operations. Thus, the net reuse of energy is 55°C when the mix is cold but only 32°C when the mix is warm on entry into a regenerative type pasteurizer. This difference of 23°C must be paid for in the forms of both heat (usually steam) and refrigeration. Of course, this simplified example does not consider that some ingredients may be inherently warm (e.g., syrups) and others cold. Furthermore, desired characteristics of the finished product, available equipment, or available ingredients may dictate the type of process that is used. Sugars, syrups, and NDM dissolve slowly in cold liquids, and incorporated air does not escape cold mixes as readily as it does the less viscous warm (35–40°C, 95–104°F) ones.

PASTEURIZATION OF THE MIX

Pasteurization of all mixes is required because this process destroys all pathogenic microorganisms, thereby safeguarding the health of consumers. Furthermore, most hydrolytic enzymes, even the natural ones of raw milk, that could damage flavor and texture are destroyed by pasteurization. Pasteurization adds little additional expense, because it is necessary to heat mix to dissolve or hydrate dry ingredients. Furthermore, homogenization can be best accomplished at temperatures near those of pasteurization.

The industry generally follows federal regulations (Table 9.1) for pasteurization standards even though some states and local health authorities have similar requirements. Each manufacturer should be familiar with the regulations that apply to the market area, taking particular note of the following:

1. whether any dairy product may be used without being repasteurized
2. maximal bacterial counts of ingredients used, even when the mix they are used in is pasteurized
3. time and temperature requirements of pasteurization
4. maximal permitted aerobic plate counts (often called Standard Plate Count) and coliform counts of the finished products
5. whether a product must be frozen on the premises where it is pasteurized.

Proper pasteurization consists in rapidly heating to a definite minimal temperature, holding at that temperature for a minimal time, then rapidly cooling to < 5°C (< 40°F).

Table 9.1. Public Health Service Recommended Times and Temperatures for Pasteurization of Ice Cream Mixes

Method	Time	Temperature (°C/°F)
Batch	30 min	69/155
HTST	25 sec	80/175
HHST	1–3 sec	90/194
UHT	2–40 sec	138/280

Source: 21 CFR 135 and Grade A Pasteurized Milk Ordinance, 1993 Revision, U.S. Public Health Service.

Pasteurization (1) renders the mix substantially free of vegetative microorganisms, killing all of the pathogens likely to be in the ingredients, (2) brings solids into solution, (3) aids in blending by melting the fat and decreasing the viscosity, (4) improves flavor of most mixes, (5) extends keeping quality to a few weeks, and (6) increases the uniformity of product.

There are two basic methods of pasteurization: batch or low-temperature long-time (LTLT), and continuous. Continuous pasteurization can be done with any of several combinations of temperature and time. Representatives of manufacturers, users and public health authorities have worked together to develop and adopt 3-A Accepted Practices for the Sanitary Construction, Installation, Testing and Operation of High-Temperature Short-time and Higher-Heat Shorter-Time Pasteurizer Systems (3-A Sanitary Standard No. 603-06 1992). These practices are endorsed by the Interstate Milk Shipper's Conference in the control of sanitary quality of grade A milk in the United States. Any pasteurizer, or any of the many other types of equipment that bear the 3-A symbol, is deemed to have been manufactured under the provisions of the accepted practices.

In the batch system the mix is usually compounded in the vat. Heat is applied, by circulating hot water through the double walls of the vat, while the ingredients are being added and blended. Once all ingredients have been added to the vat and the minimal temperature of 69°C (155°F) has been reached, timing of pasteurization is started. As soon as the minimal time of heating of 30 min has elapsed, the mix is pumped to a homogenizer, then to a continuous cooling device such as a plate heat exchanger. Only in very small operations is mix cooled in the processing vat. Heating and cooling in a vat increases the total heat treatment by a large amount, resulting in a relatively high intensity of cooked flavor in the mix. However, cooked flavor is not usually objectionable in ice cream. Furthermore, the increased hydration of proteins and stabilizers induced by the LTLT method can impart improved body and texture, increase resistance to heat shock, and reduce the time needed for aging of the mix.

Continuous pasteurization can be done at several combinations of temperature and time; see Table 9.1 for specifications for high-temperature short-time (HTST), higher-heat shorter-time (HHST), and ultra high temperature (UHT) processes. As discussed under preparation of the mix, continuous flow pasteurizers facilitate use of regenerative heating and cooling with consequent large savings in costs for energy. Most continuous type pasteurizers consist of a series of parallel plates with grooved or waffled surfaces. Heat is exchanged from warmer liquid passing in one direction of flow on one side of the plates to cooler liquid passing in the reverse direction of flow on the opposite side of the same plates (Figure 9.7). Other continuous flow heat exchangers include tube-in-tube, triple tube, and steam injection designs. With the latter type the water added in the form of steam condenses in the mix and must be removed by vaporization and condensation. Controls are necessary on such systems to ensure that the correct amount of total solids remains in the mix.

The principal outcome of pasteurization is that the mix is rendered free of microorganisms that have the potential to cause disease among consumers. To ensure that pasteurization is accomplished within the parameters of time and temperature needed to kill all potential pathogens, controls are prescribed

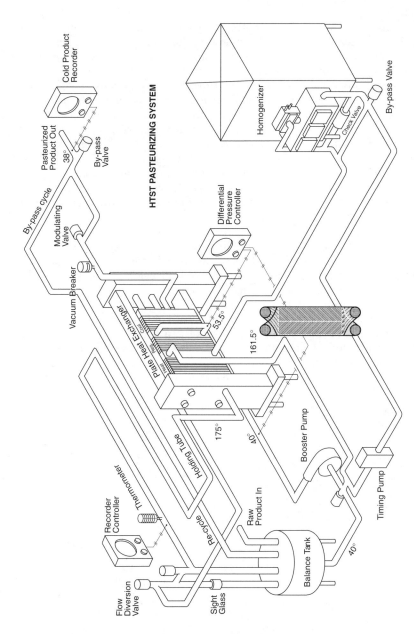

Figure 9.7. Diagram of HTST type pasteurizer with three sections of plate heat exchangers, viz., regenerating, heating and cooling. (Courtesy Borden, Inc., Dairy Division, Ogden, UT.)

by regulations. Among them is a requirement that holding tubes be designed to expose mixes to the minimal temperature for the minimal time that matches the temperature, e.g., 25 sec when the minimum is 80°C (175°F). Determinants of holding time are rate of mix flow, length and diameter of the tube, and amount of mixing of the fluid in the tube. Each continuous-flow pasteurizer must have either a positive displacement-type timing pump or a meter-based timing system to control the rate of flow through the holding tube. The meter-based system consists of a centrifugal pump, a control valve (or the pump must be of the variable speed type), and a magnetic flow meter, which uses an electric signal to control the flow rate of the product through the holding tube (Figure 9.8). The magnetic flow meter must produce a linear signal proportional to the flow through the holding tube. Also, there must be high flow and low flow alarms to signal the **flow diversion device** to close in the event the rate of flow goes above or below preset limits.

Holding tubes must be so designed (3-A Sanitary Standard No. 603-06) that the simultaneous temperature difference between the hottest and coldest product in any cross-section of flow at any time during the holding period will not be greater than 0.5°C (1°F), and the average velocity through the holding tube shall not be less than 1 ft/sec (3.05 cm/sec).

Ice cream mixes may be so viscous as to cause flow within the pasteurizer holding tube to be of the laminar type (smooth and stable). In such cases the holding tube must be sized to hold the mix twice as long as would be necessary were the flow turbulent (high rate of mixing). Goff and Davidson (1992) tested viscosities and flow characteristics of representative ice cream mixes. Many of the mixes to which they added stabilizers showed non-Newtonian flow (appar-

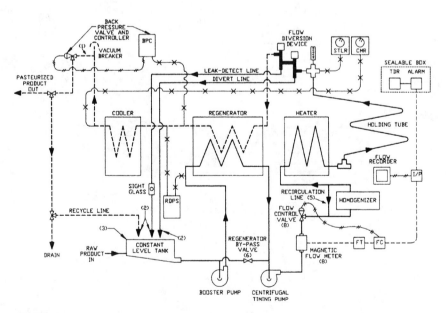

Figure 9.8. HTST pasteurizer with rate of flow controlled by a magnetic flow meter and programmable controller. (U.S. Public Health Service. 1993.)

ent viscosities decreased as shear rates increased). This type of viscous mix can create laminar flow in the holding tube, requiring extension of tube length, or can decrease the rate of pumping through the pasteurizer. Additionally, the mix nearest the wall of the holding tube, being held longer than that mix flowing through the center of the tube, may suffer in quality from the extra long application of heat.

Holding times of pasteurizers usually are determined by measuring the rate of flow of a salt solution from the upstream to the downstream end of the holding tube (3-A Sanitary Standard No. 603-03). The low viscosity of the salt solution permits turbulent flow so that mixing within the tube makes the time of passage of each portion of the solution very close to the average flow rate. Denn (1980) reported that the flow pattern is laminar for Reynolds numbers (N_{Re}) < 2,100. Turbulent flow exists when N_{Re} exceeds 4,000, but the International Dairy Federation recommends that an N_{Re} of >12,000 be maintained in HTST pasteurization tubes.

Viscosities of ice cream mixes are affected by the concentration, type, and degree of hydration of the stabilizer, carbohydrates, colloidal salts, and proteins of the mix; type of heat treatment; whether the mix is homogenized prior to holding; and the rate of shear in the holding tube. Shear rates varied from 50 to 180 sec^{-1} in experiments by Goff and Davidson (1992), and viscosities ranged from 8.7 centipoise (cP) in an unstabilized mix to 103 cP in a mix containing 0.25% carboxymethyl cellulose. The latter was measured at a low shear rate. Mixes contained 14% fat and 41% total solids.

Continuous pasteurizers are capable of heating mixes to temperatures well above those required to meet pasteurization standards. Furthermore, it is relatively easy to increase the length of the holding tube or to slow the rate of pumping to increase the holding time. Each of these three methods of increasing the heat treatment increases processing cost. Increasing temperature increases costs for heat energy; increasing holding tube length increases resistance to fluid flow, resulting in increased energy costs for pumping and, possibly, requiring a larger pump; reducing rate of flow reduces plant capacity. Reasons for higher heat treatment include the following: (1) it reduces the amount of stabilizer needed by up to 25%, (2) it improves body and texture because of the denaturation of proteins and the consequent increase in their water-holding capacity, and (3) it increases resistance to oxidation because of the exposure of reducing groups on the unfolded proteins.

An important advantage of continuous flow pasteurizers is the ability to clean them in place (CIP) by circulating rinse water, detergent, and more rinse water through them. Systems designed for CIP can be operated automatically in many installations. Thus, computer or microprocessor control is facilitated. To avoid buildup of films of denatured protein on heating surfaces, mixes should be substantially free from entrained air as they enter the plate heater of the pasteurizer.

HOMOGENIZATION

The main purpose of homogenization is to make a stable and uniform suspension of the fat by reducing the size of fat globules to less than about 2 μ. When a mix is properly homogenized, the fat will not rise and form a cream layer nor will the frozen product have a greasy or buttery appearance or mouthfeel. Homogenized fat churns very slowly in the freezer so that emulsifiers are usually required to provide the amount of controlled churning that results in a frozen product that is dry in appearance and slow to melt. Homogenization is required for any mix containing a fat or oil that is not in a relatively stable emulsion. The list of sources of such lipids includes butter, butter oil, anhydrous milkfat, fractionated milkfat, plastic cream, frozen cream, and oils from vegetable sources.

Homogenization is usually accomplished by forcing the mix through a very small orifice under suitable conditions of pressure and temperature, using a positive displacement pump to provide the pressure. Homogenizers are piston-type pumps (Figures 9.9 and 9.10) that move a constant amount of liquid through the very thin orifice of one or two valves (Figure 9.11). Therefore, homogenizers can be used as timing pumps in HTST systems. Fat globules, which must be in the liquid state, are greatly distorted as they travel at a velocity of about 12,000 cm/sec between the parallel walls of the valve and valve seat [some valves differ in design (Figure 9.12)]. The globules experience a sudden release in pressure and rate of flow as they exit the valve and impact a ring that surrounds it. This drop in pressure momentarily lowers the vapor pressure of the mix to a point at which vapor pockets are formed. The extremely

Figure 9.9. Homogenizer crankcase and pump head. Cutaway view of heavy-duty gearless model with five cylinders. (Courtesy APV Homogenizer Division, Wilmington, MA.)

Figure 9.10. Gaulin low-pressure homogenizer featuring the Micro-Gap® homogenizing valve and low-pressure cylinder for increasing homogenization efficiency. (Courtesy APV Homogenizer Division, Wilmington, MA.)

unstable environment in which the globules find themselves causes them to be disrupted. As bubbles of vapor form and collapse, shear forces known as cavitation are highly active.

Natural fat globules are coated with phospholipids to which are adsorbed other lipids and proteins. As fat globules are reduced in size during homogenization, the amount of phospholipid available to be adsorbed becomes limiting, and added emulsifiers are adsorbed to the fat. When the average diameter of the fat globules is reduced to one-half the original diameter, the number of globules increases by eight times and the total surface area is doubled. Thus, the amount of materials adsorbed increases markedly. Usual homogenization treatment reduces globule diameters about tenfold and increases total surface area about 100 times. Since proteins are adsorbed on the outer surfaces of the newly formed membranes, the amount of hydrated surface area is greatly

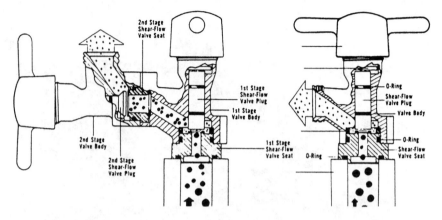

Figure 9.11. Diagram of two-stage (left) and single-stage (right) homogenizer valve assemblies. (Courtesy APV Homogenizer Division, Wilmington, MA.)

increased by the treatment. This may account for the increased smoothness of texture associated with homogenization of mixes.

Efficiency of homogenization is reduced when the homogenizing valve is worn, the pump does not deliver fluid at the designed rate (usually because the intake or discharge valve is damaged or worn), temperature of mix is low, or air is present in the mix. Inadequate homogenization can lead to churning in the freezer. Evidence of such churning can be seen as a greasy appearance on the scraper blades, elbows in pipes and surfaces of extrusion nozzles. Churning is most likely to occur with high-fat mixes formulated to be extruded in a very stiff and dry form.

To check for adequacy of homogenization, diluted mixes should be examined under an oil immersion objective of a microscope for the appearance of large globules or excessive clumping (Figure 9.13). An eyepiece micrometer can aid in determining the sizes of the globules. An inexpensive alternative is to place mix in an Erlenmeyer flask or slender bottle for several hours and then run tests for fat in the top and bottom portions of the mix. Differences in fat content greater than about 5% of the test of the mix indicate that homogenization is not adequate.

Homogenization efficiency increases with increases in temperature up to about 80°C (176°F). If homogenization precedes pasteurization by several seconds and any part of the mix is composed of raw milk, the minimum temperature for homogenization is 60°C (140°F). At this temperature milk lipase is virtually inactivated, ensuring that lipolysis will not take place as the protective membrane is stripped from the fat globules during homogenization. Furthermore, higher temperatures limit clumping of fat globules and reduce the energy needed to run the homogenizer.

Location of the homogenizer in the process line depends on several factors. When the pasteurizer is of the batch type, homogenization follows immediately after pasteurization. If the homogenizer is to function as the timing pump in

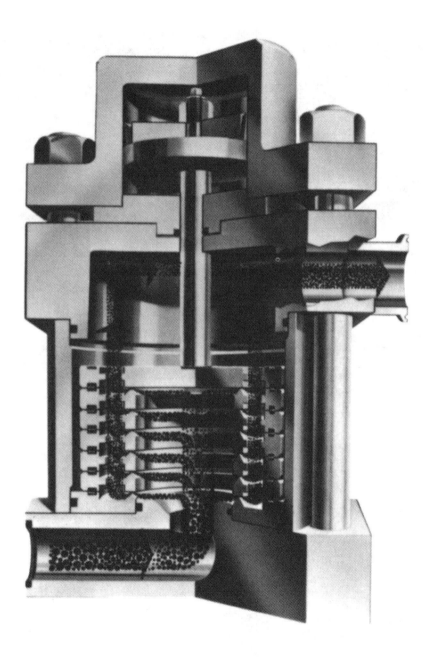

Figure 9.12. Cross-sectional view of the Micro-Gap® homogenizing valve. Mix enters the bottom of the valve assembly, is homogenized as it passes through the microgaps between the circular valve seats, and then exits at the top of the valve body. (Courtesy APV Homogenizer Division, Wilmington, MA.)

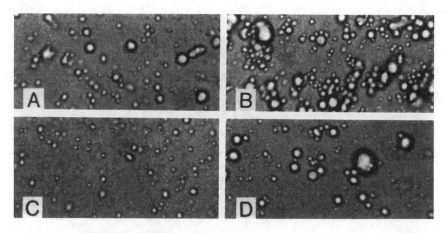

Figure 9.13. Microscopic appearance (×800) of fat globules in ice cream mix diluted 1:4 with water: (A) reference sample, 2,000–500 psi (141–35 kg/cm^2); (B) unhomogenized; (C) homogenized at 4,000–500 psi (281–35 kg/cm^2); (D) homogenized at 500 psi (35 kg/cm^2). Only the reference sample contained emulsifier (0.075%). (From Schmidt and Smith 1989.)

a continuous system, it must be located between the raw side regenerator and the heating section of the pasteurizer (Figure 9.14). Placed here, it produces a negative pressure on the raw product side of the regenerator and a positive pressure on the pasteurized side, thus preventing contamination of pasteurized mix in case of a leak in the regenerator. Furthermore, heat produced in the

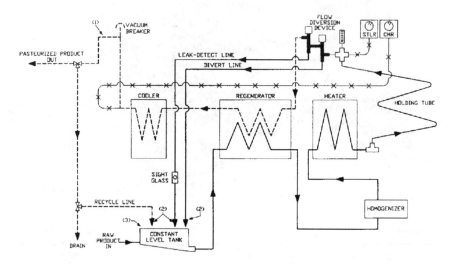

Figure 9.14. HTST pasteurizer system with homogenizer used as the timing pump. (U.S. Public Health Service 1993.)

process is recoverable, and less heat needs to be added in the heating section of the pasteurizer than if the homogenizer were to be placed downstream from the flow diversion valve that is located at the end of the holding tube.

In systems in which a second positive displacement pump serves as the timing pump, it is possible to locate the homogenizer downstream from the heating section or from the flow control device of the pasteurizer (Figure 9.15). However, since the two pumps are highly unlikely to move exactly the same amount of product, a bypass line must be installed around the homogenizer, and the homogenizer must operate at a rate greater than that of the timing pump.

Undissolved particles passing through the homogenizer in continuous flow lines tend to cause wear on valves. The most difficult of the solid ingredients to dissolve is usually the stabilizer. Placement of the homogenizer as far downstream from the blending tank as possible allows for maximal hydration of such materials. In any case there must be adequate turbulence to keep solids suspended during passage through the processing line. Wear on homogenizer valves should be checked routinely. Increased pressure is not likely to correct problems due to defective valves. Entrainment of air in the mix and leakage of air into the mix on the suction side of the homogenizer must be avoided if efficient homogenization is to be accomplished.

Since several homogenizer valve designs exist and mixes vary in their fat content and other components, pressures necessary to produce adequate dispersion of fat also vary. Commonly accepted pressures for plain mix with 10% milkfat in a two-stage homogenizer are 2,000 psi (13.8 MPa, 136 atm., 141 kg/cm^2) and 500 psi (3.45 MPa, 34 atm., 35 kg/cm^2) on the first and second stages, respectively. Schmidt and Smith (1989) reported that no emulsifier was needed in a stabilized (0.25%) mix containing 10.2% milkfat that was homogenized at

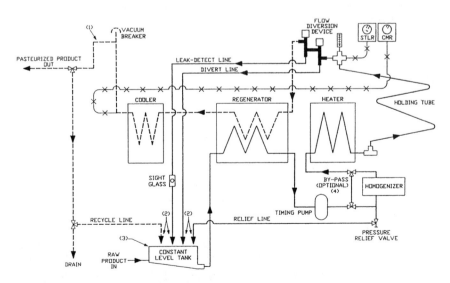

Figure 9.15. HTST pasteurizer system in which the homogenizer does not serve as the timing pump. (U.S. Public Health Service, 1993.)

Table 9.2. Approximate Homogenization Pressures[a] for Mixes of Varying Fat Contents

Fat (%)	Single stage (lb/in^2)	Two stage	
		First valve (lb/in^2)	Second valve (lb/in^2)
1–8	2,500–3,000	2,500–3,000	500
10–14	2,000–2,500	2,000–2,500	500
15–17	1,500–2,000	1,500–2,000	500
18	1,200–1,800	1,200–1,800	500
>18	800–1,200	800–1,200	500

[a]To convert lb/in^2 to Système International (SI) units in MPa, multiply by 6.895×10^{-3}.

only 500 psi. (See discussion of emulsifiers in Chapter 6.) In general, as fat content increases to 14–18%, homogenization pressures on the first stage should be reduced to prevent excessive mix viscosity (Table 9.2).

Chocolate mixes require pressures that are about 500 psi (3.45 MPa) higher than those used for plain mixes containing the same amount of fat. Cocoa contains varying amounts of bean shell that can cause wear on the homogenizer valves. Therefore, an important specification for the cocoa purchaser is a maximal amount of shell.

Under usual conditions fat globules tend to cluster upon exiting the first homogenizing valve. This is overcome by operating at a low pressure, usually 500 psi, a second valve installed immediately downstream from the first. High-fat mixes usually contain reduced amounts of NMS, therefore reduced protein. This may contribute to excessive clustering of fat globules and, consequently, to excessive mix viscosity. Other factors that promote clustering include use of nonemulsified fat, low homogenization temperature, use of only one homogenizing valve, and high mix acidity. Pressure of 500 psi on the second stage valve is generally satisfactory for cluster breakup regardless of the quantity of fat in the mix.

Minimizing homogenization pressures conserves energy and reduces costs of operation. When the practice also reduces requirements for emulsifier, costs can be reduced further.

AGING MIXES

Pasteurization and homogenization change the physical forms of the suspended solids of ice cream mixes. Pasteurization melts all of the fat, and homogenization reduces fat globule diameters. New and different fat globule membranes are formed (Figure 9.16). Hydrophilic colloids are hydrated and reduced in size. Cooling to < 4°C (< 40°F) that follows these processes causes fat to begin to crystallize. However, the mix is not ready to freeze at that point in the process. Crystallization of the fat, adsorption of proteins and emulsifiers to the fat globules, and hydration of the proteins and stabilizers need to continue for a few hours, especially if gelatin is used as the stabilizer. Sodium carboxymethyl cellulose and guar gum, commonly used stabilizers, hydrate well during the

9 MIX PROCESSING

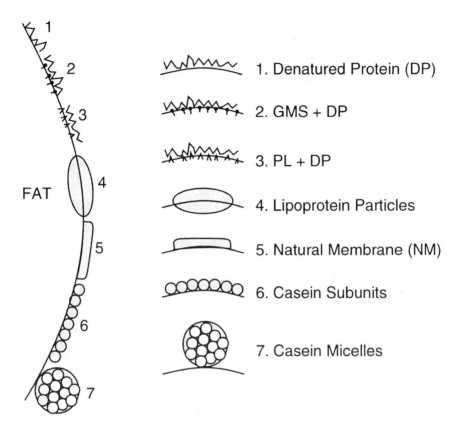

Figure 9.16. Substances in mixes compete for adsorption sites on surfaces of fat globules formed in the homogenizer. GMS—glycerol monostearate; PL—phospholipid.

processing of most mixes. However, hydration proceeds slowly for the small amount of carrageenan (Irish Moss) that is usually added with these stabilizers to prevent whey separation on long-term storage.

Cooling of mixes to 0–2°C increases the rate of crystallization of fat, increases capacities of freezers, and almost completely eliminates the possibility that microorganisms can grow in the mix. Cooling of mixes to such low temperatures is done efficiently in direct-expansion type swept-surface heat exchangers. Such cold temperatures add several days to the shelf life of mixes sold to other firms for freezing.

PACKAGING MIXES FOR SALE

Mixes are frequently prepared in a large plant for sale to ice cream retailers who freeze ice cream on the premises where sales take place. Depending on the size of the operation, mix containers range in capacity from one-half to 5 gal. Filling is done on the same equipment used for fluid milk, and the containers are firm-walled plastic, plastic-coated paper, or flexible plastic bags. Each type is distributed in cases or cartons.

UHT processing and aseptic filling of sterilized containers is an important development for mix distributors. The long shelf life, reduction in risk of unexpected spoilage, and almost complete elimination of risk of having pathogens in the product may result in significant savings that offset the higher costs of processing when UHT/AP systems are used.

FLAVORING MIXES

Most manufacturers process mixes in the plain unflavored form, choosing to add flavoring materials at the freezer. Furthermore, most flavorings are purchased in the ready-to-use form from specialists in that field rather than being prepared in the ice cream plant (Chapter 7). Flavoring materials are chosen based on consumer preferences, availability, costs, equipment needed to introduce flavoring into the product, labeling implications (such as the perceived advantage of having "all natural" on the label), stability of flavoring material, and packaging considerations. Chocolate flavorings are usually added to mixes prior to processing. Liquid and pureed flavorings are added to mix in the flavoring tank prior to freezing. Solid materials, fruits, nuts, candies, etc., are usually added to soft-frozen product as it exits the continuous freezer. Fruit and ingredient feeders (Figures 9.17 and 9.18) are made to distribute these materials at the rate chosen for the particular product. However, it is possible to add fruits and nuts to mixes before continuous freezing provided they are finely ground or chopped and are evenly distributed in the mix.

Gear-type positive displacement pumps are used to force syrups, jams, or purees through small orifices into flowing streams of soft-frozen desserts to make variegated, ripple, or swirl-type products. Depending on the distribution pattern desired, the nozzle, which usually has multiple orifices, may be rotated at speeds controlled by the operator.

In the case of batch freezer operations the most common practice is to flavor plain mix after it is placed in the freezer. Time of addition of solid ingredients depends on the amount of breakup of the materials that will produce the optimal size and distribution of the flavoring material. For example, the distribution of cookies in cookies-and-cream ice cream will vary from very large chunks to miniature pieces depending on the time and rate of agitation after addition of the cookies to the frozen mix. To control this variable the freezer should be emptied quickly and with minimal agitation once the cookies are broken to the desired size.

Addition of ingredients to pasteurized mix constitutes a **critical control point** in ice cream manufacture, because no further lethal process is given the product. Therefore, it is important that all operations at this point be done in a sanitary manner and that ingredients be free of pathogenic microorganisms. The most likely pathogens to be encountered at this point are *Listeria monocytogenes* and salmonellae.

Figure 9.17. Ingredient feeder. (Courtesy Waukesha Cherry-Burrell, Louisville, KY.)

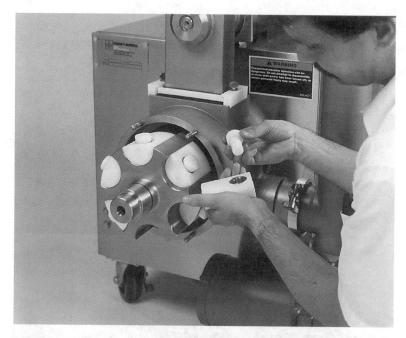

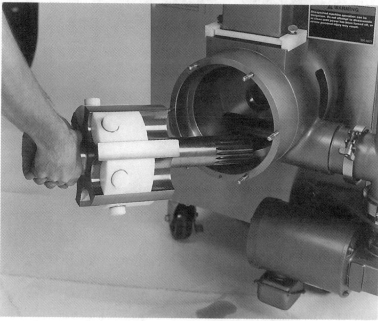

Figure 9.18. Positive displacement pump that moves semi-frozen product beneath the ingredient-dispensing auger: (a) pistons being inserted; (b) pump head being placed in housing. (Courtesy Waukesha Cherry-Burrell, Louisville, KY.)

REFERENCES

Denn, M. M. 1980. *Process Fluid Mechanics*. Prentice-Hall, Inc., Englewood Cliffs, NJ.

Goff, H. D., and V. J. Davidson. 1992. Flow characteristics and holding time calculations of ice cream mixes in HTST holding tubes. *J. Food Prot.* 55:34–37.

3-A Sanitary Standard No. 603-06. 1992. 3-A Accepted Practices for the Sanitary Construction, Installation, Testing and Operation of High-Temperature Short-Time and Higher-Heat Shorter-Time Pasteurizer Systems, revised. 3-A Sanitary Standards Committee, McLean, VA 22101-3850.

Schmidt, K. A., and D. E. Smith. 1989. Effects of varying homogenization pressure on the physical properties of vanilla ice cream. *J. Dairy Sci.* 72:378–384.

10
The Freezing Process

Freezing the mix is one of the most important operations in making ice cream, for upon it depend the quality, palatability, and yield of the finished product. Freezing consists of two parts: (1) the mix is frozen quickly while being agitated to incorporate air and to limit the size of ice crystals formed; and (2) the partially frozen product is hardened without agitation in a special low-temperature environment designed to remove heat rapidly.

The general procedure of the freezing process is easily learned since it involves only accurate measurement of the ingredients, movement of the mix into the freezer, operation of the freezer, and removal of frozen product from the freezer. However, mastering the details of freezer operation to produce a uniformly high-quality product requires considerable practice. If all operations are done manually, the several variables are not easily controlled. It is unusual that two persons will execute the details of freezing in exactly the same manner. Therefore, different people obtain different ice creams even when they use the same ingredients, formulas, and equipment. This is a major reason why programmed freezers are becoming widely used.

PREFREEZING TESTS

Before mixes are frozen they should be tested to determine whether the composition meets the specifications of the formula. Recommended methods of testing are found in Standard Methods for the Examination of Dairy Products. They are the ether extraction method for fat (15.8E,F) and the vacuum or forced draft oven methods for total solids (15.10A,C; Bradley *et al.* 1993).

FREEZING OPERATIONS

Cold, flavored ice cream mix is pumped into the continuous freezer barrel that is under pressure and is chilled with a liquid refrigerant. The mix is whipped with a dasher. Onto it are attached sharp scraper blades that contact the very smooth surface of the freezing cylinder. Their function is to scrape minute ice crystals from the cylinder wall. Removal of these crystals immediately upon

10 THE FREEZING PROCESS

their formation ensures that the product will have a smooth texture and that an ice layer will not build on the cylinder wall. Dull scraper blades increase the load on the drive motor, lengthen the time to freeze the mix, and produce a coarse texture.

Ice does not conduct heat as fast as does steel. Therefore, if ice is permitted to form a layer inside the cylinder, it acts as an insulator, slowing release of heat from the mix to the refrigerant. Furthermore, buildup of ice increases the distance heat must be transferred before it enters the liquid refrigerant and transforms the refrigerant into the gaseous state. The rate of heat transfer is a function of the difference in temperature, the thermal conductivity of the heat transfer surface, and the rate of renewal of surface films.

The temperature of the mix drops quite rapidly in the freezer cylinder. Removal of sufficient sensible heat (heat that thermometers measure) to start the mixture to freeze should take less than 1–2 min. Meanwhile viscosity decreases as rapid agitation disrupts gel structure and breaks clusters of fat globules. Rapid agitation also causes air to be incorporated. The air cells that form differ remarkably in size depending on whether the product is being made in a batch or continuous type freezer. The batch type freezer operates at atmospheric pressure, so air that is incorporated exists at the same pressure both inside and outside the freezer. However, the freezing cylinder of the continuous freezer is held under pressures up to about 100 psi (7 kg/cm^2, 690 kPa). Pressures of about 75 psi (5 kg/cm^2, 520 kPa) produce about 100% overrun with the normal mix, and the air cells in the freezer make up 15–20% of the volume of the mix. They will constitute 50% of the product volume when the pressures equilibrate to atmospheric pressure (14.7 psi, 1.03 kg/cm^2, 1 atmosphere, 101.4 kPa) outside the freezer.

When the freezing point is reached inside the freezer, liquid water begins to change to ice crystals. These crystals are practically pure water, and removal of this solvent from the mix causes dissolved materials to become more concentrated. Consequently, the freezing point of the unfrozen mix is lowered. Crystallization occurs at the expense of removal of latent heat of fusion (heat not measurable with a thermometer) as well as sensible heat. When pure water is being frozen, the temperature will not change noticeably as ice crystals are forming. However, in freezing ice cream the freezing point is continually being lowered by crystallization of water, so the temperature continues to drop but at a slow rate since both sensible heat and latent heat are being removed. As concentration of dissolved substances increases, the freezing point finally drops to a temperature at which no more ice is formed. Therefore, never is all of the water in ice cream frozen, even at temperatures of hardening. Depending on the drawing temperature, 33–67% of the water is crystallized in the freezer, and the hardening process then may account for freezing of an additional 23–57% (Table 10.1).

Another factor that can slow the drop in temperature in ice cream being frozen is the crystallization of substances dissolved in the aqueous phase. Not only do these substances lower the freezing point below that of water, but they have unique solubilities and *cryohydric points*. The latter is the temperature at which a dissolved substance separates out of solution. On separation the crystals release heat, causing a slight rise in temperature unless more heat is

Table 10.1. Approximate Percentage of Water Frozen in Ice Cream at Various Drawing Temperatures

Temperature (°F)/(°C)	Water frozen (%)	Temperature (°F)/(°C)	Water frozen (%)
25/–3.9	33	20/–6.7	59
24/–4.4	41	19/–7.2	62
23/–5.0	47	18/–7.8	64
22/–5.6	52	17/–8.3	67
21/–6.1	56	15/–9.4	90

removed than is released. Each of the dissolved substances of ice cream has its own cryohydric point. Representative ones include: lactose –0.28°C to –4.1°C depending on the isomer, glucose at –5°C, sucrose at –8.5°C to –14.5°C, disodium phosphate at –0.9°C, potassium chloride at –11.1°C, and calcium chloride at –55°C. Sommer (1951) explained that "In a solution that contains several dissolved substances, the following must be expected. When the saturation point (due to removal of water as ice) has been reached with respect to substance A, further freezing will not increase the concentration of A in the unfrozen portion, but will increase the concentration of B, C, etc. Therefore, the temperature will not be constant on further freezing, as it would be in a single component solution. There will be merely a decrease ion the rate of temperature change, after A has reached its saturation point. The same holds true as the saturation points of B, C, etc. are reached in the remaining unfrozen solution." Since ice cream is seldom, if ever, frozen to below –55°C, it is evident that all of the water will never be frozen. Therefore, with any change of temperature there is a change in the state of the water. The amount of water involved in the change of state increases as the temperature rises. A 1°C change in hardening room temperature from –20°C to –19°C involves only about one-fifth as much water as a 1°C change in dipping cabinet temperature from –14°C to –13°C. The latter, in turn involves only about one-fifth as much change in state of water as is involved in a change at freezer temperature from –6°C to –5°C. These facts are vital to the understanding of the effects that fluctuating temperatures can have on the quality of frozen desserts.

CHANGES THAT TAKE PLACE DURING THE FREEZING PROCESS

Inside the freezing cylinder the liquid mix, with its suspended fat globules and colloidal proteins, carbohydrates, and salts, is transformed into a highly viscous foam. Ice crystallizes from the continuous phase, transforming it into a thick syrup. Air cells form, and hydrophilic colloids adsorb to their surfaces, stabilizing them. Fat globules become increasingly crystalline, and some of them coalesce, forming structure that supports the foam (Figure 10.1). As the product exits the freezer, it has about one-half of its water frozen and has expanded up to about 100% in volume. The continuous phase is a thick syrup while the disperse phase consists of air cells, ice crystals, fat globules, casein micelles,

10 THE FREEZING PROCESS

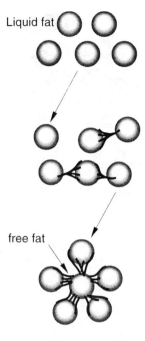

Figure 10.1. Explanation of the theory of how destabilization of fat by churning causes coalesence of fat globules during freezing of ice cream.

and other hydrocolloids. This makes ice cream a three-phase system: gaseous, solid, and liquid (Figures 10.2 and 10.3). The product of phase one is soft frozen and flowable, but the degree of stiffness varies with formula, process, freezer design, overrun, and temperature. Churning of the fat and formation of minute air cells result in a dry-appearing and relatively stiff product as it exits the freezer (Figure 10.4). Therefore, transport and packaging of the product are affected by each of these variables.

Choices of ingredients can markedly affect the physical properties of the finished frozen dessert. To assist formulators, Bakshi and Smith (1985) developed a computer program to evaluate the effects of using selected sweeteners and substituting whey solids for NMS in mixes. The program predicts freezing point and the amount of unfrozen water at typical storage temperatures. As whey solids are substituted in equal weight for NMS, the freezing point decreases because the effective molecular weight of whey solids is 235 Daltons compared with 370 Daltons for NMS (Bakshi and Johnson 1983).

Results of the freezing process can best be explained by examining the structure of the frozen product. Berger et al. (1972) illustrated the inner structure well in a series of electron photomicrographs made by the freeze-etching tech-

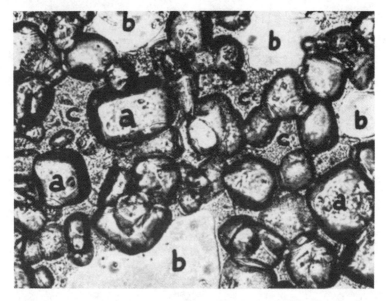

Figure 10.2. The internal structure of ice cream. (a) Ice crystals: average size 45–55 µm. (b) Air cells: average size 110–185 µm. (c) Unfrozen material: average distance between ice crystals or between ice crystals and air cells 6–8 µm; average distance between air cells 100–150 µm.

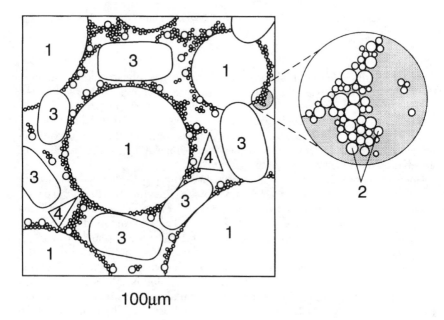

Figure 10.3. Drawing of the physical structure of ice cream: (1) air cells, (2) emulsified fat surrounding air cells, (3) ice crystals, (4) lactose crystals.

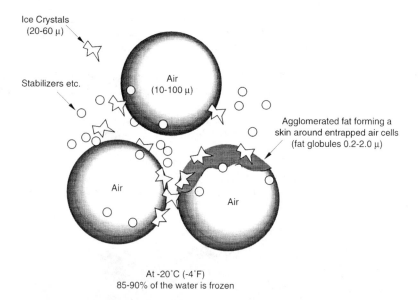

Figure 10.4. Concept of how liquid fat may spread over surface of air cells causing the frozen product to be stiff and dry in appearance. (Courtesy Danisco, New Century, KS.)

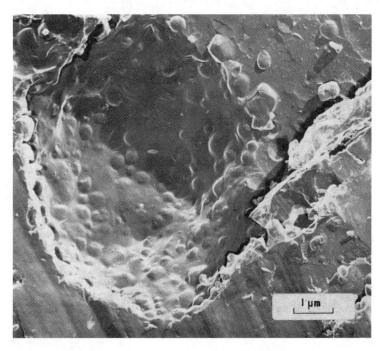

Figure 10.5. Electron micrograph of interior of a small air cell in frozen ice cream. The projections are fat globules. [Reproduced from Berger *et al.* (1972) by permission.]

nique. Figure 10.5 shows an air cell coated with fat globules. The air cell–mix interface is a continuous layer of unstructured material, coated with fat globules, which go through to the inside of the air cell. This suggests that the interfacial layer is very thin and is easily distorted. Air cells range in size from 5–300 µm, and lamellae between them measure 5–300 µm in thickness. As fat globules are broken during freezing, liquid fat spreads at the air/mix interface where it displaces protein and acts as a foam depressant. Coalesced fat in the walls of air cell lamellae limits their thinness. In ice cream containing 100% overrun in the form of 60-µm-diameter air cells, the number of air cells per gram approximates 8.3×10^6, and their total surface area approximates 0.1 M^2.

Figure 10.6 shows crystallized fat and agglomerates of casein micelles. The continuous phase or lamella between ice crystals was shown to contain clumps

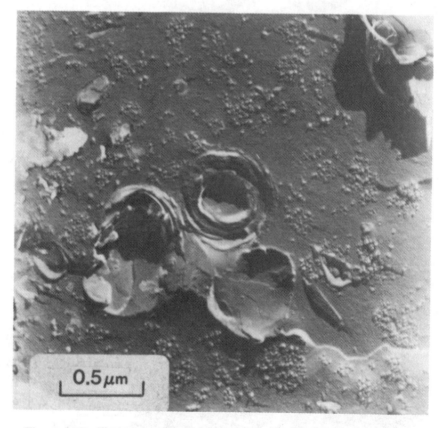

Figure 10.6. Electron micrograph of fat globules fractured in the freezing process and cemented together by liquid fat. Note the shell of high melting triglycerides. [Reproduced from Berger et al. (1972) by permission.]

10   THE FREEZING PROCESS   171

and individual globules of fat as well as casein micelles (Figure 10.7). The
number of ice crystals, if assumed to be in the form of a 40-μm cube, would be
approximately 7.8×10^6 per gram, and the total surface area about 0.08 M^2
(Campbell and Marshall 1975). At times sugars, especially lactose, crystallize
in frozen desserts. Berger et al. (1972) demonstrated microcrystals of sucrose
in ice cream (Figure 10.8).

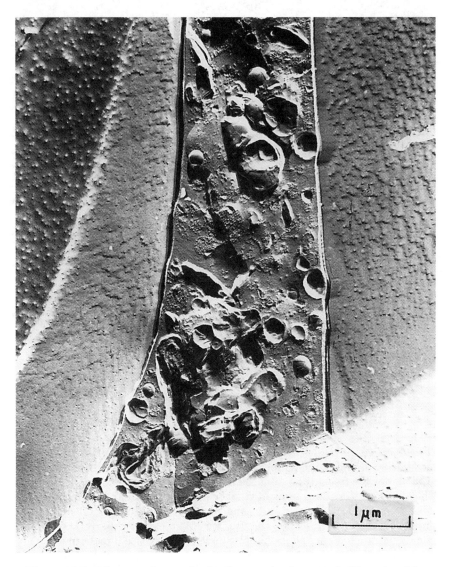

Figure 10.7.   Electron micrograph of wall separating ice crystals. [Reproduced from Berger *et al.* (1972) by permission.]

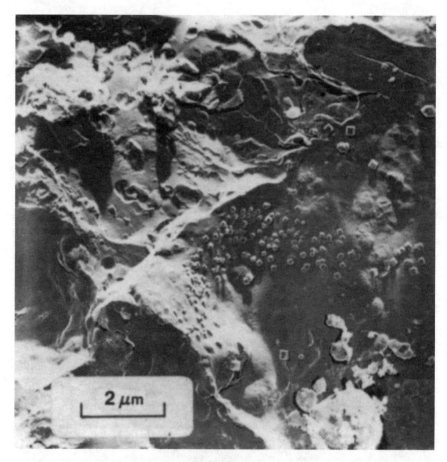

Figure 10.8. Electron micrograph of sucrose crystals in ice cream. [Courtesy Berger et al. (1972) by permission.]

REFRIGERATION NEEDED TO FREEZE ICE CREAM

The objective in freezing ice cream is to produce the maximum number of ice crystals by dropping the temperature of the mix well below its freezing point. The higher the freezing point of the mix the larger the number of ice crystals that will form in the freezer at a given temperature and the smoother the texture of the ice cream (Bakshi and Smith 1985). Tiny crystal nuclei that form there can only grow in size after ice cream exits the barrel, because no more nucleation occurs after agitation and scraping cease. Furthermore, a low freezing point (melting point) results in a relatively high amount of melting of ice as the temperature rises. When the temperature is again lowered, the crystals that form are larger than the ones that melted, giving rise to coarse texture.

Although the amount of heat to be removed from a frozen dessert mix is a function of several variables, mix composition is the major one. Heldman and Hedrick (1968) varied the sugar content of ice cream from 100% sucrose to 29% sucrose plus 71% corn syrup solids to 50% of both sucrose and corn syrup solids. Mix containing only sucrose required the least amount of refrigeration, 72 British thermal units (BTU)/lb mix, to lower the temperature to −9.4°C (15°F). As corn syrup solids content increased, so did the need for refrigeration, with the mix containing 50% CSS requiring 87 BTU/lb mix.

Mixes vary in composition, so they vary in freezing point. Furthermore, composition affects **thermal capacity** or **specific heat** in calories[1] per gram that a mix must release for the temperature to be lowered by 1°C. These variables, i.e. freezing point and specific heat, are determinants of the amount of ice that will be formed with any given amount of heat energy removed.

The freezing point of an ice cream mix is lowered by lactose, sugars, salts, and other substances in true solution. Fat and protein have no direct effect on freezing point because fat is immiscible with the aqueous phase, and proteins are very large molecules. However, as these substances are increased in concentration, there is less water in which solutes can dissolve, so the freezing point will be depressed. Freezing point depression is the difference between 0°C and the temperature at which the mix first begins to freeze.

Leighton (1972) presented a formula by which the freezing point of a simple mix of known composition can be estimated. The formula assumes that lactose and sucrose depress freezing point equally and glucose 1.9 times as much. Lactose is assumed to constitute 54.5% of NMS and 76.5% of whey solids. The salt effect is assumed to be 4.26 times % NMS. The important compositional data, therefore, are % sugar, % NMS, and % water. This simplified formula assumes negligible effect of fat, proteins, stabilizers, and emulsifiers except as they displace water. Application of this formula is facilitated by reference to the freezing points of solutions of sucrose (Table 10.2).

Table 10.2. Freezing Points of Solutions of Sucrose

Sucrose Equivalent (%)	Freezing Points (°C)	(°F)
0	0.00	32.00
5	−0.42	31.25
10	−0.83	30.50
15	−1.17	29.90
20	−1.50	29.30
25	−2.08	28.25
30	−2.67	27.20
35	−3.58	25.55
40	−4.39	24.10
45	−5.69	21.75
50	−7.00	19.40

Adapted from *J. Agr. Res.*, Vol. 56. No. 2, Jan. 15, 1938.

---

[1] A calorie is the amount of heat required to raise the temperature of 1g of water 1°C at 15°C (kcal = heat required to raise the temperature of 1kg of water 1°C).

Example: Calculate the approximate temperature at which 50% of the water is frozen in mix containing 12% fat, 11% NMS, 15% sucrose, and 0.3% stabilizer.

1. Find the % of water unfrozen in the ice cream when 50% of the water is frozen (water in mix − solids in mix)/2:

$$\% \text{ unfrozen water} = \frac{100 - (12 + 11 + 15 + 0.3)}{2} = \frac{61.7}{2} = 30.85$$

2. Calculate the sucrose equivalents supplied by sugar and NMS:

$$\text{sucrose equiv.} = (\% \text{ NMS} \times 0.545) + \% \text{ sucrose}$$
$$= (11 \times 0.545) + 15 = 21.00$$

3. Calculate the % sugars in the unfrozen water:

$$\% \text{ sugar in unfrozen water} = \frac{\% \text{ sugar} \times 100}{\% \text{ sugar} + \text{unfrozen water}}$$
$$= \frac{21.00 \times 100}{21.00 + 30.85} = 40.50$$

4. Interpolation from Table 10.2 indicates the freezing point of a 40.50% solution of sucrose is −4.52°C.
5. Calculate the freezing point depression by milk salts:

$$\frac{\% \text{ NMS} \times 4.26}{\text{unfrozen water}} = \frac{11 \times 4.26}{30.85} = 1.52$$

6. Calculate the total freezing point depression as contributed by sugars plus salts:

$$4.52°C + 1.52°C = 6.04°C$$

Therefore, the temperature of the ice cream is −6.04°C (21.13°F) when 50% of the water in it is frozen.

By adaptation this process can be used to calculate temperatures when other percentages of water are frozen or when other sweeteners are used. For example, if one knows the dextrose equivalent of corn sweeteners, a close approximation of the effect of the corn sweetener on the freezing point depression is dextrose equivalent times the freezing point depression of glucose (1.9 times that of sucrose).

Thus, corn syrup solids with a dextrose equivalent (DE) of 50 would depress the freezing point $0.5 \times 1.9 = 0.95$ times as much as sucrose. The latter number is called the sucrose equivalence factor and would be used in the equation of 2. above. For example, assume the sweetener in the above-stated problem is 12% sucrose and 7% corn syrup solids testing 52 DE. Then:

10 THE FREEZING PROCESS

One-half the water content = 29.85%

Sucrose equivalents = (11 × 0.545) + 12 + (7 × 0.52 × 1.9) = 24.91

Now, the temperature at which 50% of the water in the new mix is frozen is determined by first calculating the % sugars in the unfrozen water:

$$\frac{24.91 \times 100}{24.91 + 29.85} = 45.49°C$$

and this is translated into total freezing point depression:

$$-5.75°C + 1.52°C = -7.27°C$$

Thus, changing the sweetener increased the total soluble solids, decreased the free water, and dropped the temperature at which one-half of the water is frozen by 7.27°C − 6.04°C = 1.23°C.

Jaskulka *et al.* (1993) compared the predictive abilities of ice cream freezing point depression equations. They concluded that of the 14 equations found in the literature, most are applicable to a limited number of mix types. Only three of them produced values that did not differ significantly from measured values. They developed a method for predicting freezing points more accurately and for a wider variety of formulations using an empirical model (Jaskulka *et al.* 1995). The quadratic model showed that the main variables exert the following relative effects on freezing point depression: whey 115.07; high-fructose corn syrup (42% fructose) 110.24; NDM 109.60; dry buttermilk 98.33; 36 DE corn syrup solids 88.65; and liquid sucrose (67.3% solids) 67.63.

CALCULATING REFRIGERATION REQUIREMENTS

The previous discussion makes it obvious that the amount of heat to be extracted from an ice cream mix is affected by the composition of the mix and the temperature to which it is frozen. Table 10.3 illustrates for one mix the amount of heat energy extracted as drawing temperature decreases.

Table 10.3. Heat Absorbed from Ice Cream Mix Containing 12% Fat and 39.3% Total Solids on Cooling from 4.4°C to the Indicated Temperatures. (No allowance made for error due to friction or radiation of heat.)

Temperature	BTU/lb	kcal/kg
-1.10	5.0	2.78
-1.66	12.5	6.94
-2.22	19.5	10.83
-2.78	26.5	14.72
-3.33	33.0	18.33
-3.89	39.0	21.67
-4.44	45.0	25.00
-5.00	50.5	28.06
-5.56	55.5	30.83
-6.11	59.0	32.78
-6.67	63.0	35.00
-7.22	67.5	37.50
-7.78	70.5	39.17

Extraction of heat occurs in three steps, viz., (1) sensible heat from the liquid mix, (2) latent heat from the water as it solidifies, and (3) sensible heat from the semi-solid slush.

The following example adapted from Farrell (1953) illustrates a simplified method of estimating the heat removed from a mix containing 12% fat, 11% NMS, and 16% sugar. The mix enters the freezer at 4.4°C and is drawn at −5.6°C. Specific heat of the mix is estimated at 0.80; specific heat of the semi-frozen ice cream is 0.65; latent heat of fusion of water is 556 calories/kg; water in the mix is 60.7%; freezing point of mix is −2.63°C; and water frozen is 48%. Latent heat of fusion of milk fat is ignored because most of the solidification takes place during aging of the mix.

Therefore:

	kcal/g
Sensible heat of mix = [(4.4) − (−2.6)] (0.80) =	5.60
Latent heat of fusion = 80 (0.48 × 0.607) =	23.31
Sensible heat of slush = [(−2.6) − (−5.6)] 0.65 =	1.95
Total calories absorbed per kg mix =	30.86

In this calculation the main source of error is the variance in specific heat among mixes. Specific heat is the calories required to raise the temperature of 1 g of a substance 1°C. Sommer (1951), after reviewing considerable literature, concluded that the following specific heat values work well with ice cream: fat 0.50, NMS 0.46, sugar or stabilizer 0.35, water 1.00. The specific heat of water varies little with temperature, for example, from 1.00184 at 10°C to 1.00874 at 0°C. In comparison, the specific heat of NMS varies from 0.359 at 10°C to 0.278 at 0°C, or nearly 12 times as much as that of water.

Using the values of specific heat given above, one can calculate the expected specific heat of an ice cream mix as follows:

Mix with a High Amount of Fat

Component	Specific heat (cal/g)	Concentration (mg/g)	Calculated product
Fat	0.50	0.14	0.0700
NMS	0.46	0.08	0.0368
Sugar/stab	0.35	0.18	0.0630
Water	1.00	0.60	0.6000
		Specific heat	**0.7698**

Mix with Reduced Fat

Component	Specific heat (cal/g)	Concentration (g/g of mix)	Calculated product
Fat	0.50	0.04	0.0200
NMS	0.46	0.13	0.0598
Sugar/stab	0.35	0.18	0.0630
Water	1.00	0.65	0.6500
		Specific heat	**0.7928**

The latent heat of fusion of ice is 80 cal, so to melt a gram of ice requires 80 cal (80 kcal/kg ice). In the English system 144 British thermal units (BTU) are required to melt a pound of ice. The latent heat of fusion of milk fat is 19.5 cal/g.

Heldman (1966) presented a method for predicting the total refrigeration requirements for freezing ice cream to any temperature above 0°F. He divided the total heat required into four portions, viz., (1) sensible heat above the initial freezing point, (2) sensible heat of the unfrozen portion during freezing, (3) latent heat of ice, and (4) sensible heat of ice below the initial freezing point. Calculations confirmed that latent heat is the largest portion of the total requirement; however, sensible heat of unfrozen and frozen portions account for 8–10% of the total heat at normal freezing temperatures.

Obviously, the more heat removed from the mix prior to its entry into the freezer, the higher the capacity of that freezer given a constant drawing temperature. Uniformity in temperature and rate of flow of mix increases the probability that overrun and freezing rate will be uniform.

TYPES OF FREEZERS

Freezers for frozen desserts are designed to perform specific tasks under a variety of conditions and at varying costs. The soft-serve freezer must continue to deliver frozen product intermittently over several hours of operation. Batch freezers are designed to freeze a quantity of mix for delivery in a short time period. Continuous freezers receive mix continuously from positive displacement pumps and discharge the partially frozen product continuously.

Freezing times are affected by mechanical and physical factors and the properties of the mix. The mechanical and physical factors are

1. type and construction of freezer
2. condition of cylinder walls and blades
3. speed of dasher
4. temperature of refrigerant
5. velocity of refrigerant as it passes around freezing chamber
6. quantity of oil deposited on the outside of freezing cylinder
7. overrun desired
8. temperature at which the ice cream is drawn
9. rate of unloading freezer (batch type).

The mix characteristics that affect the freezing time are

1. composition
2. freezing point
3. methods of processing
4. kind and amount of flavoring materials.

The energy required to operate a freezer varies with mix formula, probably as it affects viscosity. Smith *et al.* (1985) found that mixes containing locust bean gum and guar gum in equal amounts required significantly more energy than mixes containing 75% locust bean gum and 25% guar gum. Guar gum binds four times as much water as does locust bean gum (Wallingford and Labuza 1983). Therefore, viscosity would be expected to be increased more by guar gum than locust bean gum.

The Continuous Freezer

The continuous freezer process was first patented in 1913 but did not become widely used until the 1930s. The process consists of continuously feeding a metered amount of mix and air into one end of the freezing chamber. As the mix passes through this chamber, it is agitated and partially frozen then discharged in a continuous stream at the other end of the chamber. This product is dispensed into packages that are placed in a hardening unit to complete the freezing process.

Capacities of continuous freezers range from about 30–1,000 gallons per hour (100 to >3,800 L/hr) per freezer barrel. Some freezers have two or three freezer cylinders mounted on a single frame and operated by the same controller. Ratings of capacities are generally based on nominal conditions such as:

1. Machine is in new or excellent condition
2. Refrigerant is clean, free of oil and noncondensable gases
3. Full fat ice cream mix is used with approximately 38% TS
4. Temperature of mix entering the freezer is 4.4°C (40°F), and it is drawn at −5.6°C (22°F)
5. Evaporating temperature of the ammonia refrigerant (saturated conditions) is −30.6°C (−23°F) or ammonia back pressure at the evaporator is 2 psi
6. Rating is stated in terms of gal/hr or L/hr at 100% overrun.

Because the rating is done under optimal conditions, the user cannot be assured that operations can be maintained at that capacity. Furthermore, characteristics of mixes are important determinants of freezer capacities.

The first continuous freezers had a positive type pump metering a constant supply of mix to a second pump that displaced two to three times the volume of the first pump. An air inlet valve was positioned upstream of the second pump so a desired amount of air could be admitted to the cylinder for whipping into the ice cream. A hold-back valve located at the distal end of the cylinder was adjusted to keep pressure on the cylinder. Instead of depending on air drawn in by vacuum, today's freezers are supplied compressed air through a regulator or a mass flow meter.

Approaches to controlling overrun, stiffness, and drawing temperature of frozen product vary among freezer manufacturers and among models within manufacturers. One method is to control the speed of the mix and air pumps with electronic mass flow meters. The controller adjusts the ratio of the quantities of air and mix. A variable frequency drive, employing a frequency inverter, adjusts the speed of the mix pump motor (Figure 10.9). Product stiffness is determined by monitoring wattage required to operate the dasher (Figure 10.10). Figure 10.11 shows such a freezer. Variable speed drives on some ice cream freezer pumps are controlled mechanically or hydraulically.

10 THE FREEZING PROCESS

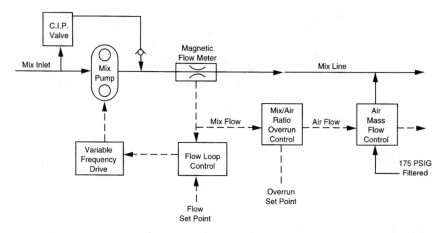

Figure 10.9. Schematic drawing showing the patented system for controlling product overrun in the Vogt Premier Series P freezer. (Courtesy Waukesha Cherry-Burrell, Louisville, KY.)

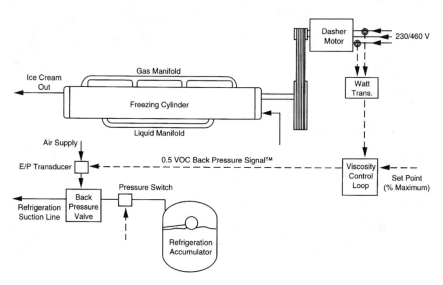

Figure 10.10. Schematic drawing of the system for controlling product stiffness in the Vogt Premier Series P freezer. (Courtesy Waukesha Cherry-Burrell, Louisville, KY.)

A major variable affecting the stiffness of ice cream exiting a freezer is temperature. However, using temperature to measure stiffness is subject to error since mix composition and overrun are also determinants of stiffness. Back pressure on the freezing cylinder and heat transfer efficiency must be optimal for control of freezing. Back pressure can be monitored and controlled

Figure 10.11. Vogt Premier Series P freezer. A programmable logic controller receives signals from sensors of cylinder pressure and current demanded to turn the dasher and uses the data to control flow rates of mix and air, which are monitored with flow meters. (Courtesy Waukesha Cherry-Burrell, Louisville, KY.)

10 THE FREEZING PROCESS

by a programmable controller (PC) that signals for changes in pressure on a product outlet valve.

Another freezer design (Figure 10.12) has both a mix pump and an ice cream pump. The semi-frozen ice cream discharges from the front of the freezing cylinder through the product pump, the speed of which is controlled to keep a constant pressure inside the freezing cylinder.

Figure 10.12. Continuous freezer with both a mix pump and a product pump, the "WS" model of APV Crepaco, Lake Mills, WI.

Mitten and Neirinckx (1986) explained: "With the two-pump arrangements, the mix or mix-air pump works against only the cylinder pressure, while the discharge or ice cream pump works against the difference between the downstream line pressures and the cylinder pressure. With the mix or mix-air pump and hold-back valve arrangement, the mix-air pump has to work against the imposed cylinder pressure plus the downstream line pressures." The latter arrangement puts considerable stress on pumps if downstream pressures are high. Furthermore, downstream pressures need to be held relatively uniform for optimal control of overrun. Changes in product viscosity can affect significantly the pressures against which pumps must work in transporting ice cream from the freezer to the filler. These changes in viscosity can arise from changes in extrusion temperature and from heat conducted through pipelines covered with varying amounts of frost. The two-pump system isolates the freezing cylinder from external pressure changes and tends to yield more constant overrun Figure 10.13.

In freezers operated manually air flow into the cylinder can be monitored as a function of pressure and flow through a meter that can be observed visually.

Some manufacturers offer freezers designed to deliver ice cream at temperatures about 3°C (5°F) lower than the normal drawing temperature. These low-temperature continuous freezers have two freezing chambers; the first being essentially the same as a conventional freezer and the second having a dasher especially designed to subject the mix to vigorous agitation. The freezer chamber is maintained under 4 to 6 times the pressure of the conventional barrel (14–21kg/cm^2, 200–300 psi, 1,400–2,100 kPa). Ice crystal size is reduced as much as 40%, e.g., from 45–55 μm to 18–22 μm. Slight decreases are observed in air cell size while the number of cells increases slightly. Air cell lamellae decrease in thickness. The product is more resistant to adverse handling than is regular ice cream.

Another approach to decreasing air cell sizes is to install a pre-emulsifying device ahead of the freezing cylinder. The dasher in the barrel cannot spin fast enough to divide the mass of air injected with the mix until the mix viscosity increases during freezing. Since viscosity increase due to freezing typically starts at about one-third of the distance from the entrance end of the chamber, most of the air is incorporated in the distal two-thirds of the chamber. By substantially dispersing the air before mix enters the cylinder, air cells are made smaller. Consequently, more and smaller ice crystals are formed, and the exiting product appears dryer than when no pre-emulsification is done. Small air cells are more stable than large ones, rendering ice cream made in this way comparatively less susceptible to shrinkage in the container. Some types of novelty products, such as stickless bars, must be extruded from the freezer in very stiff form. In addition, lowfat and nonfat products need to have very small air cells; therefore, pre-emulsification is a desirable treatment for these products.

Dashers function in ice cream freezers to carry the sharp blades that scrape ice from cylinder walls, to agitate the mix and air so a finely divided foam will form, and to partially churn the fat to help stabilize the foam. Dasher design is basically of two types, open and closed (solid), but this is an oversimplification. Displacement dashers, those with a solid core, that rotate at a high speed tend

10 THE FREEZING PROCESS

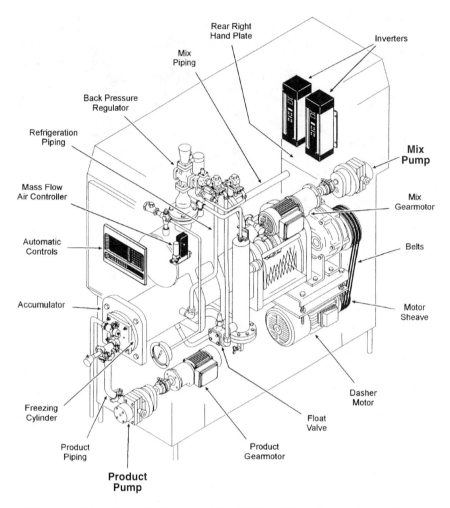

Figure 10.13. Schematic drawing of the "WS" model continuous freezer that has both a product pump and a mix pump. (Courtesy APV Crepaco, Lake Mills, WI.)

to produce a more stiff product than the open dasher, which is driven more slowly (Figure 10.14).

In early models about 80% of the volume was displaced by the dasher, and speed of rotation was high. Ice cream produced in such a freezer tends to have minute ice crystals but to be highly churned so that its melting rate is slow. This type of dasher action is desirable for producing extruded products such as ice cram bars that are to be enrobed in chocolate. Here product shape must be maintained long enough to affect hardening, and shape must be maintained when the bar is covered with the warm chocolate. However, the combination of solid dasher with a small annular space between the dasher and the freezer

Figure 10.14. Dashers designed for producing specific types of products in continuous freezers: left—closed, center—open with closed beater, right—open with beater. (Courtesy APV Crepaco, Lake Mills, WI.)

cylinder wall limit the volume of mix in the chamber. As the surface-to-volume ratio increases, so do chances of freeze-up within the cylinder.

By increasing the diameter of the freezing cylinder and reducing the displacement of the dasher, the freezer becomes much less sensitive to variations in refrigerant supply. Mix tends to act as a buffer against physical changes within the system, and the output is increased in uniformity of temperature and overrun. However, less churning is likely to occur in such freezers so that the ice cream tends toward wetness in appearance and quickness of melt.

Currently supplied open dashers displace 15–50% of cylinder volume and can be purchased with or without beaters. Solid dashers that displace 65–80% of cylinder volume are preferred for production of extruded products.

Internal structure of frozen desserts is highly dependent on freezer design and operation. The design and state of repair of the cylinder dasher and blades as well as the capacity of the freezer to carry away heat are important determinants of finished product quality. The ice cream manufacturer is advised to gain full knowledge of these parameters from the manufacturer of the freezer before making a purchase.

The continuous freezing process has the following advantages over the batch process:

1. Less stabilizer is needed because a larger amount of ice crystals can be formed in the freezing cylinder instead of in the hardening unit, where slow freezing produces large crystals.
2. A shorter aging time is needed because incorporation of air is less dependent on viscosity of the mix.
3. Smoother ice cream is obtained because the ice crystals are uniformly smaller than those obtained with batch freezing.
4. A more uniform yield is obtained with less variation among packages, especially small ones.

10 THE FREEZING PROCESS

5. Continuous freezing facilitates the making of specialties such as center molds, special shapes, combinations of flavors or colors in one package, variegated products, or individual serving-sized packages.
6. Throughput and quantity of product per worker can be greatly increased over that of the batch process.
7. The probability of contamination of product during filling is reduced as hand filling is seldom used.

The continuous process has a few disadvantages when compared with the batch process:

1. Less tolerance is available for variance in fit of many parts that must fit with minute clearances. This means that parts must be manufactured within small tolerances and handled with extreme care during cleaning and assembly of the freezer.
2. Greater training is required of operators and maintenance personnel.
3. It is easier to obtain excessive overrun.
4. Initial cost of the equipment is relatively higher.
5. An ingredient feeder is usually required for adding fruits, nuts, and other solid flavorings, whereas the batch freezer permits addition of these materials to the freezer barrel.

The Refrigeration System

Commercial continuous ice cream freezers are supplied with liquid refrigerant, usually ammonia, from the in-plant refrigeration system. This refrigerant enters the chamber surrounding the freezing cylinder (Figure 10.15) through

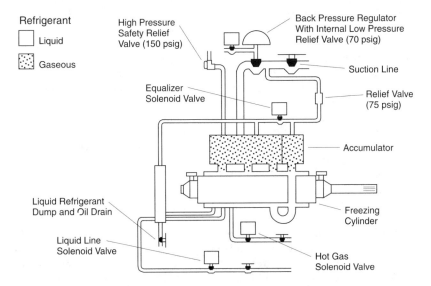

Figure 10.15. Schematic diagram showing the refrigeration system for an ice cream freezer in the "on" mode. (Courtesy APV Crepaco, Lake Mills, WI.)

an electrically controlled solenoid valve and a float valve that maintains the proper refrigerant level around the freezing cylinder. This cylinder is flooded with liquid refrigerant, which, on absorbing heat from the ice cram mix, boils and vaporizes. Vaporized refrigerant flows to the accumulator located above the freezing cylinder. There entrained liquid is separated from the gas and prevented from entering the suction line that carries the gas away from the freezer. The refrigerant is then reliquefied by compression and cooling so it can be again circulated to the freezer.

For protection against freeze-up, most modern continuous freezers have a hot gas line, equipped with a solenoid valve, to carry hot gas into the chamber surrounding the freezing cylinder. This unit can be activated manually or automatically. For example, if the ammeter, because of high demand for current, were to indicate the torque on the dasher motor to be exceeding a set point, the solenoid valve could be automatically opened so hot gas would flow into the chamber with the liquid refrigerant. The temperature would immediately rise, and defrosting of the freezer cylinder would occur.

3-A Sanitary Standards for Ice Cream Freezers

3-A Sanitary Standards for the manufacture of batch and continuous freezers are now in their fourth version (IAMFES 1990). Standard 19-04 became effective on March 20, 1990, and 19-03 was rescinded at that time. The purpose of this standard 19-04 is to describe the sanitary features of batch and continuous freezers for ice cream, ices, and similarly frozen dairy foods and equipment integral therewith. This includes pumps, equipment for incorporating air or flavoring material into the product, and mix supply tanks attached to and made as a part of the freezer. Soft-serve freezers are not covered under this standard. Permitted materials are described as are features of sanitary design and fabrication. The major material to be used is stainless steel of the American Iron and Steel Institute (AISI) 300 series. Bearings, springs, shafts, couplings, and similar parts shall be made of stainless steel of the AISI 400 series. Optional metal alloy (8% zinc maximum, 19% nickel minimum, 3.5% tin minimum, 5% lead maximum, 1.5% iron maximum, and the balance copper) is permitted for some parts that contact product, but only in freezers that are to be cleaned manually. These freezers must bear an information plate that warns against using acid cleaning compounds because the alloy is seriously corroded by acids.

Freezers that have been approved as meeting the standard are permitted to bear the 3-A symbol. Copies of the standard are available from the Dairy and Food Industry Supply Association, McLean, VA, or IAMFES, Des Moines, IA.

Operating the Continuous Freezer

The principal responsibilities of the freezer operator are (1) to regulate the amount of air being introduced into the mix to produce the desired overrun, and (2) to control the temperature of the refrigerant on the freezing chamber to give the desired stiffness to the product as it leaves the machine. These two variables need consistent monitoring by manual or microprocessor means, but changes are usually minimal once the system has been brought to a stable

10 THE FREEZING PROCESS

condition. Stability is achieved when temperatures of the equipment have been lowered to a steady state by removal of the heat stored in them and the rate of flow of mix and air have been stabilized. A source of error in overrun control is entrained air in the mix. This can result from adding incompletely melted rerun (mix that previously went through the freezer) to the mix tank, air leaks on the suction side of the mix pump, or air left over from blending operations.

To achieve optimal freezer operation mixes must consist of the intended composition and be processes as planned. Air incorporation prior to the freezing process must be minimized, and air removal may be necessary. Mix to be refrozen should be reprocessed both to optimize freezing and to assure microbiological safety. Mix should be supplied to the freezer pump at a low and constant temperature and a constant pressure.

Care and maintenance of the freezer and refrigeration system must be given priority if freezing is to be optimal on a daily basis. The following are the chief requisites for keeping the system operating properly:

1. Keep the ammonia jacket clean and free from oil, water, and nonvolatile ammonia fractions. Routinely check and drain water, oil, etc. from trap.
2. Keep the scraper blades clean and straight. Utmost care should be exercised in handling the blades to avoid bending or damaging.
3. Keep mix pumps in proper working condition. Especially check lubrication and tightness of belts and chains.
4. Make certain there is an adequate supply of refrigerant at the freezer. This requires that the entire refrigerant system be maintained.
5. Provide steady suction pressure at about 1 lb lower than the pressure at which the freezer was designed to operate. The continuous freezer requires a steady supply of liquid refrigerant. An insufficient supply or a significant rise in the suction pressure will soon show up as softness of the discharged product.

Applications of Programmable Controllers

Programmed freezers are designed for economy through the elimination of several manual operations and improvement of efficiencies. Basic functions can be performed simply by use of a manual selector switch and air-operated valves such as illustrated in Figure 10.16. This unit contains a rerun tank, which melts and deaerates mix as it is produced so it can be added to the mix vat or reprocessed as chosen. The system utilizes an air purge or blow-down feature to clear much of the material from the lines. Even greater degrees of control can be built into a system (Figure 10.17).

Automated overrun controls involve regulation of air input in a desired proportion to the amount of mix entering the freezing chamber. Control is achieved with a microprocessor. Manual settings of the controls is necessary at the beginning, but once set, the controller will maintain flow rates, pressures, and refrigeration outputs. A major change in mix composition requires the recheck of settings or preprogramming of the microprocessor. Since these automated systems do not control based on density of the mix or the finished product, variation in the amount of air entrained is sure to cause variation in the

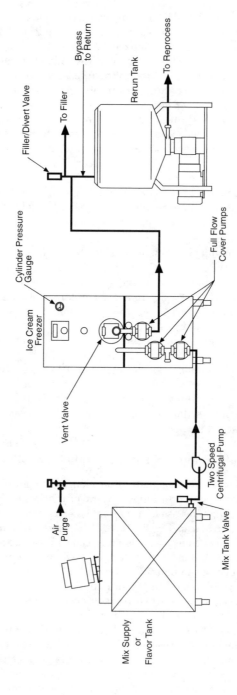

Figure 10.16. Basic programming elements for ice cream freezing systems. (Courtesy APV Crepaco, Lake Mills, WI.)

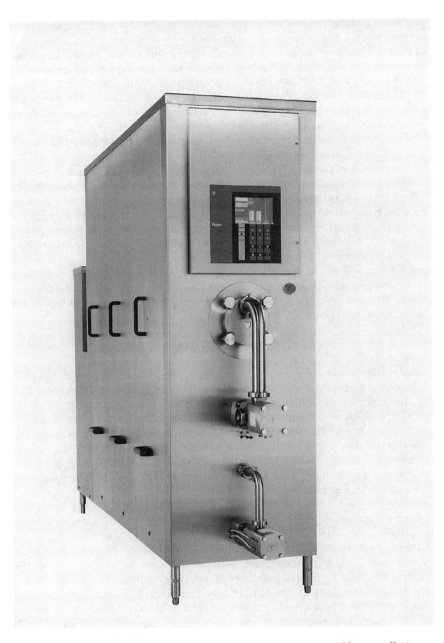

Figure 10.17. A Tetra Laval Hoyer microprocessor (programmable controller) controlled freezer. This represents units that have automatic controls for start/stop, viscosity, and overrun. (Courtesy Tetra Laval Food Hoyer, Pleasant Prairie, WI.)

overrun. The problem with measuring density in-line is that air is compressed somewhat until the product is at atmospheric pressure in the package. Measurement of package weight can be made on-line, but variability of filler performance may be significant. Furthermore, the time between introduction of air into the freezer and filling of packages with product that received that air is on the order of 1–2 min, depending on the length of the line to the packaging machine. This long lag time from control adjustment, to response, to readjustment tends to cause automated control systems to overcorrect.

Some programmable freezers can be linked to the filler machine to permit it to control the mix flow rate, hence to adjust rate of production to rate of fill. Programmable controllers (PCs) have a series of displays that permit the setting of control parameters as well as visualization of operational parameters. The following is a list of major screens of one PC: password, production data, product parameters, clean-in-place (CIP) parameters, sanitize parameters, general parameters, start pumpdown, calibrate, debug menu, communications errors, program information, hardware information, setpoint high/low. Another offers touch screen control for more than 95 formula items as well as data entry, monitoring, and control screens.

It is also possible for the controller to set the rate of operation of the ingredient feeder. Graphic displays of output of the freezer and the ingredient feeder can be produced with some PC units.

Advanced Programmable Freezers

Although it is not the purpose of this book to endorse any product of industry, it is informative to present examples of technology applied by industry. It is in this spirit that the following descriptions of recently developed freezers are presented. Four major manufacturers of high-capacity freezers were asked to provide pictures and descriptive materials of their programmable freezers. The descriptions that follow are intended to assist readers in identifying the various principles of design and operation that can be used to achieve similar results in freezing ice creams of various types. Pictures of three of the machines have been shown in Figures 10.11, 10.12, and 10.17.

The APV Crepaco model WS freezer (Figure 10.12) consists of a freezing cylinder that can accommodate open or solid dashers with cylinder volume displacements of 30% and 80%, respectively. Mix is fed into the cylinder via a sanitary rotary pump driven by a variable speed motor. The PC provides an input signal to an AC inverter that increases or decreases the frequency of the electrical current to change the motor speed, thus changing the pumping rate. The PC also monitors the speed of the product pump to empty frozen product from the freezing cylinder as pressure builds to a set point. Air input is sensed by a hot-wire type mass flow meter. The signal is transmitted to the PC, which signals an electrically operated solenoid valve to maintain or adjust the amount of air for the overrun selected. Liquid refrigerant enters the chamber around the freezing cylinder at a pressure of at least 50 psig (350 kPa). As it absorbs heat from the ice cream mix, the refrigerant boils, creating vapor that flows via the suction line to the unit's accumulator, located above the freezing cylinder, then to the compressor. A back pressure regulator controls this process.

10 THE FREEZING PROCESS

A float valve maintains proper refrigerant level around the freezing cylinder. An electrically operated hot gas solenoid valve opens when there is a need to defrost the freezing cylinder or to prevent freeze-up. Equipment downstream from the freezer can be linked for rate of operation to the freezer via the PC.

Gram Equipment Company of America offers five freezer types with six models per type. Capacities range to 1,000 gal/hr. Models equipped with a PC (Figure 10.18) enable automatic control of production and communication among computerized components of the plant. For overrun control the PC signals the air mass flow controller on the basis of mix quantity measured by the mass flow controller and the set point entered for overrun. The mix pump and product outlet pump are operated separately with the speed of the outlet pump controlled in relation to the cylinder pressure, which is measured by a pressure transducer. The PC compares the pressure to a set point and corrects possible variations. Viscosity of ice cream within the cylinder is measured by a watt meter that senses the power required to turn the scraper-dasher. The PC compares the wattage required to a set point and changes evaporation temperature of the refrigerant when needed. Dashers are available in open whipping style with 30% or 40% displacement, or in solid style displacing 80% of cylinder volume. The Gram aerator (Figure 10.19) is built as a separate unit from the freezer. It consists of the aerator unit, a mix pump, a meter-controlled air supply, and a PC to control processing and CIP cleaning. When used with the aerator the continuous freezer is used to freeze the mix/air mixture only and has no direct air supply or a mix pump.

The Hoyer model KF freezer by Tetra Laval Food (Figure 10.17) is equipped with a rotary mix pump regulated in rate by input from a magnetic flow meter. Air is precisely measured through a mass flow meter. Pressure within the freezer cylinder, normally held at 60 psig (414 kPa), is established by the speed of the ice cream outlet pump that receives its signal from the PC. Pressure required is dependent on the overrun desired, and this can be either set into the PC manually or taken automatically from the PC memory for each type and flavor of product. Refrigeration is adjusted by control of pressure on the suction side of the freezer. A float valve maintains proper liquid refrigerant level around the freezing cylinder. The refrigerant is automatically dumped into the dumping tank when there is need to defrost the freezing cylinder or to prevent freeze-up. The dasher is of the open type and displaces 15% of the cylinder volume. The PC can be tied to the ingredient feeder so output of the freezer determines the amount of ingredient dispensed into the frozen dessert. Similarly, the filler can be monitored by the PC to control output of the freezer. The PC software can store production data (Figure 10.20) for up to 50 products. If disturbances occur to the process during production, they will be indicated on the monitor. The PC can be programmed to take action to prevent damage or freeze-up when disturbances occur.

Vogt Premier freezers by Waukesha Cherry-Burrell are offered in cylinder sizes of 6-inch diameter by 24-, 48-, 72-, or 80-inch length, and 8-inch diameter by 80-inch length. Capacities at 100% overrun range from 150–800 gal/hr. Freezers are offered in single, double, and triple cylinder designs. Dashers are of the open type, displacing 44% of the cylinder volume and containing an inner beater. The mix pump is of the single lobe type. Three types of control

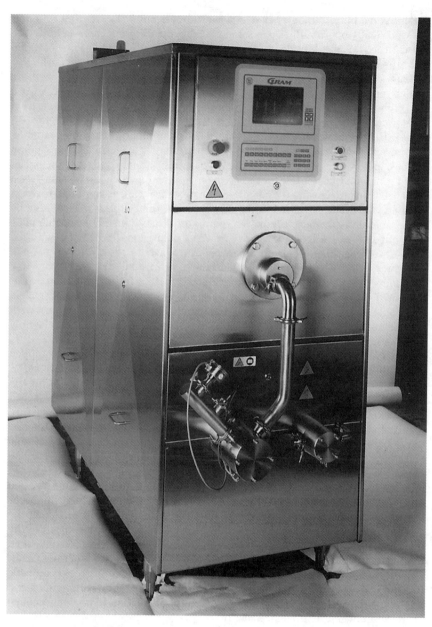

Figure 10.18. Microprocessor-controlled freezer model CF with a rated capacity of up to 1,000 gal/hr. (Courtesy Gram Equipment Company of America, Tampa, FL.)

10 THE FREEZING PROCESS

Figure 10.19. Aerating unit including mix pump, metered air supply, and PC. (Courtesy Gram Equipment Company of America, Tampa, FL.)

of mix flow, overrun, and viscosity are offered among the four series of models, viz. manual control, discrete automatic control, and PC control (Figure 10.11). In controlling up to three freezers the PC uses signals from either a magnetic flow meter or a mass flow meter. Speed of the mix pump is adjusted by means of a solid state variable frequency drive. Overrun is selected by choosing a set point, and the PC responds by signaling the air mass controller, which measures and controls the amount of pressurized air admitted to the cylinder. Cylinder pressure generally runs between 80 and 100 psig (550 and 690 kPa) and is imposed by an electronically controlled back pressure valve. The refrigeration system is of the full flooded type, and is controlled through measurements of product stiffness as determined by continuous monitoring of motor wattage required to turn the dasher. Production can be stopped for several minutes and resumed instantly with ice cream temperature, stiffness, and overrun

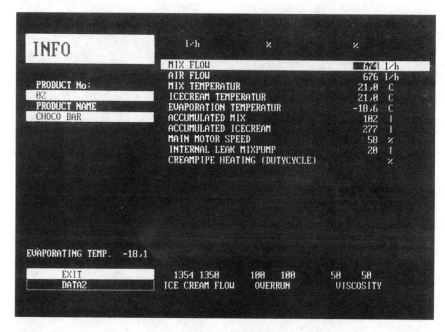

Figure 10.20. Production data readable on a screen of a PC of an ice cream freezer. By activating "INFO" the screen will switch to a production data display with information on mix flow, air flow, temperatures, accumulated mix, and ice cream values. (Courtesy Gram Equipment Company of America, Tampa, FL.)

maintained. A mix aerating device is optional with the freezer. This aerator is preferred for use in producing very stiff extruded products.

Shut Down and Cleaning of the Freezer

Once the freezer is started the refrigerant is shut off from the freezing chamber only when the machine is to be stopped. Usually the refrigerant is turned off a few minutes before the last mix is to enter the chamber. This permits the machine to warm to an extent that the rinse water that follows the last of the mix can pass through without being frozen. The rinse water temperature should not exceed 38°C (100°F) to avoid heat stress damage to the very cold cylinder. To avoid high wear due to friction, the dasher should be turned only a few revolutions during rinsing.

Once the water passing through the freezer has flushed out most of the solids, the freezer should be shut down. For manual cleaning the dasher, pumps, and other parts, depending on construction features, can be removed, disassembled, and placed in a sink or tank for washing in a warm (50–55°C) detergent solution as recommended by the manufacturer. The freezing chamber, inside and out, should be cleaned with hot detergent solution. All components should be rinsed thoroughly with water containing sufficient acid, usually

10 THE FREEZING PROCESS

phosphoric acid, to soften it and to remove mineral residues from the equipment surfaces.

CIP cleaning is possible and desirable with many modern freezers. The usual steps are (1) rinse with water at 38°C until the rinse runs clear, (2) circulate alkaline detergent (1–1.5%) solution at 65–70°C through the equipment for 20–30 min, (3) rinse with tap water, and (4) rinse with a low concentration of phosphoric acid (approximately 0.1% depending on water hardness). Drain at all low points in the system. During the cleaning cycle, the freezer should be run about 10 sec each 10 min. The entire assembled system must be sanitized with an approved sanitizer before use. Assuming that equipment surfaces are clean, factors that determine effectiveness of approved sanitizers are concentration, time of exposure, and temperature at the site of sanitization. If any one of the conditions of use is below the accepted minimum, the probability of failure is unacceptably high. It has been clearly shown that microorganisms that reside in films of soil on equipment surfaces are much more difficult to destroy than those suspended in a solution.

The Batch Freezer

The batch freezer (Figure 10.21) is a less complicated machine than the continuous freezer. Each batch of product is measured, flavored, colored, and frozen separately. This type of freezer is chosen for use in small plants and in product development laboratories. Batch freezers generally consist of a horizontally oriented cylinder of capacities ranging from 4–10 gal (15–40 L), a dasher, designed to propel the product toward the discharge port, the refrigeration unit (compressor, condenser, expansion valve, and evaporator—in this case the freezer cylinder), a cabinet, and in some models a mix supply tank. The refrigerant is typically a halocarbon type, usually R22 or R502, well known as Freons. With the concern over the effect these chlorofluorocarbons (CFCs) may have on depletion of the earth's protective ozone layer, new refrigerants are being developed. A major concern in design of equipment for their use is effectiveness of lubricants. A major supplier of refrigerants has produced a series of new refrigerants to replace R12 and R502 in both existing and new equipment (Williams 1993). Production of R12 and R502 was prohibited at the end of 1995. R22 will be phased out by banning use in new equipment beginning in 2015, and production is set to end by 2030. Venting of these CFC refrigerants to the air became illegal in 1992. Any technician servicing refrigeration systems must be certified and must have appropriate recycling and recovery equipment.

The temperature of the refrigerant should fall in the range of –23°C to –29°C (–10°F to –20°F) to provide for rapid formation of small ice crystals while permitting development of desired overrun. Freezing too slowly causes large ice crystals to form, while freezing too fast limits time for incorporation of air. The drawing temperature of ice cream from the batch freezer, –5°C to –6°C (23–25°F), is slightly higher than that used for continuous freezers. This results in batch-frozen products having larger ice crystals than continuous-frozen products. Typically, air cells are larger also. Precise overrun control is difficult with a range of 10–15% being experienced during emptying of the barrel. This

Figure 10.21. A 40-quart capacity batch freezer. (Courtesy Emery Thompson Machine and Supply Company, The Bronx, NY.)

occurs because the dasher continues to operate during unloading of the freezer. Whippability of mixes must be high because air is present at atmospheric pressure.

Although freezers for soft-frozen products, including milk shakes and frozen beverages, operate on the same general principles as batch freezers, they have

10 THE FREEZING PROCESS

some unique characteristics. They must be capable of holding the product frozen while avoiding churning and significant variation in overrun. This subject is discussed in Chapter 12.

Operation of the Batch Freezer

In preparation for use freezer parts should be examined for cleanliness and state of repair. After the operator's hands have been thoroughly washed, the freezer should be assembled according to manufacturer's instructions. Sanitization can be accomplished with hot water, but in doing so the final temperature must be 82°C (180°F). Since that heat must be removed from the cylinder during freezing, chemical sanitizers are often chosen. It is important to drain the sanitized freezer well to avoid contamination of the first batch of product. While sanitizing, the dasher should be turned no more than a few revolutions to minimize wear.

The second step in freezing is adding a measured amount of mix, flavoring, and coloring. The total volume of these components should be about one-half the capacity of the freezer. Flavoring and coloring materials must be added allowing enough time before emptying of the freezer for them to become uniformly distributed, but the order and moment of adding the materials varies with the product. For example, acid fruits and nuts should be added only after the mix has begun to thicken. Fruits may cause localized coagulation of casein because of their acidity if added before ice crystals begin to form. Nuts, fruits, cookies, and baked goods will remain in larger pieces, and candies will be less likely to dissolve if they are added late in the freezing cycle. Manufacturer's suggestions serve as general guides, but each operator should determine for the specific conditions and product what is the proper time for adding such ingredients.

The mix is run into the freezer, the dasher started, and the refrigerant turned on. This order of operation is important in the avoidance of damage to the machine. Scraper blades will dull rapidly if the dasher is operated with no mix in the cylinder. Refrigerant must never be turned on unless the dasher is in motion. Starting a dasher that is frozen to the cylinder wall can cause damage to the dasher or the drive mechanism.

The time to shut off the refrigeration may be determined by the operator peering through the peephole in the front of the machine. Experience and training are necessary to recognize the appearance of the partially frozen mix that dictates stopping the refrigerant from flowing into the chamber around the barrel. Much of the whipping takes place after the refrigerant is turned off. Some freezers have automatic shutoff controls that respond to selected amperage of the circuit that supplies electricity to the dasher motor.

Shutting off the refrigeration too soon can result in (1) an extended whipping time with possible failure to obtain desired overrun, (2) soft ice cream, and (3) coarse texture. Likewise, failure to turn off the refrigerant soon enough can result in (1) extended whipping time to obtain desired overrun, (2) stiffer ice cream with a lower temperature, and (3) smoother texture provided the product does not have to warm excessively to produce desired overrun. The main consideration, then, is selection of the optimal consistency for whipping the mix. This property will vary with composition of the mix, flavorings added, construction of the freezer, rate of freezing, and temperature in the freezer. During whipping, air bubbles form but some burst, yet there is a net gain in air content until maximal overrun is reached. With continued whipping and a rise in temperature there is a net decrease in air cells. For this reason, it is important to empty the contents of the freezer quickly. When the ice cream is drawn from the freezer, it should be stiff enough to form a "ribbon," yet soft enough to slowly settle into the container and to lose its shape in a minute or two. If a small amount of ice cream becomes too soft to package, it may remain in the barrel and be refrozen with the next batch provided the color and flavor will blend well. Pieces of residual fruit and nuts cannot be removed without taking the freezer apart, so it is best to plan the sequence of freezing starting with plain ice creams and progressing to those that contain particulates.

Filling Containers from a Batch Freezer

In preparation for filling containers it is best to prechill any container that has sufficient mass to melt the ice cream. For example, plastic containers generally have sufficient mass to carry enough heat to cause significant melting. Plasticized paperboard has less mass and heat does not penetrate the paperboard as rapidly as it does plastic. Product that melts next to the container will refreeze, forming large ice crystals. Furthermore, melting causes some loss of overrun.

The freezer should be emptied as rapidly as possible to prevent wide fluctuations in overrun among packages. Caution is in order to prevent formation of air pockets that leave the container partially filled. This problem is experienced when the ice cream is so stiff as to not flow readily. Well-frozen ice cream is desired from a quality standpoint, but some compromise in freezing temperature endpoint may be necessary to permit efficient and adequate fill.

Large containers are usually filled directly from the freezer; small containers are most properly filled by a packaging machine fed through a hopper. Otherwise it may be necessary to fill them manually after discharging the ice cream from the freezer.

REFERENCES

Bakshi, A. S., and R. M. Johnson. 1983. Calorimetric studies on whey freeze concentration. *J. Food Sci.* 48:1279–1283.

Bakshi, A. S., and D. M. Smith 1985. A computer-assisted method for evaluating ingredient substitutions in ice cream formulations. *J. Dairy Sci.* 68:1926–30.

Berger, K. G., B. K. Bullimore, G. W. White, and W. B. Wright. 1972. The structure of ice cream. Parts 1 and 2. *Dairy Indus.* 37:419–425, 493–497.

Bradley, R. L., Jr., E. Arnold, D. M. Barbano, R. G. Semarad, D. E. Smith, and B. K. Vines. 1993. Chemical and physical methods. In *Standard Methods for the Examination of Dairy Products*, 16th ed. R. T. Marshall (ed). American Public Health Assn., Washington, D.C.

Campbell, J. R., and R. T. Marshall. 1975. *The Science of Providing Milk for Man.* McGraw-Hill Publishing Co. New York.

Farrell, A. W. 1953. *Dairy Engineering.* John Wiley & Sons, Inc. New York.

10 THE FREEZING PROCESS

Heldman, D. R. 1966. Predicting refrigeration requirements for freezing ice cream. Quarterly Bull. Michigan AES. 49(2):144–154. East Lansing, MI.

Heldman, D. R., and T. I. Hedrick. 1968. Refrigeration requirements for ice cream freezing. *J. Dairy Sci.* 51:931.

IAMFES. 1990. 3-A sanitary standards for batch and continuous freezers for ice cream, ices and similarly frozen dairy foods. Number 19-04. Formulated by International Association of Milk, Food and Environmental Sanitarians, United States Public Health Service and The Dairy Industry Committee. IAMFES, Des Moines, IA.

Jaskulka, F. J., D. E. Smith, and K. Larntz. 1993. Comparison of the predictive ability of ice cream freezing point depression equations. *Milchwissenschaft* 48:671–675.

Jaskulka, F. J., D. E. Smith, and K. Larntz. 1995. Development of an empirical model to predict freezing point of ice cream mix. *Milchwissenschaft* 50(1):26–30.

Leighton, A. 1927. On the calculation of the freezing point of ice cream mixes and of quantities of ice separated during the freezing process. *J. Dairy Sci.* 10:300–308.

Mitten, H. L., and J. M. Neirinckx. 1986. Developments in frozen products manufacture. Chapter 4. In *Modern Dairy Technology, Vol. 2. Advances in Milk Products*. R. K. Robinson (ed). Elsevier Applied Science Publishers, Ltd. Essex, England.

Smith, D. E., A. S. Bakshi, and S. A. Gay. 1985. Changes in electrical energy requirements to operate an ice cream freezer as a function of sweeteners and gums. *J. Dairy Sci.* 68:1349–1351.

Sommer, H. H. 1951. *Theory and Practice of Ice Cream Making*. The Olsen Publishing Co. Milwaukee, WI.

Wallingford, L. W., and T. P. Labuza. 1983. Evaluation of water binding properties of food hydrocolloids by physical/chemical methods and in low fat meat emulsion. *J. Food Sci.* 48:1.

Williams, L. 1993. Personal communication. E. I. du Pont de Nemours and Company, Wilmington, DE.

11
Packaging, Labeling, Hardening, and Shipping

CONSIDERING THE PACKAGE

As ice cream is drawn from the freezer, it should be put into containers that give it the desired **form**, **size**, and **appearance** for convenient handling, efficient hardening, consumer appeal, consumer information, convenience, and economy. The label must reveal the ingredients in the order of highest to lowest concentration and the nutrient content in standard format.

Whereas the ice cream manufacturer is basically interested in using containers that will protect the product, will be desired by potential purchasers, and will cost a minimal amount, there are many factors that ultimately determine what will be the package that is used. In view of the importance of the subject to the dairy industry, the International Dairy Federation recently published the third edition of the Technical Guide for the Packaging of Milk and Milk Products (IDF 1995). The bulletin describes factors that determine the nature of the packages that enter and stay in the marketplace:

1. *variables that affect choice of raw materials*—such as uniformity of materials, compatibility of product and package, and efficiency of production
2. *performance of packages in storage and transport*—how they fit into the overall system of packaging and handling
3. *marketing considerations*—including suitable consumer units for storage, space savings, and stackability, ability to convey a good sales message and minimal losses or damage
4. *consumer acceptance, purchase, and use*—reasonable price, attractiveness, reliability, safety, information conveyed and clarity of message
5. *public concerns*—energy use, expenditures of nonrenewable resources, public health and welfare, and environmental contamination.

Several of the factors mentioned above are of sufficient public concern that legislation has been passed to control them. Food legislation has two basic objectives—first, to prohibit the sale of foods injurious to the health of the consumer, and second, to ensure that a product of nutritive substance and

11 PACKAGING, LABELING, HARDENING, AND SHIPPING

quality is available to the consumer. Because the majority of today's food packages are either made wholly from or coated with plastics, there is concern that the polymers are safe and that toxic substances do not migrate from them into the foods. Since plastics consist of a basic polymer plus auxiliary substances added to achieve desired function, it is required in many countries that manufacturers make a positive list of all polymers, additives, adhesive coatings, and lubricants that are permissible for use in packages for foods. In the United States that list is published in the Code of Federal Regulations.

Migration of materials from the package into food is of major concern. Regulations that address potential outcomes of migration usually address the following (IDF 1995):

1. The taste, smell, or character of the food shall not be altered after storage under normal conditions.
2. Total migration from the package, even if completely harmless and not sensorially perceptible, shall not exceed the "contamination test" limit set by national regulations.
3. The packaging material shall contain only substances included on the list of permitted ingredients, and these shall not exceed the permitted maximum concentration. Usually the manufacturer guarantees this.
4. If the list of permitted ingredients contains provision that migration of certain components from the packaging material shall not exceed a set level in the food, the packaged food will be checked against that provision. (Important in this regard are monomers of the material that was polymerized and breakdown products that arise from overheating.)
5. Fundamental hygiene shall be practiced in manufacturing the package and producing the packaging materials.

With the emphasis on environmental quality that has continued to grow in magnitude, it is important to consider packaging systems with a "Life Cycle Analysis." The concept was broken into key phases (Kool and van den Berg, 1995) as follow:

- Extraction and processing of raw materials
- Package manufacture
- Transport of raw materials to manufacturers, packages to dairies, and packaged products to the ultimate consumer
- Consumer sales and use
- Recycling or waste disposal.

The environmental impact can fall into four categories, viz. use of raw materials (particularly nonrenewable ones), energy consumption, air- and water-borne emissions, and waste disposal.

LABELING

The Nutrition Labeling and Education Act of 1990 required the U.S. Department of Health and Human Services to propose new labeling regulations by November 1991. The responsibility for these regulations as pertains to foods other than meat and poultry falls to the Food and Drug Administration. The regulations issued by the FDA made up 900 pages of the *Federal Register* of January 6, 1993. They became effective on May 8, 1994. Firms affected are those that sell more than $500,000 of all food and nonfood products and sell more than $50,000 of food per year. Packages having less than 12 in^2 of surface area and on which there is no nutrition claim are exempt from the rule. Those packages must bear an address or telephone number that a consumer can use to obtain the required nutrition information. Major provisions of the regulations as they affect labeling of frozen desserts are presented in the following.

All nutrient and food component quantities shall be declared in relation to a serving. A serving is an amount of food customarily consumed per eating occasion by persons 4 years of age or older. Furthermore, the serving size is to be expressed in common household measure. For frozen desserts this size, or reference amount, was set at one-half cup, including the volume of coatings and wafers for novelty type products. This regulation, combined with the minimum weight per gallon stipulation of the federal standard for ice cream, 4.5 lb/gal minimum weight, sets the minimum weight of a serving at 63.8 g. This is calculated as follows:

$$4.5 \text{ lb/gal} \times 453.5924 \text{ g/lb} = 2041.2 \text{ g/gal}$$
and:
$$2041.2 \div 128 \text{ fl oz/gal} = 15.95 \text{ g/fl oz}$$
since: ½ cup = 4 fl oz, 4 fl oz × 15.95 g = 63.8 g (round to 64 g)

By consulting the weight per serving declared on the container one can easily calculate the target overrun of the manufacturer. In a 1995 survey by the author's laboratory the declared weight per serving of regular vanilla ice cream in half-gallon containers averaged 67 g with a standard deviation of 2.9 g. Eight of the 14 brands surveyed declared the serving size as 65 g. The maximal net weight per serving was 73 g, which, if a mix weight of 9 lb/gal is assumed, is equivalent to the overrun shown in the following calculation:

$$\frac{[(9 \text{ lb/gal} \times 453.5924 \text{ g/lb}) \div (128 \text{ oz/gal} \div 4 \text{ oz/cup})] - 73 \text{ g/cup}}{73 \text{ g/cup}} \times 100 = 75\%$$

This formula can be used for any similar calculation by substituting the actual weight per gallon of mix and the target weight in g/cup to calculate the desired overrun, or if the desired overrun is known, to calculate the desired weight per cup.

In addition to the minimum weight per gallon, ice cream must contain at least 1.6 lb of food solids per gallon. This equates to 22.7 g of food solids per 4-fl-oz serving. Thus, the minimum TS in the mix would be 35.44% when overrun is set at 100%. The minimum fat content, unless the label contains a descriptor, is 10% in plain ice cream. This equates to a 6.4 g/serving, and the label would read 6 g of fat. The total milk solids must equal at least 20%, so the minimum NMS in this ice cream would be 10%. The estimated protein content would be 36% of the NMS, or (64 g × 0.1) 0.36 = 2.3 g, and the label would read 2 g. If the remainder of the solids of the mix is composed of 10%

11 PACKAGING, LABELING, HARDENING, AND SHIPPING

sugar, 5% corn syrup, and 0.44% stabilizer/emulsifier, the amount of carbohydrate is 15% of 64 g plus the amount of lactose in the NMS. If all of the NMS is supplied by concentrated skim milk, the lactose can be estimated as 56% of the weight of the NMS. Total carbohydrate would be [0.10 + 0.05 + (0.56 × 0.10)] 64 = 13.18, and this would be rounded to 13 g. The calcium content would be about 90 mg, and this is 9% of the Recommended Daily Allowance for men and women between the ages of 25 and 50 (National Institutes of Health, 1994).

The label must also show amounts per serving for calories, calories from fat, total fat, saturated fat, cholesterol, sodium, total carbohydrates, sugars (mono- and di-saccharides including lactose), dietary fiber, and protein. Furthermore, the % Daily Value must be given as referenced to a 2,000-calorie-per-day diet for the above components, for vitamins A and C, and for the minerals calcium and iron. A label typical of one for regular vanilla ice cream is shown in Figure 11.1. The list of nutrients covers those most important to the health of today's consumers, most of whom need to be concerned about getting too much of certain nutrients rather than too few vitamins or minerals as in the past.

At the same time as labeling regulations were changed, the Federal Standard of Identity for ice milk was dropped. This action was made possible by the adoption of descriptors to describe foods in terms of selected important characterizing ingredients. The important characterizing ingredient of ice cream is milkfat, and the descriptors used follow:

Reduced fat: 25% less fat than the reference product.
Light: 50% reduction in total fat from the reference product, or one-third reduction in calories if fewer than 50% of the calories are from fat.
Lowfat: not more than 3 g of total fat per serving.
Nonfat or fat free: less than 0.5 g of fat per serving.

Interestingly, in the calculation of the amount of fat contained, the values expressed are the free fatty acid portion, and the glycerol portion is not included. Furthermore, the trans fatty acid portion may be excluded also.

These lower fat products must not be deemed nutritionally inferior in the nutrients shown on the nutritional label; therefore, vitamin A, a fat-soluble vitamin carried by milkfat must be added. Since milkfat contains an average of nearly 40 International Units (IU) of vitamin A equivalents per gram of fat and regular ice cream is required to contain at least 6.4 g of fat per serving, the amount of vitamin A to be contained in ice creams with lowered amounts of fat would be 40 × 6.4 = 256 IU, and this is 5.12% of the U.S. Recommended Daily Allowance. The rounding rule specifies that for vitamins amounts are expressed in 2-unit increments; therefore, the label would read 6% of the Daily Value. (Readers are referred to 21 CFR 104.20 for rules on fortification of foods.)

These lower fat products must possess physical and functional properties that resemble those of the product they replace, i.e., flavor, body, texture, and appearance must be in semblance of ice cream. The same safe and suitable ingredients are to be used as for ice cream, and it is permissible to add fat analogs and water to replace fat and calories. Furthermore, it is permissible to lower the content of lactose by ultrafiltration, to add hydrolyzed dairy

Nutrition Facts

Serving Size 1/2 cup (66 g)
Servings Per Container 16

Amount Per Serving

Calories 130   Calories from Fat 70

	% Daily Value*
Total Fat 7g	**11%**
Saturated Fat 4.5g	**22%**
Cholesterol 30mg	**10%**
Sodium 55mg	**2%**
Total Carbohydrate 16g	**5%**
Dietary Fiber 0g	**0%**
Sugars 15g	
Protein 2g	

Vitamin A 6%   •   Vitamin C 0%

Calcium 8%   •   Iron 0%

* Percent Daily Values are based on a 2,000 calorie diet. Your daily values may be higher or lower based on your calorie needs:

	Calories:	2,000	2,500
Total Fat	Less than	65 g	80 g
Sat Fat	Less than	20 g	25 g
Cholesterol	Less than	300 mg	300 mg
Sodium	Less than	2,400 mg	2,400 mg
Total Carbohydrate		300 g	375 g
Dietary Fiber		25 g	30 g

Calories per gram:
Fat 9   •   Carbohydrate 4   •   Protein 4

Figure 11.1. Example of Nutrition Facts label on a half-gallon package of regular vanilla ice cream.

Table 11.1. Summary of Average Quantities Per Serving Declared on Nutrient Labels of Eight Types of Vanilla Ice Cream Packaged in Half-Gallon Retail Containers

Descriptor[a]	Calories	Fat (g)	Protein (g)	CHO[b] (g)	Sugars[c] (g)	Calcium (%)[d]
Fat free/sugar free	80	0.0	4.0	18.0	5.0	15.0
Nonfat	98	0.0	3.5	21.0	16.0	9.5
Lowfat	102	2.3	3.0	18.0	15.2	8.7
Light	110	3.3	3.0	17.2	15.0	9.2
Reduced[e]	100	4.5	3.0	13.0	3.5	9.0
Regular	134	7.2	2.4	15.6	13.8	8.3
Premium	155	9.5	2.5	15.5	13.3	7.5
Superpremium	170	18.0	5.0	20.0	16.0	6.0

[a] Numbers of samples per descriptor group: fat and sugar free 1, nonfat 4, lowfat 6, light 5, reduced fat 2, regular 16, premium 4, superpremium 1.
[b] Carbohydrate: polysaccharides with more than two saccharide units per molecule.
[c] Sum of all free mono- and disaccharides, including lactose.
[d] Calcium expressed as % Daily Value in a 2,000-calorie diet.
[e] Reduced fat ice cream sweetened with nonnutritive sweeteners.

proteins, and to utilize safe and suitable sweeteners such as aspartame and acesulfame K. In each case the product's label must identify each ingredient. Finally, these products may weigh as little as 4.0 lb/gal, one-half pound less than the whole fat variety of ice cream.

It is permissible to label some ice creams as "healthy." Such a product must contain, on a per-serving basis, not more than 3 g total fat, 1 g saturated fat, 15% of calories from saturated fat, 60 mg of cholesterol, or 360 mg of sodium (as of January 1, 1998). Additionally, it must contain at least 10% of the Daily Value of at least one of the following nutrients: vitamin A, vitamin C, calcium, iron, protein, or fiber.

The market survey of ice creams, referred to above, revealed the mean values shown in Table 11.1 for selected nutrients in variously labeled retail ice creams in half-gallon packages in the summer of 1995.

THE PACKAGING OPERATION

The packaging of frozen desserts is of two types: **bulk packaging** for the sale of dipped products, including cones, and **consumer packaging** for direct retail sale. Packaging should be done as close to the freezer as possible to limit the back pressure on the freezer cylinder exerted by friction in the pipes. Bends in the line to fillers should be minimized, because they produce several times more friction head than straight pipe of the same size. For example, in terms of friction loss elbows in 2-inch and 3-inch lines are equivalent to 7 ft and 12 ft of straight pipe, respectively (Farrall 1953). Therefore, for a flow rate of 600 gal/hr through 20 ft of 2-inch line with three elbows the friction loss in lb/in^2 is 0.7 for the pipe plus 0.74 for the three elbows (total = 1.44 psi). At the same rate of flow through a 3-inch pipe friction loss would be 0.23 lb/in^2 for 20 ft of pipe and 0.42 lb/in^2 for the three elbows (total = 0.65 psi). These numbers apply to water rather than frozen ice cream, but they serve to illustrate the point.

Bulk Packaging

Practically all bulk frozen desserts are packaged in single-service containers. Some are packaged in reusable plastic, but little use is made of steel cans. Container sizes include 5, 3, and 2.5 gal, and most are formed on site manually with a single-spindle former or mechanically with multiple spindle machines (Figure 11.2). Flattened sidewalls, bottoms, rings to form the bottom and top, and overlapping tops are shipped separately to the ice cream plant.

Important functions of filling machines are accurate and precise fill, avoidance of air pockets, maintenance of the distribution of variegates and other inclusions, and freedom from product on the outside of the container. In filling large containers the container may be lifted to near the filler head or "tooling plate," then spun and lowered as product is pumped in. Weight and stiffness of the product can be used to lower the elevator as the product fills the container. In such cases, pressure on the elevator must be adjusted when mix stiffness changes. Lids are commonly pushed on mechanically as the filled container moves away from the filler on a conveyor (Figure 11.3).

Proper function and economy of operation of fillers are truly important, but no filler is satisfactory that does not meet strict standards for sanitary

Figure 11.2. Forming machine for 1-gal (4-L) to 6-gal (24-L) sizes of bulk containers. Produces up to 780/hr as it semiautomatically attaches sidewalls and bottoms together with seam holders formed from flat metal strips. (Courtesy Sealright, Overland Park, KS.)

11 PACKAGING, LABELING, HARDENING, AND SHIPPING

Figure 11.3. Bulk ice cream filler with volumetric "bottom-up® filling." Fill accuracy is ± 1 oz for a 3-gal tub. (Courtesy T. D. Sawvel Co., Maple Plain, MN.)

construction and operation. To this end the industry has accepted 3-A Sanitary Standards (1993) for Equipment for Packaging Viscous Dairy Products (Standard Number 23-02). The standard covers materials and fabrication of equipment used to package frozen desserts. Manufacturers whose equipment meets the standard may attach a 3-A symbol to the packaging machine. The symbol assists sanitarians in inspecting the equipment by letting them know that approved materials and fabrication criteria have been applied.

Packaging for Direct Sale to Consumers

Packages for sale directly to consumers vary in size from 3 fl oz to 2 gal (88.7 ml to 7.57 L). The most common size is 0.5 gal (1.89 L). Shapes are rectangular, cylindrical, and conical. Percentages of package sizes in recent retail trade are

Table 11.2. Package size shares of the ice cream, sherbert and frozen yogurt marketed in retail stores in 1995. (Source: International Ice Cream Association).

Package Size	Product		
	Ice cream %	Sherbet %	Frozen Yogurt %
Pint	3	1.7	6.8
Quart	2	27.9	2.9
Half gallon	83.6	66.1	87.7
All other sizes	11.4	4.3	2.6

shown in Table 11.2. Conical type packages are shipped preformed and nested within each other, whereas rectangular and cylindrical types are formed at the location of filling.

In 1995 the National Conference on Weights and Measures voted to remove container size restrictions on dairy foods, including frozen desserts. This permitted states that had adopted the model "Uniform Regulation for the Method of Sale of Commodities" to drop their requirement restricting sizes to fluid ounces and other volume increments of gallons.

Since most packaging is done on continuous freezer lines, this discussion will be limited to such installations. The operation consists of forming, filling, closing, weighing and overwrapping containers. Overwrapping or, less frequently, bagging is done to limit the number of items to be handled and to protect the packages during handling. Further consolidation onto pallets is usually done subsequent to hardening. Palletizing before hardening limits heat transfer and causes course texture.

Frozen desserts packaged for carryout usually have a lower overrun than those filled for dipping. Furthermore, overrun varies with the implied quality of the product, the higher quality products having lower overrun (Table 11.3). The lower the overrun, the higher the price that has to be charged for the product. Because dipping squeezes air out of ice cream, it is necessary to have overrun of bulk ice creams 15–20% higher than that of packaged products to have the same overrun at serving.

Fillers for small packages, cups, and cones are available with rotary tables (Figure 11.4) for smaller systems or with a series of parallel lanes (Figure 11.5)

Table 11.3. Typical Overrun for Frozen Desserts

Product	Overrun (%)
Superpremium ice cream	20–40
Premium ice cream	60–75
Ice cream, packaged	75–90
Ice cream, bulk	90–100
Sherbet	30–40
Ice	25–30
Soft ice cream	30–50
Milk shake	10–15

Figure 11.4. Rotary type filler for cups and round nested containers: accommodates timed or volumetric fill of up to 120 cups/min in 6- to 16-oz sizes and 60 cups/min in quart or half-gallon sizes. (Courtesy Sealright, Overland Park, KS.)

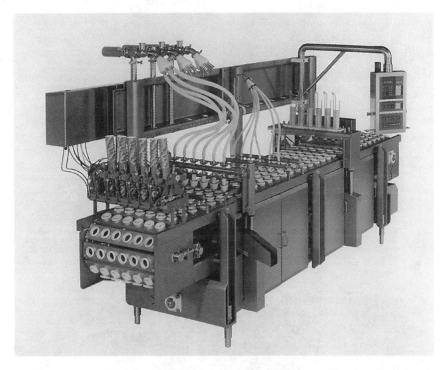

Figure 11.5. Cup or cone filler with multiple parallel lanes. (Courtesy Tetra Laval Food Hoyer, Pleasant Prairie, WI.)

for larger operations. These fillers are usually adaptable to volume and package type changes. The filler shown in Figure 11.5 is computer controlled and can be pre-programmed to fill 25 different products. It can be linked by computer with a companion freezer, ingredient feeder, and freezing tunnel to provide a fully integrated rate-controlled system. Containers in a single lane are filled simultaneously and may be topped and capped at other stations on the stainless steel conveyor.

Economy in Packaging Operations

Scheduling of production is dependent primarily on rate of sales, how much product is in inventory, production rate possible with the equipment, and the most economical length of a production run. An analysis of the problem can be done using an optimizing lot size formula, in which the number of units produced per run, the number of units expected to be shipped or sold per time unit, the cost of holding a unit in inventory (storage) per unit of time, and the cost of setting up and tearing down for a production run are factors (see, for example, Kramer and Arbuckle 1965).

Although ice cream is sold by volume, a minimal weight per gallon is required. Furthermore, it is in the interest of the producer to control weight. This is done primarily by controlling overrun. However, some variation is introduced by the filler and the package.

Getting the most out of filling operations requires the application of statistical quality control. Fill control involves examining how much variation normally occurs from one filler to another and from one container to another on the same filler. Relatively simple procedures are available for using this information (Kramer and Arbuckle 1965). Variations of statistically significant magnitude provide information that is useful for filler adjustments or for selecting filler systems. Control charts that have upper and lower control limits are quite useful at the filler.

To establish a target weight is rather easy with the help of a modern calculator that provides a mean and standard deviation from container weights entered into it. A representative sample is necessary if the values are to be useful. About 30 samples should be taken randomly after the freezing and filling systems have been observed to operate in a steady state. Each sample should be weighed and the standard deviation and mean calculated. With larger containers the variance in container weight is likely to be insignificant compared with the variance in weights of the product. In this case an overall standard deviation will suffice. With small containers it is important to obtain a standard deviation of both container and product weights and to pool the two standard deviations. To pool the values, (1) square each standard deviation, (2) sum the squared values, and (3) determine the square root of the sum. Using either of the values to ensure that at least 95% of the packages are filled with the minimal weight of product, the target weight for fill should be set above the minimal weight by 1.65 times the standard deviation.

For example, assume the minimal fill weight for half-gallon containers is set at 1,100 g and that the weights of 30 randomly selected filled containers taken over 15 min of time had a standard deviation of 30 g and a mean of 1,112 g. The target weight should be set at $1,100 + (1.65 \times 30) = 1,149.5$ g. The decision must then be made whether to adjust the filler or the overrun or both based on observations of container fullness. When a freezer and filler are working well, the weights of half-gallon containers should not vary more than 2% (22 g) from the target.

THE HARDENING PROCESS

When ice cream is drawn from a freezer and placed in containers, it is of a semi-solid consistency and is not stiff enough to hold its shape. (The exception is some extruded products.) Therefore, the freezing process is continued in containers without agitation until the temperature reaches $-18°C$ ($0°F$) or lower, preferably $-25°C$ to $-30°C$. Quick hardening is desirable because the formation of large ice crystals takes place with slow hardening. Stacking of packages before they are hardened may cause deformation of packages, loss of overrun, and surface discoloration because air is pressed out.

The time to accomplish hardening has been assumed to be the time for the temperature at the center of the package to drop to $-18°C$ ($0°F$). This hardening time for a still air operation may be as short as 30 min for 4-oz packages to as long as 24 hr for 5-gal packages. A shorter time always results in a smoother ice cream. A hardening time of 6–8 hr for 5-gal packages indicates excellent conditions in the conventional hardening room. Many operators allow 12 hr in such systems. Hardening tunnels and plate type systems remove heat at an accelerated rate.

Factors Affecting Hardening Time

Factors that affect the rate of hardening include temperature differential; conductance properties of the product, the container, and the cooling plate (if used); and distance for heat to travel. Temperatures of the product and the cooling medium determine the temperature differential. The greater this difference the faster heat will be removed; however, the higher the temperature at drawing, the longer the time of hardening. Conductance of heat is affected by composition of the product and the overrun. Since fat and air conduct heat more slowly than an aqueous phase, increasing the content of fat or air increases hardening time. The ability of the container and the overwrapping material to transfer heat can also be a significant factor. Plastic conducts heat faster than paper. (This means also that product can melt faster in plastic than in paperboard containers.)

Distance for heat to travel is affected by the size and shape of the container as well as by how much separation there is between packages. Thus, overwrapping increases this distance, as does stacking of packages. Doubling the size of the package or of the stack increases the hardening time in a conventional hardening room by about 50%. Overwrapping also affects shape and the amount of surface area per unit volume. The smaller the surface-to-volume ratio, the slower heat will be transferred.

Once heat gets to the surfaces of the containers, it needs to be removed promptly. Velocity of air movement thus becomes a critical factor unless packages are exposed to cold plates that transfer heat away by conductance. Plate type hardeners work well when packages are of the rectangular type and are of equal height when filled. Since rectangular containers constitute a major type of package in many plants, plate type hardeners work well in a large number of installations.

Experiments have shown that immersion of packaged ice cream in liquid nitrogen at $-196°C$ ($-320°F$) reduced the center temperature of product in pint packages to $-34°C$ ($-30°F$) in 5 min (Der Hovanesian 1960). However, exposure to liquid nitrogen for over 1 min made the texture "salvy" and the body crumbly. This treatment also increased the tendency for shrinkage in the packages. It was calculated that 0.56 lb of liquid nitrogen was required per pound of ice cream.

Types of Hardening Facilities

Numerous types of facilities have been designed to harden frozen desserts. These include cabinets, cold cells within storage rooms, hardening tunnels, and contact plate freezers.

11 PACKAGING, LABELING, HARDENING, AND SHIPPING 213

Cabinets are used only in small operations. They resemble retail ice cream cabinets. Air is circulated in them when the door is closed to speed heat removal. Some cabinets have refrigerant circulating through channels within each shelf; such a design speeds heat removal.

Cold cells are zones within storage rooms in which chilled air is blown over newly arrived packages as they are conveyed toward the ultimate storage area. In many installations this process takes place within a hardening tunnel. The tunnel design may be straight-through or spiral. Devices for moving packages through tunnels include conveyors, trays, and belts. Trays may form moving shelves within an enclosed area, each tray being moved stepwise (indexed) from the receiving end, where a row of freshly packaged product is pushed onto the shelf, to the discharge end, where hardened product is pushed off simultaneously (Figure 11.6). Between these points trays typically make two passages through the machine, being exposed to very cold air (−30°C or lower) moving at high velocity.

Another type of tunnel hardener employs trays attached to a continuously moving chain drive. The conveyor unit gradually inclines and declines within the tunnel and accepts and delivers products at a common point situated at the front of the freezer (Figure 11.7). The spiral type hardening tunnel typically

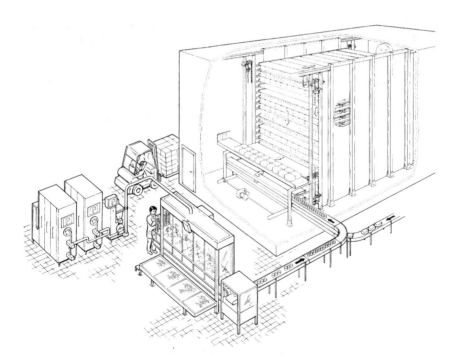

Figure 11.6. Tray type freezing tunnel capable of providing different dwell times for different products, separating flavors, and unloading product independent of loading order. (Courtesy Tetra Laval Food Hoyer, Lake Geneva, WI.)

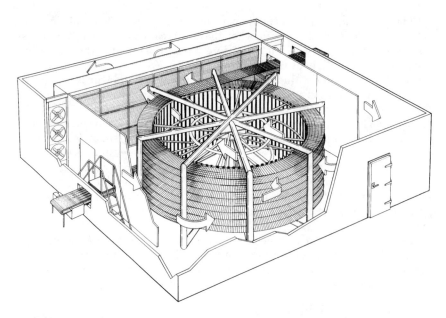

Figure 11.7. Hardening tunnel that uses a chain drive to move packages along a spiral track through the freezing zone. (Courtesy Northfield Freezing Systems, Northfield, MN.)

receives soft product at the bottom of the spiral and discharges hardened product at the top. Operating speeds can be selected to fit manufacturing speeds. Time within a tunnel hardener may vary from 40–160 min.

Special tunnels have been constructed to freeze stickless novelties such as the Klondike Bar and BonBons. Product is extruded onto stainless steel plates that are moved mechanically through a cabinet where blasts of frigid air chill the products. As the novelties exit the freezer a hammer-like device taps the steel plate to loosen the hardened piece. It is then swept to another conveyor for passage through an enrobing machine, drying, and packaging.

Contact plate freezers chill product rapidly by conducting heat from packages. They have the additional advantages of requiring minimal floor space and operating with high efficiency. Because they typically have no fans, they do not have to remove the heat of electric motors from the unit. Packages of the same height are contacted on both top and bottom by plates containing circulating refrigerant. Hardening is usually completed within 2 hr. A disadvantage of contact plate hardeners is that containers need to be of the same vertical dimension. They work especially well for half-gallon rectangular containers.

Small firms may elect to use the hardening room to both harden and store product. In this case compromise is necessary either in the speed with which hardening is done or in the costs of keeping the product very cold and moving air rapidly within the room.

Decisions must be made about the size of the hardening unit and the storage unit. The rate of product throughput should determine the capacity of the

11 PACKAGING, LABELING, HARDENING, AND SHIPPING

hardening unit. Factors that influence storage capacity include maximal demand expected, number of different products and container shapes and sizes marketed, capacity of production equipment, time each product can be permitted to stay in storage for reasons of quality, and interest on investment costs. The size and shape of the container as well as the depth of stacking on pallets or shelves are significant factors determining the capacity of storage rooms. For example, rectangular containers stacked together contain 7 gal/ft^3. The same cubic volume holds 5.1 gal of 3-gal cylindrical bulk containers and 4.5 gal of nested type 0.5. gal containers.

Precautions to Observe in the Operation of Hardening and Storage Rooms

The most commonly used refrigerant in ice cream plants is anhydrous ammonia. Only in small operations are chlorofluorocarbon (Freon) refrigerants used. Ammonia is a dangerous chemical. It is toxic to humans and animals and is able to penetrate packages readily. Therefore, ammonia leaks from refrigeration systems must be treated as serious occurrences. The U.S. Code of Federal Regulations (29 CFR 1910.120) requires training for personnel responsible for facilities using ammonia. There must be an emergency facility response plan based on Hazardous Waste Site Operations and Emergency Response (HAZWOPER) regulations, which are the responsibility of the U.S. Occupational Safety and Health Administration (OSHA). Necessary training includes worker protection requirements, materials handling procedures, compliance with evacuation and containment regulations, responsibilities of the on-scene incident commander, components of the incident command system, personnel roles, facility emergency response planning, facility emergency response audits, the Emergency Planning and Community Right-to-Know Act, spill reporting, and decontamination programs. Obviously, the need is great to protect from hazardous incidents related to escape of ammonia. Although the cost of prevention is not insignificant, the costs of an incident can be much higher.

It is not within the scope of this book to treat this subject further, but managers and operators must realize that responsibility rests on them. They must be fully informed of the regulations and they must have taken the necessary steps to follow them.

Besides the protective clothing and masks that must be provided for workers in the area of ammonia usage, several other protective measures must be taken in conjunction with walk-in type freezer operations, whether they be refrigerated with ammonia or another chemical. These include the following:

1. Provide a fail-safe latch on the inside of the freezer door(s). However, it is also necessary to provide an alarm system inside the hardening facility in case someone gets trapped inside. The system should attract help from the outside when regular plant workers are absent.

2. Keep an ax and sledge hammer in a prescribed location inside the door. These can be used as a last resort to break through the door in case assistance cannot be obtained.

3. Avoid use of wood for doors and other components. Moist wood tends to give off odors and is difficult to clean and sanitize.

4. Ensure that the defrost cycles are effective in removing frost and that they do not last so long as to cause temperature of the product to rise significantly.

5. Avoid placing ice cream packages together until hardened unless the apparatus used operates on the principle of removing most of the heat by conductance.

6. Avoid placing newly frozen products close to packages that are already hardened. Heat from the warmer product will migrate to the colder one and cause heat shock.

7. Have all packages dated, and use the first-in, first-out principle of product rotation.

8. Minimize temperature fluctuations. Fluctuating temperatures produce temperature gradients within the frozen product. These fluctuations drive diffusion processes, and they result in defective concentration and mobility gradients.

9. Maintain inventory control.

HANDLING, STORING, AND SHIPPING

The ice cream hardening facility is not a pleasant place in which to work. The very cold temperature and high velocity of air flow make it necessary for workers to wear heavy insulated clothing, footwear, and head covers. Most firms rotate their personnel frequently to prevent frostbite or discomfort and to maintain high efficiency. Completely mechanized handling is possible. In such systems product is conveyed into the hardening device and emerges on a conveyor from which it is mechanically palletized. Containers may be secured on pallets with plastic film that is wrapped mechanically around the containers. The pallet is placed onto a rotating table and is turned as the roll of wrap is moved vertically to allow the total area of the sides to be covered. The pallet is then moved via forklift truck or conveyor to a forklift type elevator that is programmed to place the pallet at a predetermined location in storage.

Storage/retrieval systems are available that enable one worker to move palletized product into and out of up to seven levels of racks (Figure 11.8). Pallets can be stored and retrieved in multiples on either side of a single aisle. Storage racks running perpendicular to the single aisle have been constructed up to 75 ft long. Pallets are raised or lowered to these storage racks by a vertical carriage and are positioned in the racks by an 8-wheel rack vehicle. Both of these electro-mechanical devices are computer-controlled, and positioning is done with the aid of infrared scanners that can position the carriage and rack vehicle within one-eighth inch of any reflective locator pad. The rack vehicle receives single pallets from a forklift on the vertical carriage. One drive in the rack vehicle raises and lowers a platform, and another drive moves the vehicle horizontally.

An inventory system integrated with the hardware provides correct product identity and first-in, first-out inventory control. The location and date of storage of each pallet is stored in computer memory so that management of inventory and proper rotation of product are readily and efficiently accomplished. The operator of the lift is sheltered in an air-conditioned cab equipped with a

11 PACKAGING, LABELING, HARDENING, AND SHIPPING

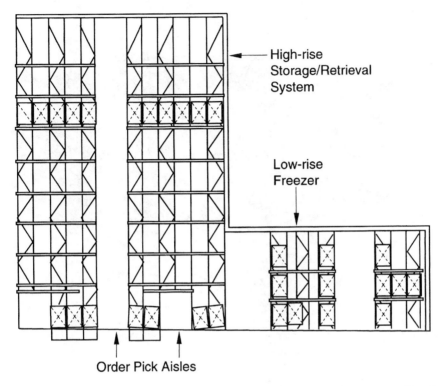

Figure 11.8. Concept drawing of a seven-story storage/retrieval system capable of moving product into and out of storage by computer controlled-vehicles. (Courtesy Woodson, Inc., Lancaster, PA.)

telephone. The computer can be networked with management, production, and sales so that all information is available at each location.

Such storage and retrieval systems are useful where inventories are large (Figure 11.9). However, palletized products will need to be broken down into order size lots either at the storage warehouse or later in the distribution chain. In any palletized system it is possible to use gravity flow racks at floor level to feed products to truck loadout positions. There the pallets can be loaded directly onto trucks or broken into order size lots for shipment to buyers. Gravity flow racks have wheels or rollers on which pallets roll slowly to a stopping device. The temperature in the loadout room can be higher than in the storage room, thus permitting personnel a more worker-friendly environment.

Each package should show a product code indicating time of production and lot number in the event it is necessary to trace the product to its origin. All data about each lot must be recorded and retained until there is minimal risk that any of the product remains available for consumption. Inventory control requires a record of product identity, location in storage, and date of production. Quality control and risk management require records of composition (by test),

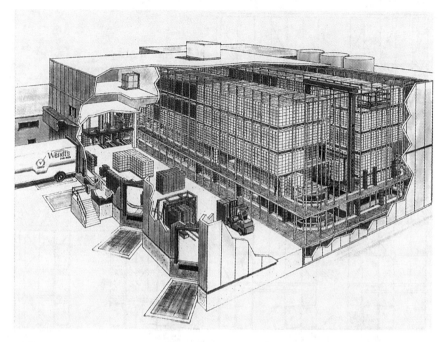

Figure 11.9. Storage/retrieval system in a facility designed for a large inventory. (Courtesy Woodson, Inc., Lancaster, PA.)

ingredients, microbial test results, and overrun. Management must establish a maximum storage time during which the product can be held without deteriorating to an unacceptable level of quality. This time will vary with the product—for example, nonfat ice creams have an inherently shorter shelf life than full fat products. Also affecting this time will be conditions of storage and the expected conditions of distribution, sales, and handling at the final destination. Personnel of the firm must be able to detect defects of each type of product and interpret the cause(s). Furthermore, it is important to have someone capable of predicting the proper storage time for each product.

Monitoring of temperatures during processing, freezing, and storage is highly important. Even the temperatures of stored ingredients must be monitored. Control within a plant is accomplished more readily than in the distribution system. To monitor temperatures on trucks and in facilities that do not have built-in temperature recorders, it is possible to obtain and use temperature monitors that preserve a record on a computer chip (Figure 11.10). The monitor is placed in a reading device (PC interface) once a record has been made. It is important to place monitors in locations where the optimal temperature is most likely to be compromised. Shippers often place recorders at the front, center, and rear of trucks. Vehicles must be designed, constructed, and maintained to avoid heat shock to products.

The temperature of frozen desserts in the distribution system should never

Figure 11.10. Device for monitoring temperature in sensitive locations. The monitor cell contains a sensor, power supply, and memory components. It is queried by a computer for deviations from the desired temperatures after shipment is completed. (Courtesy Ball Electronic Systems Division, Broomfield, CO.)

get above the desired temperature in the retail cabinet. Therefore, truck bodies must have been well insulated, cooled prior to loading, and kept cold by mechanical refrigeration. The refrigeration system must be operated optimally during distribution. Exposure of frozen product to warm temperature during off loading into store freezers must be limited to a few minutes to avoid heat shock and development of large ice crystals.

Shipping with Dry Ice

On special occasions it may be necessary to ship small amounts of frozen desserts under conditions that require dry ice (carbon dioxide) as refrigerant. This lightweight refrigerant has a high cooling capacity and leaves no residue as it vaporizes. Heat removed from the materials around it causes the vaporization. The recommended practice is to saw the ice into appropriately sized pieces and to wrap them in paper to slow vaporization.

Some disadvantages to the use of dry ice include its relatively high cost; its very low temperature (–109°F, –78°C), which may freeze the product so that it takes a long time to soften enough to be consumed; and its potential for burning the handler. The extremely low subliming point of dry ice makes it easy to freeze fingers without feeling the cold sensation until damage has been done. Dry ice should always be handled with insulated gloves.

QUALITY IS THE GOAL

The highest quality product is unlikely to be purchased unless care is taken to package it in convenient, safe, attractive, appropriately labeled, and economical packages of the desired size. The package must stand the rigors of handling, be identified by lot number, and protect the product from tampering, contamination, and chemical reactions, especially oxidation. The product should be hardened quickly and efficiently and stored for a minimal time at –30°C or lower. Delivery must be in clean, cold trucks by drivers who see that heat shock is prevented in transferring the product from storage room to truck and from truck to a freezer at the destination. Unless these conditions are met, the producer cannot be confident that the buyer will be back for another order. Efficiency and economy are important, but they pale in comparison to the importance of high quality and perceived value of the purchased product.

REFERENCES

3-A Sanitary Standards Committee. 1993. 3-A Sanitary Standards for Equipment for Packaging Viscous Dairy Products, Number 23-02. International Association of Milk, Food and Environmental Sanitarians, United States Public Health Service and Dairy Industry Committee. IAMFES, Des Moines, IA.

Der Hovanesian, J. 1960. Quick hardening of ice cream by liquid immersion. *Ice Cream Rev.* 43(9):98.

Farrall, A. W. 1953. Hydraulics and pumping. In Dairy Engineering. 2nd ed. John Wiley and Sons, N.Y.

11 PACKAGING, LABELING, HARDENING, AND SHIPPING 221

IDF. 1995. Technical Guide for the Packaging of Milk and Milk Products. 3rd ed. Bulletin of the International Dairy Federation N°300/1995.

Kramer, A., and W. S. Arbuckle. 1965. Getting greater economy in your filling operation. *Ice Cream World* 72(12):10, 14–15.

Kool, J., and M. G. van den Berg. 1995. Environmental constraints on packaging of dairy products. In Technical Guide for the Packaging of Dairy Products. 3rd ed. Bulletin of the International Dairy Federation N°300/1995.

National Institutes of Health. 1994. NIH Consensus Statement. Optimal Calcium Intake. Vol. 12, No. 4. June 6–8.

12
Soft-Frozen Dairy Foods and Special Formulas

SOFT-SERVE PRODUCTS

The demand for soft-frozen dairy desserts has steadily grown for many years as the number of fast-food and drive-up or drive-through facilities has increased. The introduction of soft-frozen yogurt also tended to increase sales of soft-frozen products. Consumers have commonly called these products ice cream, and now that descriptors are used to specify general levels of fat in ice cream, the consumer is usually correct to use this name. However, frozen products other than ice cream are marketed in the soft form, and this chapter reveals differences and similarities among this general class of frozen desserts. Common among all of these products is that they are soft-frozen on the premises and are consumed soon after being served.

The term "frozen custard" is frequently used erroneously to refer to all soft-serve products. The federal definition of frozen custard specifies that it shall contain not less than 1.4% egg yolk solids. Custards may be served soft or hard frozen.

FREEZERS FOR SOFT-SERVE AND SHAKES

Few changes in the basic design of soft-serve freezers have occurred since they were first introduced. Although advances in electronics and mechanics have allowed them to be made smaller, they still operate in basically the same way: they receive a cold liquid mix from either an integrated or remotely located storage vessel, feed it into a cylinder surrounded with refrigerant, freeze it while beating in air and scraping a thin layer of frozen product from the cylinder wall, and dispense it through a dispensing head into a container.

Several functional types of soft-serve freezers are available. These include models for the floor and the counter top; single- and multiple-flavor models; models for milk shakes, soft-serve, or both together; low-overrun machines; and frozen drink machines. Examples include a single-cylinder model (Figure

12 SOFT-FROZEN DAIRY FOODS AND SPECIAL FORMULAS

12.1), a floor model with two cylinders from which two flavors can be dispensed separately or in combination as in a twist (Figure 12.2), and a single-cylinder shake dispenser that permits the operator to dispense any of four flavors by activating an internally housed pressurized syrup dispenser (Figure 12.3).

Capacities of the evaporator and mix reservoir generally range from 10–20 qt, and compressors are sized from ½ to 1 hp per dispenser head on the machine. The freezer cylinder capacity ranges from 2–4 qt. The dasher(s) is driven by a separate motor rated at ½ to 1 hp. The larger freezers require 208–230 volts and 1- or 3-phase circuits; whereas smaller units run on 110-volt circuits. It is important that there be an adequate and steady power supply to prevent failure of the major electrical components. Most units are available in either air- or water-cooled versions. Ventilation or adequate water for cooling are,

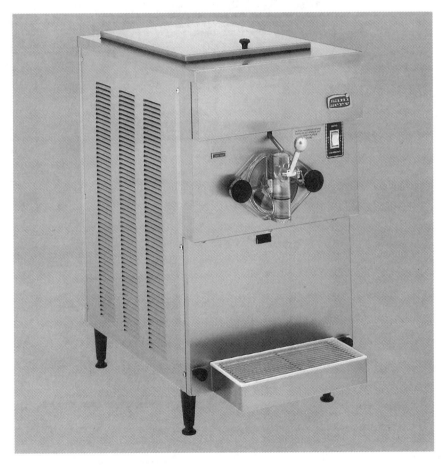

Figure 12.1. Single-cylinder, counter model soft-serve freezer of 20-qt capacity. (Courtesy SaniServe, Indianapolis, IN.)

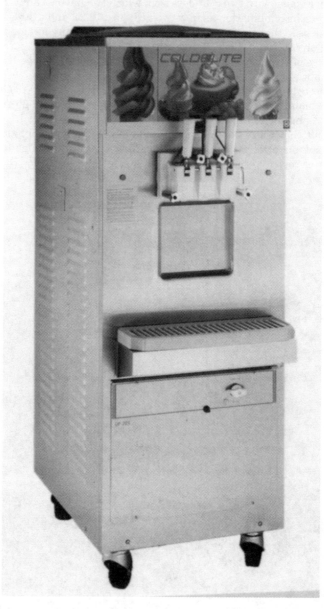

Figure 12.2. Double cylinder, floor model soft-serve freezer with center spigot to dispense "twist"—a combination of the two flavors being dispensed through the two outside spigots. (Courtesy Coldelite, Lodi, NJ.)

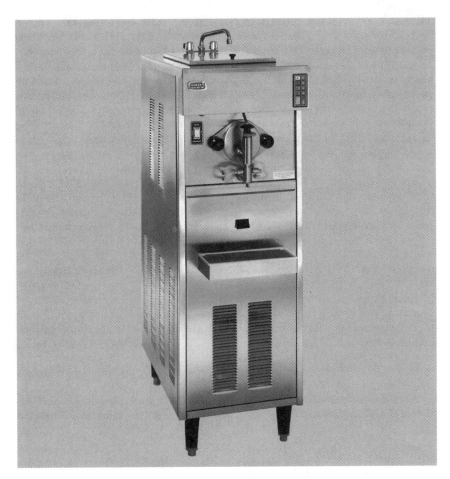

Figure 12.3. Single-cylinder floor model shake dispenser equipped with four pressurized syrup dispensers to flavor the shakes. (Courtesy SaniServe, Indianapolis, IN.)

therefore, important for proper operation. Another consideration related to efficient operation is locating freezers out of direct sunlight.

Because temperatures of dispensing are higher for shakes and frozen beverages than for soft-frozen desserts, motors are smaller in freezers designed for shakes and beverages. In contrast, motors for frozen custard machines are considerably larger than those of shake machines. Frozen custard mixes contain more total solids and are frozen to lower overrun than are the typical lowfat, light, and nonfat soft-serve mixes. Therefore, more power is required to turn the dasher and to deliver the product at the lower drawing temperature demanded by the lower freezing point of the typical frozen custard mix.

Freezers used to make soft-serve ice cream or milk shakes are designed to dispense product on demand. Thus, dispensing may occur either in rapid succession or at infrequent and varied time intervals. Sometimes the frozen product remains in the barrel for several hours and is subjected to agitation and successive refrigeration cycles. To maintain high quality under such circumstances requires specially formulated mixes and a well-insulated freezer barrel that will minimize the frequency that refrigerant is needed. Churning is the most frequently observed defect. Efficient homogenization and proper emulsification are the usual corrective actions for overcoming this problem. Product that is obviously churned should be emptied from the freezer. The freezer should then be cleaned and sanitized and a new batch made. Soft-serve products are usually drawn at –7°C (19°F) and shakes at –3°C (27°F) with some variance due to differences in freezing points of mixes.

A patented injection device can be attached to certain soft-serve freezers to apply flavoring both internally and externally (Figure 12.4). The patent-holding firm offers 22 or more concentrated flavors for use with the machines. The fat-free swirl is composed of the flavoring plus sodium alginate, starch, and sugar.

CLEANING AND SANITIZING SOFT-SERVE FREEZERS

Managers and operators must continually emphasize to employees the great importance of proper cleaning and sanitizing of freezers. The fact that much of the help in retail ice cream stores consists of inexperienced persons and that employee turnover is typically frequent means that training is usually needed on a consistent basis. Freezer manufacturers typically provide specific instructions for cleaning and sanitizing, and these should be followed closely unless they prove to be inadequate.

The following procedure for **DAILY CLEANING** is adapted from a publication of the West Virginia Department of Agriculture, Consumer Protection Division.

1. Turn off refrigeration and turn beaters on to expel remaining product.
2. Rinse freezer with about 2 qt (2 L) of cool tap water.
3. Prepare 2 gal (8 L) of chlorinated alkaline detergent solution at 52–55°C (125–130°F) at the concentration recommended by the detergent manufacturer (usually 1–1.5%).
4. Remove the hopper cover and mix tube assembly. Pour about 1 gal (4 L) of detergent solution into the hopper and brush thoroughly as the solution flows into the freezer cylinder. Be sure to brush the mix feed tube.
5. Run dasher for 30 sec, then draw off the detergent solution. Rinse with warm water.
6. Remove the freezer door and disassemble all freezer parts. Brush all parts, including the hopper cover and feed tube assembly, using the remaining detergent solution.
7. Rinse all parts thoroughly. Lubricate seals and valves as directed by the manufacturer and with approved lubricant before reassembly of the freezer.
8. Just prior to use, sanitize the freezer with a solution of 200 ppm (mg/L) of hypochlorite solution. Drain completely but do not rinse with unchlor-

Figure 12.4. Soft-serve freezer with attachment to deliver specially formulated flavoring as a swirl in soft-serve frozen desserts. (Courtesy Coldelite, Lodi, NJ.)

inated water, because that action may reintroduce contaminants to the machine.

THE HEAT TREATMENT FREEZER

One manufacturer offers a soft-serve type freezer that needs minimal daily cleaning because the major components and the mix in the freezer are heated to destroy microorganisms on a daily schedule (Figure 12.5). Components that are hand-washed and sanitized each day are those that cannot be subjected

Figure 12.5. Double-cylinder "heat treatment" soft-serve freezer. Dispensers for syrups and condiments are at counter level. (Courtesy Taylor, Rockton, IL.)

to the heating to 66.1°C (151°F) for 30 min. The machine must be completely disassembled and cleaned every 14 days. Temperatures are monitored in the mix hopper, the freezing cylinder, and the glycol coolant. They can be read from an LCD display, and failure of the machine to provide required temperatures of cooling or heating within prescribed times will result in "hard lock-out" so that automatic operation of the freezer is not possible until the freezer has been disassembled and cleaned. Failure to disassemble and clean within 14 days will also result in hard lock-out. If a heating cycle has not been initiated within 24 hours of the previous cycle, a "soft lock-out" will occur, and a heating cycle or full cleaning must be initiated before operations can be resumed. The heating cycle is completed within 4 hours.

Types of system failure are retained in the memory of the microprocessor, and these can be accessed by inspectors or operators. This type of freezer has the advantages of reducing cleaning time, saving mix, and enhancing safety in operations where novice employees are used frequently. At the same time the capital investment is increased and the cost of heating may offset the cost of disassembly and cleaning.

SOFT-SERVE DESSERTS CONTAINING PARTICULATES

The latest innovation in soft-serve marketing has been made possible by development of dispensing machines that extrude product in the flowable state. This enables the soft-serve maker to freeze the product, add inclusions such as fruit, nuts, candies, and dough particles, then hard-freeze the product for delivery to the store where it is to be sold. The hard ice cream, yogurt, or sorbet is tempered in a special cabinet. The prefilled bags or cartridges are then placed in the extrusion cabinet for serving to customers. Three to four different flavors may be dispensed from the current machines. The dispensing machine (Figure 12.6) has minimal parts to be cleaned.

This development facilitates the sale of premium ice creams with mix-ins in locations where it would normally be difficult. The machines are highly energy efficient because they do not agitate; they simply dispense the prefrozen product.

SOFT-SERVE MIX COMPOSITION

The way soft-serve products are marketed makes it possible to use formulas that differ considerably from formulas for hard-frozen products. Although some soft-serve ice creams are hardened after being packaged or formed into novelties, the principal forms in which they are marketed are as soft-frozen products and shakes.

Soft-serve mixes are unique in composition, stability, and whippability. A fat content below about 4% increases risks of having a coarse or icy texture and weak body. A fat content above about 12% is associated with significant risk of churning in the freezer and a greasy mouth coating. The NMS content varies inversely with fat content and may be as high as 14% for a lowfat

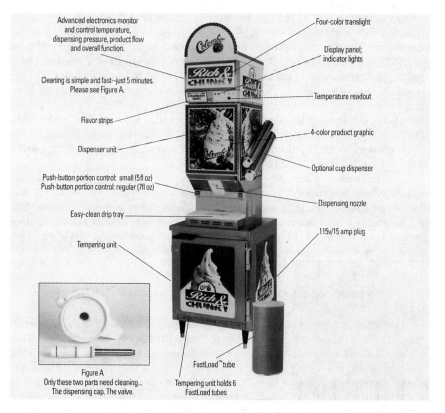

Figure 12.6. Dispensing machine for soft-serve frozen desserts containing fruits, nuts, and other particulates. (Courtesy Sealright Co., Overland Park, KD.)

formula. The sugar content ranges from 13–15%, which is somewhat lower than for regular ice cream. The amount of corn sugar or fructose used to replace sucrose must be limited to 25% to avoid too low a freezing point. Stabilizers and emulsifiers are used in amounts ranging from 0.2–0.4% and from 0.1–0.2%, respectively, to provide smoothness, desirable whipping properties, melt resistance, and firmness. Calcium sulfate may be used at about a 0.1% rate to enhance dryness and stiffness. Some typical formulas for soft-serve products are shown in Table 12.1.

Typical formulas for full fat ice cream used as soft-serve contain 2–3% less sugar than for regular ice cream. Frozen custards may be of moderate or high-fat content and are required to contain 1.4% egg yolk solids for the plain flavors and 1.12% for bulky flavored products. Typical formulas for soft-serve frozen custard have approximately the following composition: Fat 5–10%, NMS 11%, sugar 14%, egg yolk solids 1.4%, and stabilizer/emulsifier 0.4%. Total solids generally range from 32–37%.

Soft products are usually drawn from the freezer at −6.7 to −7.8°C (18 to 20°F). Fat separation, increase in sizes of ice crystals, and lactose crystallization

12 SOFT-FROZEN DAIRY FOODS AND SPECIAL FORMULAS

Table 12.1. Typical Formulas for Soft-Serve Ice Creams

Constituents[a]	%						
Milk fat	3.0	4.0	5.0	6.0	3.0	6.0	10.0
NMS	14.0	14.0	13.0	12.5	14.0	13.0	11.0
Sugar	10.0	11.0	12.0	12.0	14.0	13.0	12.0
CSS	4.0	4.5	4.0	4.0	—	—	3.0
S/E	0.5	0.5	0.4	0.4	0.5	0.5	0.4
TS	31.5	34.0	34.4	34.9	31.5	32.5	36.4

[a]NMS-nonfat milk solids; CSS-corn syrup solids; S/E-stabilizer plus emulsifier

are likely defects that result from cycling of the freezer to maintain temperature during extended holding times.

Overrun of soft-serve products ranges from 30–60%, depending on the TS content. The higher the TS content, the higher the overrun may be while maintaining desirable body and texture characteristics. Suggested limits for overrun and serving temperature vary among manufacturers (Table 12.2).

Over the past five years reduced and lowfat ice creams, known as ice milk until 1994, have constituted about two-thirds of the sales of soft-serve products while ice cream containing at least 10% fat has accounted for from 12–14% of sales (The Latest Scoop, International Ice Cream Association). Frozen yogurt has now surpassed regular ice cream in amount sold in the soft-frozen form. The U.S. Department of Agriculture started keeping records of frozen yogurt sales in 1989 and discontinued record keeping for mellorine (vegetable fat substituted for milk fat) type products in 1990. This reflects the relative importance of these products in the current marketplace.

With the change in definitions of ice cream and discontinuance of the Standard of Identity for ice milk has come the introduction of new lowfat ice cream formulas. However, this change is likely to have a minimal effect on formulations for soft-serve products. Unless there is an overt effort to inform the buyer of the caloric and fat content of soft-serve products, consumers are most likely to make choices of when and where to purchase the products on bases other than fat content. In grocery stores, where labels are required on each package the new labeling regulations are expected to affect consumer choices significantly.

Milk shake mixes fall outside the Standards of Identity for ice cream, and are not defined in the regulations of many states. However, the word milk in the name implies that they should contain at least 3.25% fat as does market milk by definition. Compared with ice cream, they are low in fat, high in NMS

Table 12.2. Suggested Limits of Overrun and Serving Temperature for Soft-Serve Products

Product	Overrun	Serving Temperature (°F)
Soft-serve	30–60	18–22
Milk shake	40–60	26–28
Sherbet	30–50	19–21
Italian ice	30–40	19–21

and low in sugar. The total solids content is characteristically 25–30%. For the best quality of milk shakes overrun should be kept below 20%. This means that it is not desirable to use emulsifier in these products. Two representative formulas follow:

Constituents	%	
Fat	3.5	4.00
NMS	12.0	14.00
Sugar	10.0	8.00
Corn syrup solids	—	3.50
Stabilizer	0.4	0.35
TS	25.9	29.85

Milk shake base is made with total solids in the range of 32–36%. It may be frozen in the same way as ice cream with 50–60% overrun, placed in bulk packages, and hardened. The hardened ice cream is dipped into a blender to which syrup and milk are added to make the shake. The following are typical formulas:

Constituents	%		
Fat	3.5	4.0	6.0
NMS	13.0	12.5	12.5
Sugar	12.0	13.0	13.0
Corn syrup solids	4.0	5.0	4.0
Stabilizer	0.4	0.4	0.4
TS	32.9	34.9	35.9

Malted milk formulas are typically slightly higher in fat but lower in sugar than milk shake formulas. They are characterized by the added malt syrup and malt base or cocoa. The product is served in a flowable but highly viscous soft-frozen state. Some typical formulas follow:

Constituent	%			
Fat	4.0	6.0	4.00	5.00
NMS	13.0	12.0	12.00	12.00
Sugar	10.5	10.5	10.00	10.50
Corn syrup solids	4.5	4.5	—	4.50
Malt syrup solids	—	—	0.75	0.75
Cocoa	—	—	3.00	3.00
Malt base	3.0	3.0	—	—
TS	35.0	36.0	29.75	35.75

Gelato is an Italian style ice cream that is very rich in fat and high in TS. It carries abundant rich flavor and has very low overrun, if any (0–10%). It is often flavored with liqueurs and various combinations of fruit. The low overrun and high solids provide the distinctive body and texture and desirable release of flavor. A typical formula contains 18.0% milk fat, 7.5% NMS, 16.0% sugar, and 4.0% egg yolk solids. Note that no stabilizer or emulsifier is recommended in this formula.

Vegetable fat frozen desserts, commonly known as **mellorine**, are formulated with same composition as soft-serve ice cream except that the milkfat is replaced with a blend of cottonseed oil and soybean oil or with coconut fat. Safflower and canola oil are also possible additives that are high in polyunsaturated and monounsaturated fatty acids, respectively. To provide good emulsification the rate of emulsifier addition is increased from the usual 0.1% used with milk fat to 0.15%. Another additive that enhances emulsifying properties is buttermilk solids. This byproduct of the churning of cream is high in the phospholipid lecithin from the membranes of milkfat globules. The usual usage rate of dry buttermilk is 1%. Since vegetable oils contain no vitamin A, some states require the addition of 8,400 USP units of vitamin A per gallon when the fat content is 10%.

13
Sherbets and Ices

A sherbet is a frozen foam made from water, nutritive sweeteners, fruit or fruit flavoring, fruit acid, milk solids, stabilizer, and coloring. Sherbets contain 1–2% milkfat (no other fat permitted) and at least 1% NMS with the total milk solids between 2 and 5% (21 CFR 135 1995). This acidic food, when characterized by a fruit flavor, has a minimal titratable acidity of 0.35% calculated as lactic acid. The final weight per gallon must be at least 6.0 lb; therefore, the overrun is less than 50%.

Compared with ice cream, sherbets have the following characteristics:

- higher fruit acid content, and a tart flavor
- lower overrun, ranging from 25–50%
- higher sweetener content (25–35%), therefore a lower melting point
- coarser or more icy texture and a more cooling feeling to the consumer
- less richness of taste due to the low milk solids content.

Ices or water ices have essentially the composition as sherbets except that they contain no milk solids and no egg ingredient other than egg white. They are frozen with from 0 (still frozen bars) to 30% overrun.

Of the total frozen desserts produced annually in the United States, about 1.5 billion gallons, sherbets comprise about 3.3% and water ices 3.7%. Sherbets make up about 1% of the soft-serve products. In Canada, where total annual production is about 366 million gallons, sherbets and water ices comprise about 1% and 4.8%, respectively. These products are in greatest demand in the summer months. The most popular flavors of sherbet, orange, pineapple, raspberry, and lime, constitute 80–85% of the flavors produced. Orange is by far the most popular flavor in most markets, commanding 29% of sales in 1994–95, with over 6.4 million gallons sold. Rainbow sherbet followed close behind with sales of nearly 6.2 million gallons. Others in the top five were raspberry (11%), lime (9%), and pineapple (8%). In all, 10 flavors were listed in the latest survey of sherbet sales by flavor (IICA 1995).

THE COMPOSITION OF SHERBETS AND ICES

The fruit-flavoring ingredients specified for ice cream can also be used for sherbets. Natural and artificial flavors are often added as supplemental flavorings. Minimum amounts of fruit or fruit juice (including weight of the water used to reconstitute dried or concentrated products to their original moisture content) required by type of sherbet are: citrus, 2%; berry, 6%; and other, 10% in relation to the weight of the finished sherbet. A typical Ingredients label may read as follows: water, milk (including nonfat milk and cream), sugar, corn syrup, dextrose, orange puree (water, natural flavoring, concentrated orange juice, orange pulp, gum tragacanth, yellow 6, citric acid), high fructose corn syrup, whey, citric acid, mono- and diglycerides, polysorbate 80, guar gum, locust bean gum, and pectin.

A recent survey of nutrition labels for orange sherbet revealed characteristics among five regional brands as shown in (Table 13.1).

Water ice has the same flavoring and sweetening provisions as fruit sherbets. Nonfruit sherbets or ices differ from fruit sherbets and ices mainly in the flavor-characterizing ingredients. The optional characterizing ingredients include ground spices, infusion of coffee or tea, chocolate or cocoa, confectionery, distilled alcoholic beverages (in an amount not to exceed that required to provide the flavoring), or any other natural or artificial food flavoring (except any having a characteristic fruit or fruitlike flavor).

Sweeteners

In general, the sugar content of sherbets and ices is about twice that of ice cream. It is important to have the correct sweetener content to obtain the desirable flavor, body, and texture. An excess results in a soft and sticky product and a deficiency causes the product to be hard and crumbly. It is also important to have the sherbet dippable at the same temperature at which ice creams are dipped, about $-13°C$ ($8°F$).

When sherbets are made with sucrose as the sole source of sweetener, they tend to develop a hard crust on the surface caused by crystallization of the sugar. Replacement of 20–25% of the sugar with corn sugar (dextrose) lowers the freezing point and lessens the chance for the defect (Day et al. 1959; Ross 1963; Turnbow et al. 1946). Corn syrup solids may function even better for this purpose, because they, having a much higher molecular weight, lower the freezing point much less than does corn sugar. The amount of corn syrup solids that can be substituted favorably for sucrose is about one-third. Amounts of sugar added to these products with fruits or with ice cream mix must be factored into the formula.

Table 13.1. Characteristics of orange sherbets as revealed by nutrition labels in 1995.

Characteristic	Range	Average
Weight per serving (g)	83–86	85
Calories/serving	110–150	130
Fat/serving (g)	1	1
Carbohydrate/serving (g)	25–34	29
Sugar/serving (g)	20–34	28
Protein/serving (g)	0–1	1
Calcium (% of Daily Value)	2–6	4

Stabilizers

Most of the stabilizers used for ice cream can be used successfully in sherbets and ices. Amounts of stabilizers commonly used in ices and sherbets are as follows: carboxymethylcellulose gum 0.20%, guar gum 0.20%, pectin 0.18%, algin products 0.20%, locust bean gum 0.25%, gelatin (200 Bloom) 0.45%. Not all stabilizers work well in the acidic pH of these products. Because sherbets contain some milk solids that bind water, they need slightly less stabilizer than do ices. Surprisingly, despite the low content of fat in sherbets, emulsifiers provide a valuable function in them.

Sherbet stabilizers are varied in composition to meet the goals established by the manufacturer in obtaining smoothness of texture at an acceptable cost. A higher cost stabilizer for sherbets contains, in order of prominence, locust bean (carob bean) gum, mono- and diglycerides, guar gum, and pectin. It is used at a concentration of about 0.3%. To enhance smoothness, the drawing temperature should be −6°C (21°F). A relatively low-cost stabilizer/emulsifier for production of a more coarse textured sherbet contains about 40% mono- and diglycerides, 25% guar gum, 25% cellulose gum (CMC), and 10% pectin. The use concentration should be about 0.4% and the drawing temperature can be as high as 5°C (23°F).

Ices have a low total solids (TS) content compared to ice cream; this means they have a greater tendency for sugar solids to separate and for the body to become crumbly than does ice cream. For this reason sherbets and ices need more stabilizers than do ice creams. This is especially true of frozen water pops that contain no sugar. A formula recommended by one supplier of stabilizer calls for 95.3% water, 4.16% Lite Pop Base No. 2, 0.24% Lite Pop Flavor, and 0.3% anhydrous citric acid. The Lite Pop Base is composed of sorbitol, locust bean gum, aspartame, and guar gum.

Acidification

Citric acid is the most commonly used acid in sherbets and ices and is usually added as a 50% solution. The amount of acid needed depends on the fruit used, the sugar content, and consumer preferences. A general rule is that the titratable acidity should be 0.36% at 25–30% sugar and should be increased about 0.01% for each 1% increase in sugar above 30%.

PREPARATION OF ICES

Formulas for ices are usually calculated on the basis of 100-lb lots. A desirable base or stock mix can be made with 21–25 lb of sucrose, 7–9 lb of corn sugar, and 0.4 lb of stabilizer. Water makes up the remainder of 80 lb of this base mix. The balance of the 100 lb is to be flavoring, coloring, acid, and additional water, the amount of which depends on the flavor (Table 13.2).

This base is prepared by slowly adding the dry ingredients to at least part of the water, taking care to avoid creating lumps. Heating will be necessary to facilitate solution if gelatin, agar, or certain gums are used as stabilizer.

13 SHERBETS AND ICES

Table 13.2. Representative formulas for ices.

Ingredients	Amount (lb)	Amount (lb)
Cane sugar	23.0	16.0
Corn sugar	7.0	10.0
Stabilizer	0.3	0.4
Fruit juice, color, citric acid solution, water	67.9	73.6
Total	100.0	100.0

Pasteurization is optional, but homogenization is not useful. The base is cooled before other ingredients are added. Aging is necessary only if gelatin or agar is used in the stabilizer, and then 12–24 hr is desirable.

This base mix is then ready for the flavoring and coloring materials. To each 80 lb of mix is added enough flavor, color, and water to make a total weight of 100 lb. The flavor and color mixture is made from the following ingredients:

1. *Fruit juices.* The amount varies between 15–20% of the finished ice, depending on the intensity of the flavor. Fruit seeds should be avoided.
2. *Flavoring.* The amount of fruit needed for flavoring varies from 5–20 lb for each 100 lb of finished mix, depending on the variety of fruit and how it is prepared. Although fruit extracts and artificial flavors may not provide as desirable a flavor as fruit juices, they are often needed to fortify the flavor and to produce a consistently uniform product.
3. *Coloring.* Approved food coloring should be selected to provide as near the natural color as possible while meeting the expectations of consumers as may be determined with a sensory panel.
4. *Acid solution.* To obtain the desired tart flavor, the fruit acids, citric or tartaric, should be used. Less desirable substitute acids are saccharic, phosphoric, or lactic. It is common practice to use 50% solutions of citric or tartaric acids made from equal weights of acid crystals and water. The amount of this concentrate to use varies from 4–10 oz (120–300 ml). The amount depends on the acidity of the fruit juice and the amount of sugar in the final mix. The final titratable acidity should range from 0.35–0.50% expressed as lactic acid.

PREPARATION OF SHERBETS

The base or stock mix described for ices combined with milk to replace part of the water can be used for any flavor of sherbet. The formula should be varied to produce the type of body, texture, and dipping characteristics desired. Limits for the major ingredients to compose 80 lb of base mix range as follow: sucrose 10–30 lb; corn sugar 7–9 lb or corn syrup solids 10–22 lb; stabilizer 0.4–0.6 lb; milk 30–55 lb; water to make a total of 80 lb.

Another approach to making the sherbet base is to combine ice cream mix with sugar, corn syrup or corn sugar, stabilizer, and water to make 80 lb. In this case the amounts of sweeteners and stabilizer in the ice cream mix must be considered in the calculations. Tables 13.3, 13.4, and 13.5 show proof sheets in which ice cream mix is used in three mixes that provide a wide range of textural and flavor release characteristics (Day et al. 1959).

The base mixes are prepared by adding the dry ingredients to the liquid ones. Stabilizer and NDM, if used, can be mixed with granular sugar to aid dispersion in the liquid. Regulations usually do not require the mix to be

Table 13.3. Sherbet: Smooth Texture; Chewy, Heavy Body[a]

Ingredients	Amount (lb)	Fat (lb)	NMS (lb)	Sugar (lb)	TS (lb)
Sugar	9.0	—	—	9.00	9.00
Corn syrup solids 42 DE, 96.5% TS	22.0	—	—	15.92	21.23
Ice cream mix (12% F, 11% NMS, 15% sugar)	17.5	2.1	1.92	2.62	6.65
Stabilizer	0.4	—	—	—	0.40
Fruit puree (5 + 1)	15.0	—	—	2.50	4.75
Water plus 300 ml acid and color	36.1	—	—	—	—
TOTAL	100.0	2.1	1.92	30.04	42.03

[a]Acidity 0.57%; freezing point −3.1°C (26.4 °F).

Table 13.4. Sherbet: Medium Smooth Texture; Medium Firm Body[a]

Ingredients	Amount (lb)	Fat (lb)	NMS (lb)	Sugar (lb)	TS (lb)
Sugar	11.0	—	—	11.00	11.00
Corn syrup solids 36 DE, 96.5% TS	10.0	—	—	6.30	9.65
Ice cream mix (12% F, 11% NMS. 15% sugar)	17.5	2.1	1.92	2.62	6.65
Stabilizer	0.4	—	—	—	0.40
Fruit puree (5 + 1)	15.0	—	—	2.50	4.75
Water plus 300 ml acid and color	46.1	—	—	—	—
TOTAL	100.0	2.1	1.92	22.42	32.45

[a]Acidity 0.55%; freezing point −2°C (28.4°F)

Table 13.5. Sherbet: Coarse Texture; Medium Firm Body[a]

Ingredients	Amount (lb)	Fat (lb)	NMS (lb)	Sugar (lb)	TS (lb)
Sugar	17.0	—	—	17.0	17.0
Corn sugar (dextrose)	7.0	—	—	5.6	6.75
Ice cream mix (12% F, 11% NMS, 15% sugar)	17.5	2.1	1.92	2.62	6.65
Stabilizer	0.4	—	—	—	0.40
Fruit puree (5 + 1)	15.0	—	—	2.50	4.25
Water plus 300 ml acid and color	43.1	—	—	—	—
TOTAL	100.0	2.1	1.92	27.72	35.05

[a]Acidity 0.55%; freezing point −3.1°C (26.4°F)

13 SHERBETS AND ICES

pasteurized if the milk products have already been pasteurized. Heating may cause some coagulation of casein in sherbets when gum arabic is used as a stabilizer. This means that the milk should be pasteurized separately when this stabilizer is used. Sherbet mixes are seldom homogenized, but they must be cooled. Furthermore, if gelatin is used as a stabilizer, aging for 12 to 24 hr is needed.

To this 80 lb of base is added 20 lb of the appropriate mixture of fruit juices, flavoring, coloring, citric acid solution, and water as described for making ices. Because citric acid may cause precipitation of proteins, it is sometimes added to the batch freezer as ice crystals are beginning to form.

FREEZING ICES AND SHERBETS

Ices can be still frozen in molds to make pops or they can be frozen while agitating in the same way ice cream is frozen. However, the rate of wear on the scraper blades is high because of the lack of fat to lubricate the metal surfaces that contact each other. Therefore, scraper blades must be sharpened frequently to maintain the capability to produce small ice crystals.

Sherbets are frozen with overrun in the range of 25–50%. The small amount of milk solids in sherbets improves their lubricating properties over those of ices.

The refrigerant of the batch freezer should be at –23°C (–0°F) or lower. This comparatively low temperature hastens freezing so that overrun can be kept low. Too high overrun in ices favors drainage of syrup. Furthermore, to avoid having high whippability, gum type stabilizers may be selected rather than those, such as gelatin, that tend to increase whippability. Another method of preventing excessive overrun is to overload the freezer, using 6 gal of mix in a 40-qt freezer.

Ices and sherbets should be of the same firmness at dipping cabinet temperature as is ice cream. If the overrun is kept at 30–35% and the sugar concentration at 28–32%, firmness should be suitable for dipping at the usual cabinet temperature of –13 to –16°C (3 to 8°F).

DEFECTS

Although consumers undoubtedly have a variety of preferences for flavor, body, texture, color, and appearance of sherbets and ices, several defects are objectionable to practically all consumers. For example, consumers want the product to appear natural and appetizing. The color should remind the eater of the flavor shown on the label, and the flavor should be true to the color and the label statement.

Common flavor defects are unnatural or atypical flavor; excessive or insufficient flavoring, sour (acid), improperly sweetened (too little, too much, or unnatural); and metallic or oxidized flavor. Terpenes of citrus fruit tend to cause bitterness. To avoid these defects requires selection of high-quality ingredients, especially fruits, juices, and flavorings, and protection of the ingredients and finished products from prolonged storage and exposure to odorous substances.

Selection of desirable artificial flavors should be given special attention during product development. Each shipment of citrus flavors should be checked for flavor in their product use dilution to ensure that defects are not present. Because of the small amount of milk solids used in sherbets, defects from these components appear much less frequently than they do in ice cream products.

As with ice cream, the most frequently observed textural defect in sherbets and ices is coarseness or iciness. Some consumers prefer the type of sherbet that freezes initially with a slightly coarse texture because it can be especially light and refreshing. Others prefer a velvety smooth texture. Nevertheless, either of the types can become offensively coarse and icy. The following steps are recommended to reduce this defect: (1) set the sugar content at 28–32% with about one-fourth of this amount, by weight, being corn syrup solids or corn sugar; (2) carefully select a stabilizer and use it at the rate proven by test in the formula; (3) draw the product from the freezer in a firm condition and harden it quickly; (4) protect the frozen product from temperature fluctuations; and (5) market the product promptly.

A crumbly body indicates an insufficient amount of or improper stabilizer. When the body is too firm, the overrun may be too low or there may be insufficient sugar in the mix. A weak or snowy body is indicative of having whipped too much air into the product. Stickiness suggests too much sugar or stabilizer in the formulation.

Surface encrustation sometimes appears because some of the sucrose crystallizes. The liberated water may evaporate or may freeze into large ice crystals. The usual solution to the problem is to increase the concentration of stabilizer and/or to lower the freezing point by adding more sugar.

"Bleeding" or settling of syrup to the bottom of the container is more of a problem with sherbets and ices than ice cream. The internal structure of the foam of ice cream is stabilized to a much higher degree by abundant proteins and partially churned fat than is the structure of sherbets or ices. To prevent bleeding, one should avoid excessive overrun, provide sufficient stabilizer, hold the sugar content to less than 32%, and keep the temperature cold, i.e., below –20°C (–4°F) until tempering it to be served. Temperature abuse is the most important cause of body and texture defects in frozen desserts.

REFERENCES

Day, E. A., W. S. Arbuckle, and D. J. Seely. 1959. Quality sherbets. *Ice Cream Field* 73(2):28, 30, 78–79.

IICA. 1995. Sherbet flavors by groups. A Nielsen Marketing Survey. Int'l Ice Cream Assn., Washington, DC.

Ross, O. E. 1963. Sherbets for tomorrow's market. *Ice Cream Field* 81(4):48, 72, 74, 76, 78.

Turnbow, G. D., P. H. Tracy, and L. A. Raffetto. 1946. *The Ice Cream Industry*. 2nd ed. Wiley, New York.

14
Fancy Molded Ice Creams, Novelties, and Specials

Since early in the history of the ice cream industry, small manufacturers with imagination and initiative have realized the potential for making profit while serving consumers by making and marketing unusual ice cream and frozen dessert products. The word *novelty* conveys the qualities of freshness, uniqueness, and cleverness in creation and marketing. Novelties have included quiescently frozen bars, special combinations of ice cream with flavors and confections, cup items, and fancy molded items. They can be made of many types of frozen desserts, including ice cream with its various fat contents, frozen yogurt, sherbet, puddings, tofu, sorbet, gelatin, and fruit ices. To these are frequently added chocolate, baked items, such as wafers and cakes, and numerous kinds of fruit.

Much of the creative work has been done with considerable manual labor, but recent advances in equipment have made possible mass production at relatively low costs. Some firms have dedicated one or more plants solely to this kind of enterprise. Because the investment in equipment is high, the industry has tended to concentrate into a few plants strategically located to serve large population centers and where there is easy access to a major highway system.

Factors that determine success in frozen novelty production are ingredients used in the mix, efficiency and effectiveness of the equipment, proper operation of the freezers, and control during the operations. Major areas of control include portion control, rate of freezing, stick insertion, extraction from the molds, yield of coating, wrapping or bagging, packing, sealing cartons, and storage.

Most of the products discussed in this chapter are produced as individual servings. The following are some reasons why they are in demand in the marketplace: (1) their quality can be especially high because they are frozen rapidly as small individual units, (2) they are easy and convenient to serve, and (3) they can take on many unique forms, flavors, and compositions for variety in eating.

For the first 50 years of production novelties were marketed principally to children. However, in the 1980s novelties broke out of their "moderate price/moderate quality" mold when superpremium novelties transformed the entire frozen desserts industry. New, indulgent, up-scale, adult-oriented novelties, including Chipwich and Dove Bar, drove the market upward. Sales increased some 30% during the decade, then tended to flatten at the end of that period, accounting for over 34% of all frozen dessert sales. By 1993, novelties accounted for a $3.4 billion market according to FIND/SVP, a New York-based research firm. Bars (with and without sticks) account for more than 85% of all novelties sold. Sandwiches follow bars with about 5% of sales. Cones and miniatures have about 2% of the total share.

PRODUCTION SYSTEMS

Novelties can be formed in either of two ways, by molding or extruding. In the molding method (Figure 14.1) ice cream that has been lightly whipped and partially frozen in the continuous freezer is transferred to refrigerated molds (Figure 14.1a). Sticks are inserted and frozen into the product (Figure 14.1b). The frozen novelty is then lifted from the mold (Figure 14.1c) and dipped in a coating (Figure 14.1d).

The extrusion method involves flavoring a mix (Figure 14.2a) and freezing it (Figure 14.2b) to a temperature of about $-6°C$ ($21°F$). A mix is chosen that will freeze into dry and stiff extrudate that can be cut with an electrically heated taught wire (Figure 14.2c). Sticks are inserted (Figure 14.2d) and the product is dropped onto trays that pass through a freezing tunnel (Figure 14.2e) refrigerated to $-40°C$ ($-40°F$), where they are hardened to about $-15°C$ ($5°F$). Finally, they are dipped in the coating (Figure 14.2f).

SPECIALTY EQUIPMENT

Versatile equipment that can produce ice cream novelties in virtually any shape is available. Such machinery provides an automated method for producing individual ice cream and ice portions, slices, chocolate-coated bars, and fancy stick products.

Extrusion Processes

In the extrusion production operation, ice cream is drawn from a freezer at about $-5.5°C$ ($22°F$), pumped through an extruder nozzle, and sliced into portions by a cutting assembly that includes an electrically heated wire. The extruder may take on a horizontal (Figure 14.3) or vertical form (Figure 14.4). If a stick item is desired, the stick is inserted in the extruded ice cream (Figure 14.5). The pieces are formed on or drop onto carrier plates and pass through a freezing chamber at $-41.7°C$ ($-43°F$) with rapid air circulation for flash freezing (Figure 14.6). Each piece is removed from the carrier plate as it emerges from the freezing chamber. Portions to be coated with chocolate or other coating are then transferred to an enrober, then through a chill tunnel to set the coating. Some candy bar type extrusion systems are more than 300 ft long, but they are capable of producing more than 70,000 bars per hour.

14 FANCY MOLDED ICE CREAMS, NOVELTIES, AND SPECIALS 243

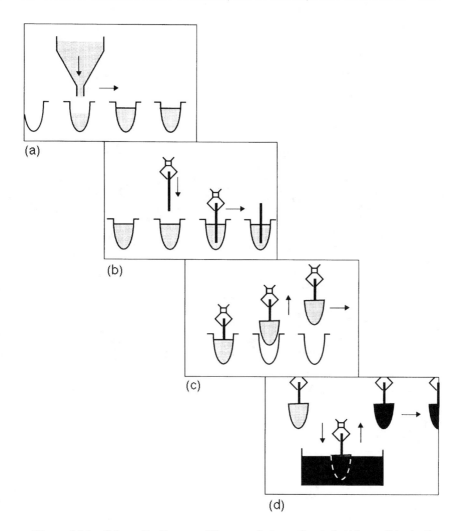

Figure 14.1. Schematic diagram of the manufacture of coated stick novelties by the molding method.

Capabilities exist for extruding single-flavor, multi-flavored, decorated, chocolate-coated, and dry-coated stickless slices; coated and uncoated ice cream sandwiches; single-portion and large horizontally extruded fancy decorated logs; bite-sized chocolate-coated miniatures; and more. Regular and boat-shaped cones can be produced with the addition of cone-filling equipment. The external contour of the slice may be almost any desired shape as dictated by the shape of the extruder nozzle. Complex extrusions in which multiple flavors or colors are extruded require the use of multiple freezers. By placing different extrusion nozzles inside each other, intricate designs, such as faces with eyes, nose, mouth, and ears, can be formed (Figure 14.7).

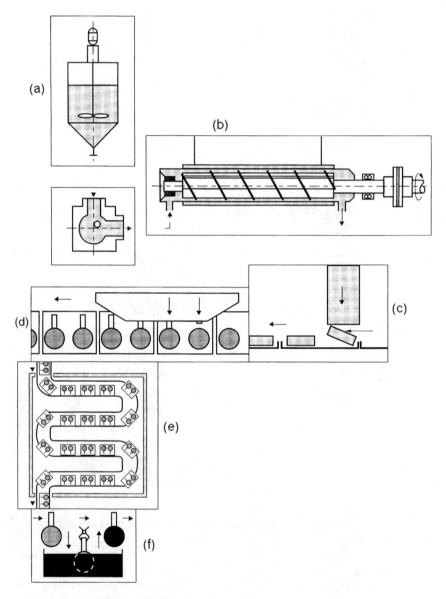

Figure 14.2. Schematic diagram of the manufacture of coated stick novelties by the extrusion method: (a) mix tank, (b) continuous freezer, (c) extruder, (d) stick inserter, (e) hardening tunnel, (f) coating tank.

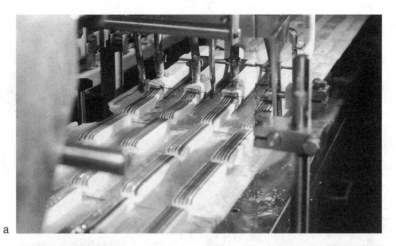

Figure 14.3. A horizontal extruder with (a) nozzles for layering flavors such as caramel, and (b) devices for sprinkling nuts, onto slices of ice cream. (Courtesy Gram Equipment of America, Tampa, FL.)

Figure 14.4. A vertical extruder that can prepare large or small slices with one or more ice creams. To ensure a clean right-angle cut, the cutter travels with the tray during the final stage of cutting. (Courtesy Gram Equipment of America, Tampa, FL.)

Figure 14.5. Sticks being inserted into extruded ice cream. (Courtesy Gram Equipment of America, Tampa, FL.)

Figure 14.6. Freezing tunnel with trays for hardening extruded novelties—the APV Glacier EXCEL plate extrusion system. (Courtesy APV Glacier, Austin, TX.)

Figure 14.7. Some of the novelties that can be produced by the extrusion process. (Courtesy APV Glacier, Austin, TX.)

Molding Novelties

In the molding type of production operation mix is pumped into molds that are immersed in or sprayed with chilled brine or glycol. The mix may or may not have been whipped and partially frozen. These machines are of the rotary (Figure 14.8) and straight through (line) types. After the mix has been partially frozen, sticks are inserted and freezing is completed in the molds. The molds then progress to a section where they are briefly exposed to heat (warm brine or water) to loosen the bar, and an extractor picks up the novelty by the stick and passes it to the next station. This station can be an enrober, decorator, or packaging apparatus. Individual packaged items are placed typically in bags or boxes, which may be packed in cartons. Because they typically are very hard when packaged, it is unnecessary to transfer them through a hardening tunnel before sending them to cold storage.

Principles and Guidelines for Frozen Dessert Novelty Equipment is a publication that resulted from a joint project of the International Ice Cream Association and the Dairy and Food Industries Supply Association in cooperation with the Milk Safety Branch of the U.S. Food and Drug Administration. The document, revised in 1990, sets forth criteria for the design and construction of equipment used to produce novelties. The document is not intended to be a regulation but rather is to be used in conjunction with other documents that deal with product safety.

The principles and guidelines are intended to provide design criteria that will result in equipment that:

1. is easily exposed for cleaning and inspection
2. can be cleaned and sanitized efficiently
3. avoids materials that may deteriorate
4. isolates the product from contact with contaminants
5. contains no areas where contaminants or vermin may be trapped
6. inhibits undesired microbial growth in ingredients
7. prevents spillage of ingredients and packaging materials
8. prevents unnecessary air movements and formation of aerosols.

ICE, FUDGE AND CREAM STICK ITEMS

Items commonly frozen quiescently on sticks include water ice, fudge, and creams. Modern equipment makes possible the production of a wide variety of styles (Figure 14.9).

A formula suitable for producing 320 dozen ice sticks (4 fl oz) has the following ingredients: Water, 100 gal; sugar, 160 lb; corn sugar or corn syrup solids, 40 lb; stabilizer, 4 lb; citric acid, 1 gal; flavoring, about 33 fl oz (depending on supplier's instructions). This produces about 120 gal of mix. The procedure for mix preparation involves mixing the stabilizer with a portion of the sugar, adding the stabilizer, sugar, citric acid, and flavoring while agitating, and continuing agitation until the sugar is completely dissolved and the stabilizer

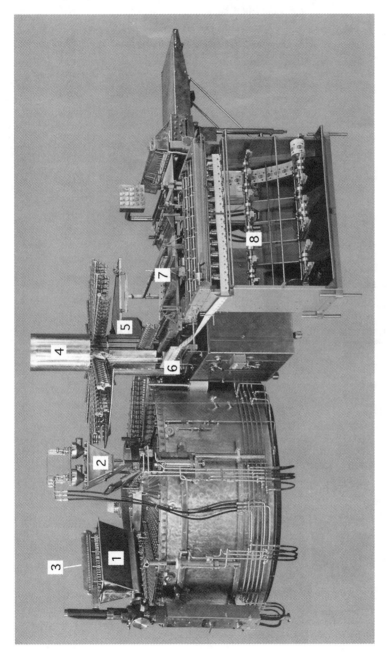

Figure 14.8. Rotary type machine for making molded novelties. Inset: Line drawing of molding machine—(1) filler for product, (2) multi-flavor filler, (3) stick inserter, (4) "remover" for extracting frozen pieces from molds, (5) chocolate dip tank, (6) depositor of bars for wrapper, (7) wrapping machine, (8) wrapper dispenser. Molding tables range in width from 6 to 26 cups. (Courtesy Gram Equipment of America, Tampa, FL.)

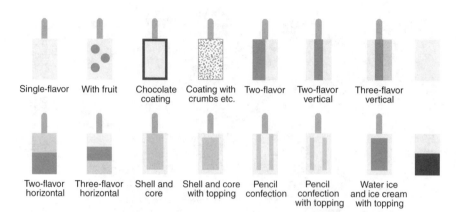

Figure 14.9. Examples of styles of novelties that can be produced in some novelty freezers. Courtesy, Gram Equipment of America, Tampa, FL.

Table 14.1. Formulas for Making Fudge Bars Containing Varying Amounts of Milkfat

Ingredient	Amount (%)		
	Nonfat	3% Fat	5% Fat
Milkfat	—	3.0	5.0
Sugar	15.0	15.0	15.0
Corn syrup (62 DE, 80% TS)	5.0	5.0	4.5
Fudge powder	3.5	3.5	3.5
Nonfat milk solids	11.0	10.0	9.0
Whey solids	1.5	1.0	1.0
Stabilizer	0.5	0.4	0.4
Water	63.5	62.1	61.6

suspended. Molds should be filled to 3/16 inch of the tops to allow for expansion during freezing.

Fudge bars are produced with varying concentrations of fat. The formulas in Table 14.1 are representative of three of these.

For cream-on-a-stick items a special filler attachment may be used so that in one operation one freezer may be used to freeze the ice cream to 100% overrun for the center while a second freezer is used to freeze the sherbet or ice to 10 to 15% overrun for the shell.

ICE CREAM BARS

The usual procedure for producing ice cream bars on a stick is to freeze the ice cream in a regular freezer to the desired overrun and a flowable consistency. Molds are then filled with the partially frozen product. The process continues

14 FANCY MOLDED ICE CREAMS, NOVELTIES, AND SPECIALS

as previously described with the frozen and extracted bars being transferred to the chocolate coating station.

CHOCOLATE COATINGS

Along with chocolate-flavored ice cream and water ices, there are frozen desserts to which the chocolate is applied as a coating or as inclusions in the ice cream. Combinations of vanilla ice cream with dark chocolate coating have a long history of popularity. Such coatings should have a true chocolate flavor, melt readily in the mouth (melting point near 33°C) without a waxy feel, solidify rapidly with minimal drip during application, adhere well to the ice cream, and form a thin layer that resists cracking or breaking during handling. A major requirement of the cocoa for such a coating is that it be finely ground. Furthermore, it should be low in moisture (less than 1%) and have no residual lipase activity to avoid the tendency for the lauric acid to saponify, producing a soapy taste. This means that residual lipase from contaminating bacteria of either the cocoa or added oils must be nil. For prevention of oxidized flavors, cocoa butter contains a relatively high concentration of antioxidants that many substitute fats do not contain. Fortunately, cocoa powder contains antioxidants too.

The quantity of fat in the coating varies with the product, but 55% is considered minimal. Replacement of cocoa butter with vegetable fats is done for two reasons, viz., economy and function. Most countries have unique labeling regulations for these modified products.

The usual composition of coatings for frozen desserts is vegetable fat, cocoa, sugar, milk (including skim milk and buttermilk), lecithin, and flavors.

To the extent that cocoa butter is removed from the chocolate liquor, it should be replaced with an oil high in lauric acid. Coconut and palm kernel oils meet this specification. A fat chosen to replace cocoa butter should have virtually no solid fat at 35°C. If more than 3% is present, a waxy mouthfeel will likely result. Additionally, the amount of solid fat should be virtually constant in the range of 20–25°C, and in this temperature range the fat should be hard and brittle.

Flexibility of the coating is important to provide resistance to cracking. Soft coconut oil makes coatings flexible, while hardened coconut and palm kernel fats reduce flexibility but cut the tendencies for the coating to drip as bars are removed from the enrobing bath. The logical choice, then, is a blend of the three. The addition of 5–10% of nonlauric oil such as peanut, sunflower, or soybean, provides the protection from smearing often required during packaging. Furthermore, these non-lauric oils do not cause a soapy flavor if they become hydrolyzed. Finally, 0.5% lecithin added to the coating reduces the tendency for viscosity to increase as ice melts producing an aqueous phase in the enrobing vat.

Depending on the ingredients combined with the chocolate liquor or cocoa there are different names for the products used for coatings. In the United States if there is no substitution for the cocoa butter, the name is chocolate coating. When a vegetable fat is substituted for part of the cocoa butter, this

Table 14.2. Formulas for Light and Dark Chocolate Coatings

Ingredients	Light milk chocolate	Dark chocolate
	%	
Natural process cocoa	6	—
Dutch process cocoa	—	10
Sugar	28.5	33
Lauric fat[a]	55	56.5
Nonfat dry milk	10	—
Lecithin	0.5	0.5
Vanillin[b]	0.03	0.03

[a]Usually the fat added is palm kernel oil or coconut oil.
[b]Flavoring is optional.

is indicated on the label, e.g., "milk chocolate with vegetable fat coating." When sweetener is added, sweet is used as a modifier of the name cocoa or chocolate, e.g., "sweet cocoa with vegetable fat coating." In the case of milk chocolate and vegetable fat coating, sweet is not a part of the name, because milk chocolate is defined as a sweetened product.

It is important to recognize that the amount of cocoa butter and the flavor-carrying nonfat component of these chocolate products vary. Approximate quantities of nonfat cocoa solids in the major ingredients follow: chocolate liquor, 45%; cocoa, 85%; sweet cocoa and sweet chocolate, 6.8%; and milk chocolate, 4.6%. The formulas in Table 14.2 are suggested by a major cocoa supplier for making chocolate coatings.

Coatings should be processed at a temperature of 35–39°C (95–105°F) if they are lauric acid based. Storage temperature is about 5°C higher than the processing temperature.

Application of Coatings

Several factors determine the temperature and procedures for coating ice cream novelties:

1. The higher the temperature, the less the amount of coating deposited on the ice cream. However, melting of the product must be avoided.
2. The lower the temperature of the ice cream, the warmer must be the temperature of the coating to obtain minimal thickness.
3. High overrun promotes rapid melting at the surface; therefore, coating temperature must be adjusted when overrun is changed significantly.
4. The higher the coating temperature, the longer the time for solidification. Drying time can be adjusted by changing the ratio of soft oil to hard fat. The chocolate is kept in an electrically heated container with a water jacket controlled with a thermostat (Figure 14.10).
5. Fast dipping is important, because coating thickness is proportional to residence time in the coating basin.
6. Moisture must be excluded from the coating to avoid increases in viscosity that will lead to thickening of the coating.

14 FANCY MOLDED ICE CREAMS, NOVELTIES, AND SPECIALS

Figure 14.10. The coating of ice cream bars in a thermostatically controlled dip tank. (Courtesy Gram Equipment of America, Tampa, FL.)

Coatings for molded novelties should have a slightly higher sugar (34 vs. 29%) and lower fat (55 vs. 60%) content than coatings for extruded novelties. Furthermore, the lauric fat needs to be somewhat harder for coatings used in dipping than for extruding.

Sometimes coatings are sprayed onto surfaces of cones. This coating needs to be quite thin, so the formula calls for about 65% fat, the remainder being 6% cocoa, 20.5% powdered sugar, 8% nonfat dry milk, and 0.5% lecithin. Coatings for ice cream cakes and pies can be more viscous than those used for cones, because the excess is usually removed by forced air or rapid vibrations. A typical formula contains 12% lightly alkalinized cocoa, 37% powdered sugar, 50.5% lauric fat, and 0.5% lecithin.

OTHER SPECIAL PRODUCTS

Continuous freezers can be fitted with special discharge nozzles and used in pairs or trios to produce combinations of flavors in the same package. Split flavor nozzles can be used to produce fancy centers (Figure 14.11). Ice cream cakes and pies (Figure 14.12) can be made with special freezer attachments. Cakes can also be made by decorating layers of ice cream cut from bulk packages by pulling a wire through the tempered ice cream. Finished cakes are usually inserted into a stiff package to support the form of the cake during handling and delivery. An aufait can be made by placing a layer of fruit between two

Figure 14.11. Ice cream log with fancy center produced by continuous freezer.

layers of ice cream. Iciness of the layer of fruit can be prevented by mixing the fruit with sugar and gelatin.

Ice cream pie shells can be made by forming the ice cream about one-half-inch thick between two pie pans. The crusts can be filled with a gelatin-fruit-sugar mixture. A more common approach is to use graham cracker mix combined with shortening to make the shells in 10-inch-deep aluminum pie tins.

Figure 14.12. Ice cream cakes.

Figure 14.13. Spumoni.

Figure 14.14. Cups containing cones filled high with very stiff ice cream and striped with chocolate. (Courtesy Gram Equipment of America, Tampa, FL.)

Ice cream can be added directly from the continuous freezer or can be ladled from bulk containers that have been filled directly from a batch freezer. To permit shaping of the top, the ice cream can be made stiff or the pies partially hardened. Whipped cream, colored to fit the occasion, makes an excellent decorating material for ice cream pies and cakes.

Layered pies are special items. For example a "mud pie" is made with a layer of mocha chip spread on the bottom of a graham cracker crust; after it has hardened, the top is filled with coffee ice cream. The top is covered with a layer of high-quality chocolate coating, and a decorative circle of frosting is run around the edge. Crushed peanuts are sprinkled on top.

Spumoni is a fancy ice cream that is usually made in a cup-shaped form in pint or quart size. The outside layer is frequently vanilla ice cream. In the bottom of this shell is placed macaroon or chocolate mousse, and this is topped with tutti-frutti mousse. In serving it is cut into wedge-shaped pieces like cake (Figure 14.13).

Cups of ice cream with fancy decorations can be produced in large quantities with the equipment shown in Figures 14.14 and 14.15. The cups may contain baked cones.

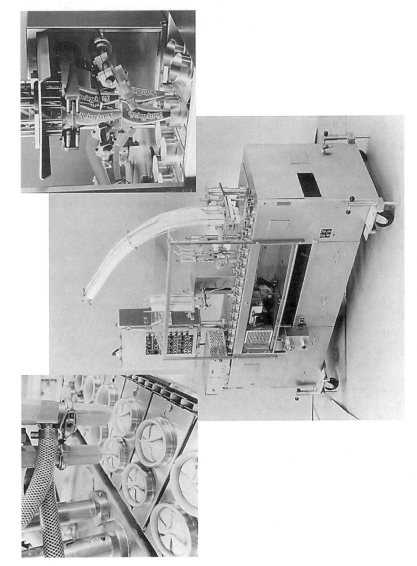

Figure 14.15. An in-line, two lane filler capable of filling cones or cups with plain or variegated ice cream. (Courtesy Norse DairySystems, Columbus, OH.)

15
Defects, Scoring, and Grading

Quality and price are two major factors that determine the volume of current sales of frozen desserts and a firm's future success. Quality is an important variable in establishing price, so it is necessary to understand the causes and the remedies of defects in quality. Defects result from faults in flavor, body, texture, melting characteristics, color, package, microbial content, and/or composition. In their efforts to ensure high quality, manufacturers assume an ideal of perfection for each of these quality attributes. The ideal product should possess a typical, fresh, clean, pleasant, and delicate flavor; have a close, smooth texture; possess moderate resistance of body; melt slowly into a liquid with the appearance of the original mix; have a natural color; have any particulates, ripples, or other inclusions evenly and liberally distributed; and have a low bacterial count. Additionally, the product must meet the compositional specifications implied by the name of the product and the requirements imposed by the ingredient and nutritional labels. This ideal standard can be embodied in a score card, which is useful for rating products on their relationship to that ideal. The American Dairy Science Association Committee on Evaluation of Dairy Products has such a score card for ice cream, and it is discussed later.

The scoring of ice creams, sherbets, ices, and other frozen desserts presents difficulties not encountered with some other dairy products. In the first place, the ideal flavor is more variable than that of some other products, because flavor quality depends on the intensity of the flavoring material, the blend of ingredients, the qualities of each ingredient, and the effects of the process on the flavor.

FLAVOR DEFECTS

Defects in flavor of frozen desserts are conveniently placed in the following categories:

1. Dairy ingredients of poor quality—sour (acid), oxidized, stale, lipolyzed, unclean, and excessively cooked or scorched

15 DEFECTS, SCORING, AND GRADING

2. Sweetener(s)—unnatural, excessive, or deficient
3. Flavoring—unnatural, excessive, or deficient
4. Blend—unpleasant balance of ingredients
5. Storage—stale or absorbed flavor.

The *acid* flavor is imparted by dairy ingredients made from sour milk or cream. The cause is growth of lactic acid bacteria that are most likely to grow in raw milk that has been inadequately refrigerated. There is a possibility that souring can occur in milk held a long time in dairy plants, but temperatures above 4°C (40°F) are necessary, and the contaminating bacteria must be able to produce lactic acid. The common spoilage bacteria in liquid pasteurized dairy products, such as concentrated milk, are proteolytic and lipolytic rather than fermentative. Because bacteria counts must reach into the millions before significant lactic acid is produced in milk, it is highly undesirable that sour products ever reach the dairy plant.

Cooked flavor is the "flavor of assurance" in ice cream, and, provided it is mild, the flavor is not objectionable. The rationale for the preceding statement is that heat breaks disulfide bonds (—S—S—) in whey proteins, exposing sulfhydryl (—SH) groups that are considered to contribute significantly to the cooked flavor. If the cooked flavor has dissipated beyond recognition, there is good reason to believe that the —SH groups have been oxidized back to the original —S—S— linkage. When cooked becomes strong enough to be described as "scalded milk, caramel-like, scorched, or burnt," there is reason to be concerned. The main contributors of such high-intensity cooked flavors are dried products, especially high-heat types. Of course, the relatively high temperature and long time of pasteurization of ice cream mixes compared with those used for milk mean that cooked flavors are likely to be stronger in ice cream than in milk. Those manufacturers that give extra holding time to their ice cream mixes with the intent of denaturing proteins so they will hold maximal water take the risk that cooked flavor will be excessive for at least a segment of consumers. The advantages that offset this potential defect must be weighed in the balance when the decision is made to use such a treatment.

Lacks freshness is a mild form of "old ingredient," stale, or lipolyzed flavors. When the term is correctly applied, either a defect has not developed to the point of being positively identifiable in the finished mix, or an ingredient that may have had an identifiable defect has been used in low enough concentration that identification is not possible.

Ice creams that have the *old ingredient* defect suffer from staleness, oxidation or fermentation of one or more ingredients. Often this defect is not readily tasted initially but develops as the product melts and is expectorated. The mouth is left with an unclean feeling.

Oxidized flavor sensations vary through a series of intensities and types, including papery, cardboardy, metallic, tallowy, fishy, and painty. The flavor is generally noted as the ice cream starts to melt and it persists after expectoration. Copper, iron, and ultraviolet (UV) light catalyze oxidation in dairy foods. The metal-induced form tends toward puckery and astringent and involves lipid oxidation, whereas the light-induced form tends toward cardboardy and involves protein oxidation. The latter has little chance to occur in frozen desserts, because their ingredients are seldom exposed to UV light.

Lipolytic rancidity is caused by the hydrolysis of the ester bonds that hold fatty acids onto the glycerol moiety of milkfat. The flavor defect is known as *lipolyzed* or *rancid*. The natural lipase of milk can catalyze the reaction if the protective fat globule membrane is broken by foaming and churning of raw milk. Prevention measures for this type of fat degradation are usually applied at the farm; however, the pumping and agitation of milk or cream before pasteurization, especially if temperatures are allowed to rise above 4°C, will increase the rate of release of fatty acids. After pasteurization, the source of the catalyst is lipase-producing bacteria. These are usually postpasteurization contaminants, especially *Pseudomonas fluorescens*. Frequently these bacteria produce heat-stable lipases that can cause cream and butter to become rancid if these products are held a long time before use. These microbial enzymes are not significantly active at temperatures used to store ice cream, so ice cream is not expected to develop lipolyzed flavor during storage. However, use of rancid butter or cream in a mix is likely to impart the flavor to the finished product. Taste receptors adapt slowly to lipolyzed flavor. The sensation may be similar to the pungency of blue cheese if the short-chain fatty acids, such as butyric, predominate, or to soapiness, if the longer chain ones, such as lauric, are highly abundant. Rancidity typically leaves an unclean and unpleasant aftertaste. Products containing perceptible free fatty acids are highly undesirable and should not be marketed.

Salty flavor is unusual in ice cream, but some formulas call for salt, making it possible that too much may be added. Additionally, overuse of NMS, dried whey, or salted butter can cause a product to taste salty. Saltiness is quickly perceived on tasting a sample.

Whey flavor often reminds analysts of graham crackers. It is imparted to ice cream when poor-quality whey solids are added or when excess whey is used. Federal standards permit up to 25% of the NMS of a formula to be replaced with whey solids, and this may be too much if the quality of this component is not satisfactory. Dry whey stored for an excessive time may be oxidized, rancid, or unclean. Whey used in ice cream should be light in color, freeflowing, absent of lumps, and clean tasting.

Egg yolk is the characterizing ingredient in French vanilla or custard ice cream. Thus, the *egg yolk flavor* is desired if it is clean and pleasant. However, overuse of eggs can lead to undesirable flavor notes in some products, and eggs are subject to development of off-flavors similar to those of milk products, especially oxidized and lipolyzed flavors. When used at concentrations up to 1%, egg yolk solids are compatible with most ice cream flavors. The high emulsifying properties of egg yolk solids (high phospholipid content) recommend them for use in vegetable fat type frozen desserts.

Stabilizer/emulsifier components may impart off-flavors, because they are prone to oxidation if not kept in a dry and cool environment. The high concentrations of these ingredients used in products low in fat and in ice cream novelties make these products the most likely ones to exhibit this defect.

Food solids from nondairy sources should be considered potential sources of off-flavors. Each laboratory should have a table of expected or potential off-

15 DEFECTS, SCORING, AND GRADING

flavors for each ingredient, a description of the ingredient, an expected storage life, and optimal storage conditions. Every shipment of ingredients should be examined on receipt and periodically thereafter if they are held longer than about 20% of their expected shelf life.

Flavoring System

Frozen desserts have a wide variety of flavoring ingredients, and these have been considered extensively in Chapter 7. Some general principles apply to the evaluation of the quality of flavorings within the product. Usually the flavoring contains highly volatile components that are detected soon after a sample is tasted.

Lacks flavoring is perceived as bland, flat, or lacking bouquet. The cause may be addition of too little flavoring, low strength of the flavoring, or masking of the flavor by other ingredients.

Lacks fine flavor is sensed as a slightly harsh or coarse flavor, and the flavoring system usually is not in balance with the ingredients. The exact cause is often obscure.

Too high flavor is not often observed, because flavorings are usually the most costly of ingredients. There is a high economic incentive to avoid overuse of flavorings. The analyst experiences sharpness of certain flavor notes when excess flavoring is used.

Unnatural flavor is too frequently observed in frozen desserts when manufacturers attempt to lower costs by using "cheap" flavors. Various sensations characterize this category of flavor defect, but the most frequent cause is addition of artificial flavoring. Artificial flavors may be suggestive of butter, candy, cherry, coconut, custard, lemon, maple, marshmallow, nuts, smoke, vanillin, and numerous other substances, depending upon circumstances. The practice of using fruit flavorings "with other natural flavors" (commonly referred to as WONF) can be associated with unnatural flavors. With vanilla the use of synthetic vanillin can cause a sharp, harsh flavor. Even when only pure vanilla extract is used, the source of vanilla beans, the way they are fermented, and the grinding and extraction treatments are sufficiently variable to cause evaluators to sometimes use the term "unnatural" to describe the flavoring.

It is important to recognize that not all consumers have the same taste preferences. Some persons actually prefer imitation flavors over natural ones in certain products. There is no substitute for having an analysis of the preferences of the target group of consumers to determine the best flavor profile for a specific product. The producer is most unwise to select flavorings and establish their rates of usage based solely on taste preferences of the "boss."

SWEETENER SYSTEM

As with flavorings, the sweetener system is characterized by defects in quantity and quality. *Lacks sweetness* is sensed as flatness or blandness, and severe deficiency of sweetener can lead to body and texture defects. A high ratio of corn syrup solids to sucrose can cause the syrup flavor.

Too sweet results from overuse of sweetener, especially nonnutritive types. The product is likely to take on a candy-like taste. When ice cream is too sweet, other flavors tend to be subdued. Furthermore, too much nutritive sweetener is often associated with low freezing point and rapid melting.

Syrup flavor is described as malty or caramel-like. Corn syrups and maltodextrins may impart off-flavors when they are inadequately refined or have been fermented. Use in high concentrations causes the syrup flavor to dominate other flavors. Syrup tends to mask delicate flavors such as vanilla. Cooked flavor enhances syrup flavor. Together they may engender the sensations of "toasted coconut" or "marshmallow." High syrup flavor is often associated with gummy or sticky body.

BODY AND TEXTURE DEFECTS

It is difficult to separate clearly the attributes of frozen desserts that characterize body and texture. *Body* has been defined as the "quality of the whole" as it applies to mouthfeel. *Texture*, then, is the "quality of the parts that make up the whole" (Bodyfelt et al. 1988). The desired texture of ice cream is smooth, creamy, and homogeneous. The desired body is firm with substantial feeling of solid matter within the foam. At dipping temperature the dipper should move readily through the product without revealing stickiness or gumminess. Arbuckle (1940) demonstrated many of the body and texture defects with the light microscope (Figure 15.1).

Defects of Body

Crumbly is the term used to describe a brittle or friable structure that tends to fall apart when dipped. The appearance of a severely affected sample is of dryness and openness like freshly fallen snow. The defect accompanies high overrun, low total solids, and insufficient stabilizer.

Gummy is the opposite of crumbly and may be described as pasty or sticky. When dipped, a gummy sample tends to curl up behind the dipper leaving an uneven, wave-like surface. Such ice cream is often slow to melt. The likely causes are overstabilization, very high solids, especially sweetener, or low overrun.

Shrunken product does not contact fully the sides of the container (Figure 15.2). The defect results from the collapse of air cells, allowing the whole structure to get smaller. Weakness of the air cell lamellae and large changes in pressure on the air cells are responsible for this defect. Contributing factors can be high overrun, low solids, and changes in external pressure such as occur during transport over mountains. Fluctuations of temperature accentuate the defect by allowing the supporting structure of ice to melt and refreeze, causing pressures on air cells to change and puncturing some of them. The physical effect of freezing to unusually low temperature, especially hardening with dry ice, can cause shrinkage.

Heavy body is manifest as soggy, doughy or putty-like. This characteristic may be desired in superpremium products for which the manufacturer wishes

15 DEFECTS, SCORING, AND GRADING

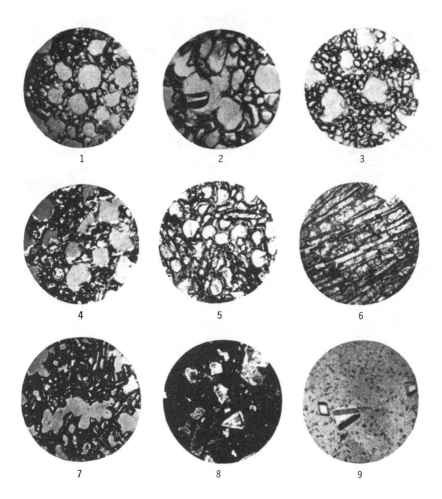

Figure 15.1. Micrographs of ice cream showing body and texture characteristics. (1) Close, smooth; (2) coarse, open; (3) short, fluffy; (4) soggy; (5) coarse, icy, because of fluctuating temperatures; (6) coarse, icy, flaky, because of slow freezing without agitation; (7) coarse, icy, because of heat shock to the surface; (8) lactose crystals from sandy ice cream; (9) lactose crystals.

to provide the impression of high solids and high value. Causes of heavy body are high total solids, overstabilization, or low overrun.

The opposite of heavy body is *weak*. This characteristic is sensed as little resistance to compression, fast melting, and low viscosity of the melted product, reminding the analyst of skim milk rather than cream.

Defects of Texture

Buttery texture in ice cream causes a greasy mouthfeel and may give the appearance of coarseness of texture due to churning of the fat in the mix. The

Figure 15.2. Shrinkage in lowfat ice cream.

defect is caused by inadequate homogenization and/or overwhipping in the freezer. Mixes that are high in milkfat, overstabilized, or agitated excessively in the freezer suffer the greatest intensity of butteriness.

Coarse texture is the most frequently observed defect in ice cream. In fact, all ice creams will eventually become coarse as ice crystals grow during storage (Figure 15.3). This grainy or icy feeling is accompanied by unusual coldness when the defect is pronounced. Factors that increase the probability of coarseness developing are: low solids, low freezing point, high drawing temperature, dull scraper blades, inadequate stabilizer, insufficient aging with some stabilizers, slow hardening, long storage time, and high and variable storage temperature. At a temperature common to dipping cabinets, −15°C (5°F), ice crystals doubled in size in 200 days (Figure 15.3; B. Liang and R. W. Hartel, University of Wisconsin, Madison, personal communication). Mean sizes and storage times were 35.8 µm on day 1, 52.9 µm on day 22, and 64.6 µm on day 200. The ice cream formula was typical of high quality product, viz., 12% milk fat, 11% NMS, 16.5% sweetener, 0.2% stabilizer, and 0.1% emulsifier for a total solids content of 39.8%. Hardening was at −40°C for 24 hours.

Obviously, with so many potential causes of coarseness, identification of the needed remedy can involve significant study.

Milkfat is an important ingredient influencing the development of coarseness. The tendency for ice crystals to grow decreases in the following order: nonfat, lowfat, light, reduced fat, regular, premium, and superpremium ice creams. Fat globules mechanically obstruct the growth of ice crystals and lubricate the mouth; therefore, the more fat the less risk the producer has that coarseness

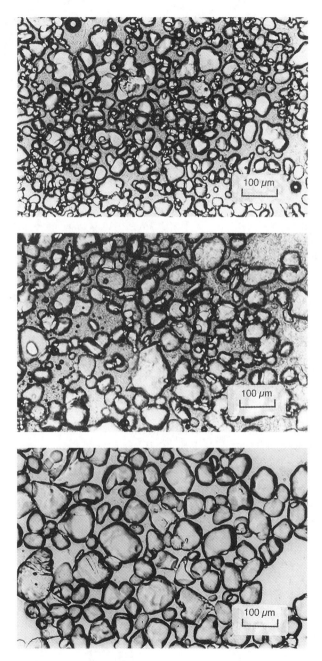

Figure 15.3. Photomicrographs of ice crystals in ice cream held at −15°C. Mean ice crystal sizes, calculated as the equivalent circular diameter derived from the projected area of each crystal, were 35.8 µm, 52.9 µm, and 64.6 µm, respectively, on days 1 (a), 22 (b), and 200 (c). (Courtesy B. Liang and R. W. Hartel, University of Wisconsin, Madison.)

Table 15.1. Effect of Nonfat Milk Solids on the Internal Structure of Ice Cream[a]

Concentration of NMS (%)	Average ice crystal size (μm)	Average air cell size (μm)	Cell wall thickness (μm)	Textural quality
9	56	177	165	Slightly coarse
11	52	188	149	Medium smooth
13	39	158	116	Smooth
15	32	103	124	Very smooth

[a]Arbuckle (1940).

will become apparent to the consumer. Arbuckle found the average size of ice crystals to be 83 × 61 μm when fat was 10% and 47 × 38 μm when fat was 16%. He found that NMS content was an even more important factor in determining textural quality (Table 15.1). As NMS content was increased in the gelatin-stabilized mix, both ice crystals and air cells decreased in size, but walls of air cells became thinner. The increased protein content of the unfrozen syrup was most likely responsible for the increased strength of the air cell lamellae. Increasing the sugar content causes a smoother texture, because it lowers the freezing point, increases the amount of unfrozen material, and decreases the amount of free water (sugar binds some water). At 12% sugar in ice cream ice crystals averaged 68 × 51 μm, while at 18% sugar the average size was 49 × 36 μm (Arbuckle 1940).

As overrun is increased, ice crystals and air cells become smaller in size (Table 15.2). However, there is the counterbalancing effect of weakening of the structure because of thinning of the unfrozen material among the air cells and ice crystals.

Arbuckle (1940) calculated the average effects of increases in concentrations of ice cream constituents on reductions of ice crystal size as follows: 1% fat, 4.4 μm; 1% NMS, 4.3 μm; 0.5% fat plus 0.5% NMS, 5.8 μm; 1% sugar in 11% NMS, 2.8 μm; 1% sugar in 13% NMS, 2.4 μm. Each 10% increase in overrun within the normal range caused an average decrease in ice crystal size of 1.6 μm.

Stabilizers reduce ice crystal size as they bind water of the mix and increase viscosity of the unfrozen portion. Emulsifiers control aggregation of fat globules causing structure to be formed to stabilize air cells and to impede growth of ice crystals; thus, they contribute to ice cream smoothness.

Large air cells that compress considerably during whipping are usually asso-

Table 15.2. Effects of Overrun on Sizes of Ice Crystals, Air Cells, and on Thickness of Intervening Unfrozen Materials[a]

Overrun (%)	Ice crystal size (μm)	Air cell diameter (μm)	Unfrozen material (μm)
85	63 × 51	165	11
100	54 × 47	142	10
115	50 × 44	109	9
130	50 × 43	104	7

[a]Arbuckle (1940).

15 DEFECTS, SCORING, AND GRADING

ciated with the *fluffy* defect. The cause is high overrun, and the product melts slowly because of the insulating effect of the air. Fluffiness can be expected when the total solids content is less than one-third of the overrun; for example, 33% total solids and an overrun of 100%.

Sandy texture is one of the most objectionable defects in a frozen dessert. The gritty mouthfeel results from crystallization of lactose from the concentrated syrup within the frozen foam. High concentrations of lactose predispose ice cream to lactose crystallization, and the reaction is speeded by fluctuations in storage temperature. Lactose crystals can be differentiated from ice crystals by their comparatively slow rate of dissolution. Prevention involves reducing the lactose content of the mix, dividing the air cells into very small units, and minimizing storage time and fluctuations of temperature.

COLOR

In high-quality ice cream the color reminds the consumer of the ice cream flavor name. Furthermore, the color should be uniformly distributed unless there is intent to variegate or otherwise mix in compatible flavors. Operators must take care not to allow succeeding batches to intermingle in the package. The products that become mixed between successive batches can be recovered and rerun in a flavor such as chocolate, provided the rerun is adequately diluted and its flavors are compatible with the chocolate or similar flavor.

The color of frozen desserts must have both the desired hue or shade and the proper intensity. If the intensity is too high, the product may appear to be artificially flavored even when it is not. If color is lacking, many consumers will think too little flavoring has been added. An inappropriate hue or shade gives the impression of artificiality.

There should be clear lines of demarcation between colors in variegates, in products containing inclusions, and in any product that contains multiple flavors (neopolitan and certain novelties).

PACKAGE

Color of the frozen dessert is important, but color and design of the package are of paramount importance because of their impact on the potential buyer. The very best ice cream in a poor package may never be tasted by the consumer because of unwillingness to purchase it. Once the package has been designed to be attractive and indicative of the product, properly labeled, printed, and delivered to the ice cream producer, the objectives become: (1) fill it adequately, (2) keep it clean, (3) deliver it undamaged, and (4) keep it cold. The defects sometimes seen in packages are: slack-filled, bulged, misshaped, improperly sealed, soiled, and ink-smeared.

MELTING QUALITY

Melting rate has the greatest significance to the consumer when the product is being eaten from a cone or stick. If the product melts too fast, a messy situation often ensues. A fast-melting product is undesirable also because it tends to become heat shocked readily. Low freezing point is the primary cause of rapid melting, environmental conditions being about equal. However, slow rate of melting can also be indicative of defective ice cream. Ice cream with a desirable melting quality (Figure 15.4F) begins to show definite melting within 10–15 min of having been dipped and placed at room temperature. Melting product flows readily and forms a homogeneous fluid with the appearance of the unfrozen mix and with little foam.

The environment for testing melting quality should be clean, well lighted, and about 20°C (70°F). The ice cream should be dipped without introducing significant heat and should be placed in a flat bottom dish such as a petri dish.

Defects of Melting Quality

Curdy melt means that irregularly shaped curd particles are seen floating in the melt (Figure 15.4B). Causes include protein destabilization by acids, imbalance of salts (high proportion of calcium and magnesium vs. phosphates and citrates), and overstabilization, especially with certain gums.

Does not melt means the ice cream retains shape after 15–20 min in a glass

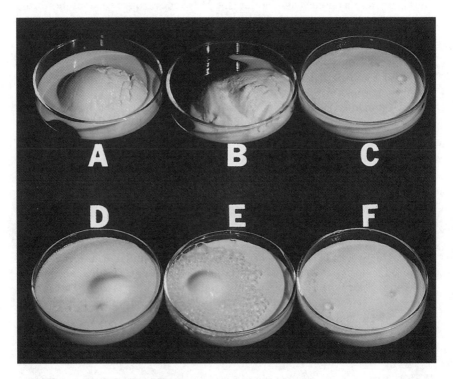

Figure 15.4. Selected melt down characteristics of ice cream: (A) *does not melt,* (B) *curdy,* (C,D,E) *foamy,* (F) *no criticism.*

15 DEFECTS, SCORING, AND GRADING

petri dish (Figure 15.4A, B). The common causes are overstabilization and high overrun.

Flaky means that light, feathery particles float in the melting product.

Foamy means that large air bubbles retain their shape and presence in the melting product (Figure 15.4C, D, E). Highly surface-active substances (emulsifiers and egg yolk) are responsible for the high stability of the foam.

Wheying off means that watery fluid appears, and in it are often seen curd particles. The usual cause is protein destabilization as with curdiness.

Low viscosity means the ice cream is thin in appearance like milk. The cause is most often low solids in the mix.

EVALUATING FROZEN DESSERTS

The evaluation of frozen desserts is no easy task. The analyst must be experienced, mentally alert, and the senses physically unimpaired. The environment must be clean, free of odors, well lighted, comfortable of temperature, convenient, and well equipped. A table or computer should be conveniently placed so results can be easily recorded.

Samples must be tempered at least overnight to a temperature of −15° to −13°C (5° to 8°F). The tempering cabinet should have a uniform temperature throughout. This is most easily ensured by placing all samples at the same height in the freezer. If samples are too cold, they are difficult to dip, and the cold tends to numb the mouth and tongue of the analyst. Samples that are too warm cannot be properly evaluated for body and texture attributes.

In taking samples it is usually wise to remove the surface to a depth of about 1 cm unless the surface needs to be examined for a special purpose. Product at the surface may contain absorbed flavors and is subjected to the greatest chances of heat shock. Individual foam type plates are suitable for taking samples for tasting. Metal or plastic spoons are favored over wooden because they do not impart flavors. About 2 oz of sample is proper for each analyst to examine. Dippers are the best sampling instruments, and they always should be removed from the product. Never should the analyst remove a sample with the personal spoon.

The accepted sequence of examining samples is as follows:

1. Examine the container for fill, shape, printing flaws, soil, and seal or closure.
2. Check the color intensity, hue, and distribution. Cut through samples containing inclusions to determine the quantity, quality, and distribution.
3. "Feel" the product as it is scooped to determine the extent of gumminess, heaviness, fluffiness, or weakness of body. Place a scoop of sample in a petri dish to observe later for melt down characteristics. Set a timer to indicate when 15 min have elapsed to examine samples for melting qualities. If foam or paper plates are used, more time will be needed for the test. Heat must be transferred into the sample to induce melting.

The main avenue of entry for heat is through the plate. Heat is transferred from air to the sample quite slowly.
4. Examine the body and texture of a small portion (about ½ teaspoon) taken into the mouth. Melting of the sample means that the sensations in the mouth are constantly changing and that the analyst must practice introspection (concentrate on the sample, shutting out all outside distractions).
5. Turn attention to the flavor. Usually this requires that a second sample be tasted. Again, the sensory characteristics change rapidly as the sample melts, so the analyst must keep mental notes of these characteristics and integrate them into thoughts that lead to a description in the end. The analyst must remember that cold dulls the senses and that it is important to allow the mouth to warm between samples as well as to allow each sample to melt completely during tasting. Sweetness suppresses other taste sensations and is fatiguing to the analyst. Fat can coat the taste receptors, lessening flavor perception. Each of these factors argues for rinsing the mouth between samples and taking time for the taste receptors to become refreshed between samples.
6. Check the melt down dish(es) for melting qualities.
7. Record accurately and clearly the results of the analyses.

SCORING METHODS

Ice cream is not sold on the basis of score, but statements of score are useful in industry and academia to help convey judgments of the quality of frozen desserts in relation to a described ideal. The American Dairy Science Association (ADSA) Committee on Evaluation of Dairy Products has established a scoring guide and a score card for use in industry and in collegiate contests in the evaluation of ice cream. The score card in Figure 15.5 is an adaptation of the ADSA score card. The highest marks that can be given a product in each of the score card categories go to the product that meets the standard of the ideal ice cream. No defect is recognized or marked when the "perfect" score is awarded.

The ideal flavor is fresh, clean, and delicate with adequate flavoring to give unmistakable identity and a natural characteristic with creamy, rich aftertaste. Flavor, sweetener, and dairy ingredients must be in balance. The ideal body is sufficiently firm to give the sensation of abundant solid matter in the product, yet not so firm as to restrict easy dipping at the usual temperature. The product does not stick to the dipper or break apart when dipped. The texture of the ideal ice cream is velvety smooth and creamy. Neither ice crystals nor air cells are large enough to be detected by the tongue. The mouth is not coated with fat or any other substance on expectoration of the sample. The mouth is not made uncomfortable by the low temperature when the product is eaten at a leisurely pace.

Reliability of scores placed on ice cream depend heavily on the judge's concept of the ideal as well as an ability to recognize slight deviations from the ideal. The latter is accomplished by considerable practice with other knowledgeable analysts and by tasting products with known types and intensities of defects. The collegiate dairy products evaluation contests sponsored by the ADSA and by industry have been responsible for making hundreds of industry profession-

ICE CREAM SCORE CARD

PRODUCT: _____   DATE: _____

FLAVOR: _____

SAMPLE NO.

CRITICISM		1	2	3	4	5	6	7	8	9	10
FLAVOR 10	SCORE▶										
NO CRITICISM = 10	**FLAVORING SYSTEM**										
	LACKS FINE FLAVOR										
	LACKS FLAVORING										
UNSALABLE = 0	TOO HIGH FLAVOR										
	UNNATURAL FLAVOR										
NORMAL RANGE = 1 10	**SWEETENERS**										
	LACKS SWEETNESS										
	TOO SWEET										
	SYRUP FLAVOR										
	PROCESSING										
	COOKED										
	DAIRY INGREDIENTS										
	ACID										
	SALTY										
	LACKS FRESHNESS										
	OLD INGREDIENT										
	OXIDIZED										
	METALLIC										
	RANCID										
	WHEY										
	OTHERS										
	STORAGE (ABSORBED)										
	STABILIZER/EMULSIFIER										
	NEUTRALIZER										
	FOREIGN										
BODY & TEXTURE 5	SCORE▶										
NO CRITICISM = 10	COARSE/ICY										
	CRUMBLY										
	FLUFFY										
UNSALABLE = 0	GUMMY										
	SANDY										
NORMAL RANGE = 1–5	SOGGY										
	WEAK										
COLOR, APPEARANCE & PACKAGE 5	SCORE▶										
NO CRITICISM = 5	DULL COLOR										
	NON-UNIFORM COLOR										
	TOO HIGH COLOR										
UNSALABLE = 0	TOO PALE COLOR										
	UNNATURAL COLOR										
NORMAL RANGE = 1–5	DAMAGED CONTAINER										
	DEFECTIVE SEAL										
	ILL-SHAPED CONTAINER										
	SOILED CONTAINER (DIRT)										
	SOILED CONTAINER (PRODUCT)										
	UNDER FILLED										
	OVER FILLED										
MELTING QUALITY 3	SCORE▶										
NO CRITICISM = 3	CURDY										
	DOES NOT MELT										
UNSALABLE = 0	FLAKY										
	FOAMY										
NORMAL RANGE = 1–3	WATERY										
	WHEYED OFF										
BACTERIAL CONTENT 2	SCORE▶										
	STANDARD PLATE COUNT										
	COLIFORM COUNT										
TOTAL 25	**TOTAL SCORE OF EACH SAMPLE**										
	TOTAL SOLIDS (%)										
	FAT CONTENT (%)										
	NET WEIGHT (LBS/GAL)										
	OVERRUN (%)										

SIGNATURE OF EVALUATORS: _____ _____ _____

Figure 15.5. A modified version of the ADSA ice cream score card (Bodyfelt *et al.* 1988).

als able to evaluate reliably the sensory qualities of frozen desserts and other major dairy foods. The Dairy and Food Industries Supply Association has been the major sponsor of the contests. Other cooperating organizations include the International Ice Cream Association. Many firms within the states entering teams have contributed product, time, facilities, and money to support these contests. The abilities of persons trained in these contests to identify and score defects are usually far superior to the abilities of consumers. This makes these trained persons valuable to the manufacturer who should desire to have defects detected and corrected before they become objectionable to consumers. However, this is not to infer that the preferences of trained sensory analysts should be sole determinants of the formulas used to produce frozen desserts. Consumer taste panels and focus groups are vital components of the product development program of ice cream manufacturers.

A defect is considered "slight" when recognized by the experienced judge or the "connoisseur" but not by most consumers. It is "definite" when detectable by many consumers and "pronounced" when detectable by most consumers. Professionals in the Dairy Foods Division of the American Dairy Science Association have developed a scoring guide for vanilla ice cream (Table 15.3). Defects that are likely to be observed are listed and the recommended score for intensities of slight, definite, and pronounced are provided. Scores allotted for the sections of the scoring guide are based on the ranges needed to rate the category. Thus, flavor scores range from 1–10 in consideration of the need to score both the intensity of the defects and their relative degree of undesirability or the variance from the ideal. For example, the defect *too high flavor* gets small point deductions. It is only a minor defect, because most consumers do not object to getting an extra amount of something good. In contrast, the *old ingredient* defect is highly undesirable and gets a much lower score.

Body and texture scores range from 1–5 as do the scores for color, appearance, and package. Melting quality is given 3 points and bacteria content 2 points in the suggested score card. This makes a total of 25 points. Compositional tests are also very important determinants of quality; however, they are usually treated as "within limits" or as indicating an "out of standards" or illegal product. The manufacturer should establish specifications for each product and ensure that representative samples are evaluated regularly by trained personnel. Furthermore, the data generated should be used effectively in quality improvement.

ICE CREAM CLINICS

Manufacturers should find it valuable to conduct ice cream grading clinics to compare their products with those of the competition and to give the employees responsible for the products valuable insights into the quality of the job they are doing. The following steps can be taken to organize a clinic:

1. Select a competent taste panel to evaluate and criticize (constructively) the products.

Table 15.3. ADSA Scoring Guide for Sensory Defects in Vanilla Ice Cream

Ice Cream	S	D	P
Flavor[a,b,c]			
Acid	4	2	U
Cooked	9	7	5
Lacks Fine Flavor	9	8	7
Lacks Flavoring	8	6	4
Lacks Freshness	8	7	6
Lacks Sweetness	9	8	6
Metallic	6	4	2
Old Ingredient	6	4	2
Oxidized	6	4	1
Rancid	4	2	U
Salty	8	7	5
Storage	7	6	4
Syrup Flavor	9	7	5
Too High Flavor	9	8	7
Too Sweet	9	8	7
Unnatural Flavor	8	6	4
Whey	7	6	4
Body/Texture[a,d]			
Coarse/Icy	4	2	1
Churned/Greasy	4	3	2
Crumbly	4	3	2
Fluffy	3	2	1
Gummy	4	2	1
Sandy	2	1	U
Soggy	4	3	2
Weak	4	2	1
Color, Appearance/Package[d]			
Dull color	4	3	2
Nonuniform color	4	3	2
Too high color	4	3	2
Too pale color	4	3	2
Unnatural color	4	3	2
Soiled container	3	2	1
Product on container	4	3	2
Underfill/Overfill	4	3	2
Damaged container	3	2	1
Defective Seal	2	1	0
Ill-Shaped containers	4	3	2
Melting Quality[e]			
Does not melt	3	2	1
Flaky	3	2	1
Foamy	3	2	1
Curdy	3	2	1
Wheying off	3	2	1
Watery	3	2	1

S = Slight, D = Definite
P = Pronounced, U = Unsalable

[a]Source: American Dairy Science Association (1994).
[b]Source: Adapted from Bodyfelt et al. (1988).
[c]"No criticism" is assigned a score of "10." Normal range is 1–10 for a salable product.
[d]"No criticism" is assigned a score of "5." Normal range is 1–5 for a salable product.
[e]"No criticism" is assigned a score of "3." Normal range is 1–3 for a salable product.

2. Chose a regular time, preferably at least once per month, to conduct the clinic.
3. Have an employee, preferably one who is not involved in the clinic, collect samples of both the firm's and competitor's products from retail outlets. For some clinics choose the oldest samples in the display cases.
4. Transfer samples in refrigerated cases to the clinic site and temper them there to −15 to −13°C (5 to 10 °F).
5. Place samples in randomly numbered paper bags or otherwise disguise them to avoid prejudicing the participants.
6. Place samples in random order.

The steps in examination of the samples follow:

1. Dip into a petri dish a 50-ml (about 2-oz) sample for melt down tests.
2. Instruct each participant to use two spoons: one for dipping, one for tasting. Alternatively, provide a dipper for each sample and place it beside the container on a large plate.
3. Numerical scoring is not necessary; score by quality class as excellent, superior, good, fair, or poor.
4. Provide for criticisms to be collected to provide insight into how to correct observed deficiencies. Tabulate and summarize the data, making results available to persons responsible for the products.
5. Follow up on the defective products to ascertain whether planned improvements have been made.

Suggested rules for evaluating frozen desserts are as follows:

1. Be mentally alert and physically well.
2. Know the score card, scoring guide, and "ideal" product.
3. Be able to recognize typical product defects and the range of their expected intensities.
4. Have samples properly tempered.
5. Secure representative samples and protect them to prevent changes in quality until examined.
6. Taste the sample, observing the texture and body characteristics.
7. Taste again, observing the sequence of flavors.
8. Practice introspection—the shutting out of all distractions.
9. Be honest with yourself, making the sample tell you what it is like instead of you assigning to the sample some defect you expect to find.
10. Avoid making facial expressions or being influenced by the expressions of others who may be present.
11. Rinse your mouth occasionally with water or cleanse it with an excellent sample.
12. Never eat strongly flavored foods just prior to judging.
13. To become an expert, practice, practice, practice!

REFERENCES

Arbuckle, W. S. 1940. A microscopical and statistical analysis of texture and structure of ice cream as affected by composition, physical properties, and processing methods. Missouri AES Res. Bull. 320.

Bodyfelt, F. W., J. Tobias, and G. M. Trout. 1988. *The Sensory Evaluation of Dairy Products*. Van Nostrand Reinhold, New York.

16
Cleaning, Sanitizing and Microbiological Quality

Consumer confidence in the quality of frozen desserts is one of the greatest assets of an ice cream-producing firm and of the industry in general. One of the world's best professors of dairy foods science, Professor W. H. E. Reid,[1] used to teach his students that **"the most important asset of a milk and milk products firm is the goodwill and confidence of the consuming public."**

It is not possible to provide high-quality frozen desserts for consumers unless soil and microorganisms are removed from the surfaces that touch the ingredients and the products that ultimately go into the consumer package. Therefore, it is important to discuss the topics of water suitability, detergency, sanitization, cleaning systems, and microbiology of frozen desserts.

Factors that determine the ultimate quality of ice cream include (1) clean, healthy employees who are well trained and have a conscientious attitude, (2) ingredients of unquestionably high quality, (3) desirable formulation and composition of mixes, (4) proper processing methods, (5) excellent equipment, (6) **immaculate cleanliness and effective sanitary procedures**, and (7) proper storage and distribution methods.

Although the frozen status of ice cream and related products relieves the manufacturer of worry about microbial spoilage of the finished product, it does not eliminate the risk of transmission of pathogenic microorganisms or toxins via frozen desserts. During the past decade more than one ice cream manufacturer has been required to recall frozen desserts from distribution channels because items were found to contain viable cells of *Listeria monocytogenes*, a pathogenic bacterium that can cause severe human illness and deaths among susceptible persons, especially infants, pregnant women, and the elderly. Another situation arose in which an ice cream manufacturer with a multi-state

---

[1]Reid was Professor of Dairy Manufacturers at the University of Missouri where he was the advisor and teacher of both W. S. Arbuckle and R. T. Marshall, authors of this book.

distribution system had to recall a large quantity of ice cream because it was found to contain a serotype of *Salmonella*.

The entire food industry has begun to follow the recommendations of the U.S. Food and Drug Administration (FDA) in doing hazard analysis followed by carefully planned control of critical points in the process. This is called the Hazard Analysis Critical Control Point (HACCP) approach to ensuring the safety of finished processed foods. The procedure calls for the processes to be completely documented in flow diagrams. Potential hazards are identified; for example, points are examined at which the lethal pasteurization process could fail or microorganisms could enter the product after the lethal process is completed. Once hazards are identified, control measures are decided. The method of documentation by which these controls have been exercised on each lot of product is then described. Follow-up to ensure continued use of the HACCP plan is vital.

Microbiological quality of frozen desserts is commonly monitored periodically by performance of the Standard Plate Count (SPC) and the coliform count on finished products. Standards for the coliform count are almost uniformly 10/g of pasteurized ingredients, such as concentrated milk, and of finished product. There are no federal standards for the SPC/g, but some health authorities set a maximum of 50,000/g.

Manufacturers should have no problem in meeting either of the recommended limits. In fact, there is strong reason to demand that the processing operation produce frozen desserts with *no* detectable coliform bacteria in 1-g samples. This group of nonsporing bacteria is readily killed by pasteurization, and, if cleaning and sanitizing are done as well as is currently possible, numbers of coliform bacteria introduced after pasteurization should be practically nil. Numbers of total aerobic bacteria in today's fresh raw milk and cream are generally well below 100,000/ml. Furthermore, only a few hundred per milliliter of these bacteria normally exist in the spore form, and it is mostly the spore-forming bacteria that are able to survive pasteurization. This means that aerobic bacteria counts (SPC/g) of ice cream should be $< 500/g$ and that coliform counts should be $< 1/g$. Any time a viable coliform bacterium is found in ice cream, the chance exists that pathogenic bacteria have entered with the coliform. Although the risk of disease transmission is extremely low when very small numbers of microorganisms enter the product after pasteurization, it is important that the processor realize that the risk does exist. Just passing the regulatory limit does not eliminate the producer's risk.

The greatest potential source of contaminating microorganisms in packaged ice cream is unsanitary equipment. The source, however, that may provide the greatest threat to safety is the careless and untrained employee. Managers must make sure that each employee dresses in a clean uniform, wears hair restraints, washes and sanitizes hands, disinfects footwear on entry into the process area, and refrains from touching any product contact surface without properly sanitizing the hands or the gloves being worn on the hands. Proper sanitary practices are essential to the ice cream plant. No one should be allowed to enter the processing environment without knowledge of the required sanitary procedures and without the required dress and personal hygiene measures taken.

CLEANING EQUIPMENT

Water quality is an important factor in the cleaning of equipment. The salts of calcium and magnesium that dissolve into water as it trickles through the soil are carried to the ice cream plant in the form of *water hardness*. To perform satisfactorily, hard water must be softened either in a water softener or with detergents. Hardness in water is expressed as the amount of calcium and magnesium salts present in milligrams per liter (the same as parts per million—ppm). Soft water contains less than 60 mg/L, medium hard water 60–120 mg/L, hard water 120–180 mg/L, and very hard water >180 mg/L. *Temporary hardness* (the bicarbonate fraction) is precipitated by most alkaline materials and heat. *Permanent hardness* (mostly the chlorides and sulfates) is not precipitated by heat. These constituents precipitate on addition of alkali or application of heat, clog boilers and leave deposits on equipment, especially where heat is applied. The films they leave on product contact surfaces can adsorb milk proteins, fat, and minerals, forming *milkstone*. Therefore, it is imperative that water for cleaning be softened. Chemical softening with detergents can be economical if hardness is low, but water softeners are recommended when hardness is high.

Soils deposited on dairy equipment differ with the location in the system. Heated surfaces tend to have the hard-to-remove films that require very hot water and strong detergents for removal. Air in the mix and high acidity increase this tendency. Areas where protein can precipitate or fat can form a film usually require alkaline cleaners; whereas films that contain lots of minerals need to be exposed to acid type detergents.

Practices that minimize soil deposition and maximize ease of removal of surface films include: (1) use of minimum time and temperature of heating, (2) cooling of product heating surfaces before and during the emptying of processing vats, (3) rinsing of films from equipment before they dry, (4) keeping soil moist until cleaning is begun, and (5) rinsing with warm—not hot—water. The ease or difficulty of removal of soil is determined by the nature of the surface on which it exists. Pores in rubber parts, crevices in insufficiently polished surfaces, and holes in corroded surfaces protect soil and microorganisms from the actions of detergents and sanitizers.

FUNCTIONS OF DETERGENTS IN THE DAIRY

The two types of detergents used to clean equipment in frozen dessert plants are alkaline and acid. However, there are several ingredients that can be selected to formulate these detergents. The five classes of these ingredients are alkalis, phosphates, wetting agents (surface active agents), chelating agents, and acids. The following are the major functions performed by formulated cleaners:

1. *Emulsifying:* causing fats and oils to be suspended in the cleaning medium
2. *Dispersing:* Breaking clumps of soil into small individual particles
3. *Dissolving:* Making both organic and inorganic soil constituents soluble
4. *Peptizing:* Hydrolyzing proteins, thus increasing their solubility
5. *Rinsing:* Assisting rinse water to remove the solution and its suspended components from the surface
6. *Saponifying:* Combining the alkaline cleaner with residual fat to form a removable soap
7. *Sequestering:* Removing or inactivating hard water components by forming a soluble complex (among substances that perform this function, the inorganic ones, e.g., polyphosphates, are called sequestrants, whereas the organic ones, e.g., etheylene-diaminetetraacetic acid, are called chelators)
8. *Wetting:* Lowering of the surface tension of a cleaning solution to increase its penetrating power.

Harper (1972) wrote that a good detergent causes the solution to penetrate the soil to be removed by strong surface action; displaces soils from the surfaces by saponifying fats, peptizing proteins, and dissolving minerals; disperses soil components by deflocculation and/or emulsification; and prevents redeposition of the dispersed soil by providing good rinsing properties.

MAJOR DETERGENT COMPONENTS AND THEIR FUNCTIONS

The major detergent components are described as follows:

Alkalis. The strongest of the alkaline detergent ingredients is sodium hydroxide (NaOH). It saponifies fat excellently but performs few other cleaner functions. Solutions of NaOH are often used to clean heavy deposits of protein and fat from surfaces of heat exchangers. A moderately strong alkaline detergent that inhibits corrosion is sodium metasilicate ($Na_2SiO_3 \cdot 5H_2O$). It emulsifies and deflocculates well. When used with water-softening chemicals, sodium carbonate (Na_2CO_3) is a good alkaline detergent ingredient for washing equipment by hand. When used in combination, baking soda ($NaHCO_3$) and sodium carbonate form sodium sesquicarbonate, an effective alkaline detergent. The effectiveness of alkaline detergents against proteinaceous films is improved as much as 40% by the addition of hypochlorite.

Phosphates. The general utility cleaner that softens water by precipitating hardness ions is trisodium phosphate ($Na_3PO_4 \cdot 12H_2O$). The polyphosphates, on the other hand, soften water without precipitating hardness components. These detergent ingredients include tetrasodium pyrophosphate ($Na_4P_2O_7$), sodium tripolyphosphate ($Na_5P_3O_{10}$), sodium hexametaphosphate ($NaPO_3)_6$, and sodium tetraphosphate ($Na_6P_4O_3$). Having the appearance of glass in the dry state, the latter two are called glassy phosphates. They rinse freely from surfaces, and emulsify, peptize, deflocculate, and suspend. Sodium tetraphosphate is the most effective of the polyphosphates.

Surfactants. Alkyl aryl sulfonates are the anionic wetting agents commonly used in dairy cleaners. Some of the best emulsifying agents are the nonionic wetting agents, and, since they do not ionize, they work well combined with either anionic or cationic wetting agents.

Chelating agents. Both salts of organic acids, such as sodium citrate, and ethylenediaminetetraacetic acid are effective chelators of metal ions in detergent solutions. They improve wetting of soils and penetration of water into the soil. They tend to adhere to particles of soil, suspending or emulsifying them.

Acids. The most commonly used acid for cleaning dairy equipment is phosphoric, and it is usually combined with a wetting agent. Although it can corrode metals, it is relatively inexpensive. More corrosive is hydrochloric acid, but an inhibited form of this inexpensive acid is sometimes used to remove heavy mineral scale. Nitric acid is less corrosive than hydrochloric acid because it produces protective oxide films on stainless steel. Gluconic acid, an organic type, is far less corrosive than the mineral acids and is therefore favored for many applications. The best practice is to prevent the buildup of significant amounts of mineral deposits by using soft water and detergents that suspend minerals in the alkaline cleaner. To further decrease the possibility of needing a strong and corrosive acid detergent, it is desirable to use acidified water for the final rinse of equipment.

PRINCIPLES OF CLEANING

The first principle to remember in regard to cleaning is that it is not possible to kill residual microorganisms with chemical sanitizers if they are protected in films. Planktonic (suspended) bacteria are much more easily inactivated by dairy sanitizers than are sessile (adsorbed) ones. Even more protected are bacteria residing in cracks and crevices. This is why stainless steel surfaces of dairy equipment are polished to a fine number 4 finish (compare Figure 16.1a, b). Even this finish provides plenty of space for soil and bacteria (Figure 16.1c) to find protection from the lethal effects of sanitizers if they are not dislodged and carried away or at least suspended by the detergent solution.

No single procedure or set of conditions will work equally well in cleaning all of the equipment in a frozen desserts operation. Already it has been stressed that water quality, soil type, and type of detergent are important variables. It is now important to show that **temperature, physical action**, and **time** affect cleaning markedly. If a cleaner is suited to the soil and the water, and if it is of sufficient concentration, then these three physical factors become relevant variables. Decreasing the magnitude of any one of them requires that at least one of the others be increased in magnitude. Higher temperatures promote faster rates of reaction and decrease viscosities. Decreasing viscosity increases turbulence, a very important force. Within the temperature range of 32–85°C (90–185°F), an increase of 10°C (18°F) will produce approximately double the cleaning rate. Milkfat is partially solid at temperatures below 33°C; therefore, cleaning solutions should be at least this warm.

Physical force is necessary to dislodge firmly adsorbed films. Brushing does this well. In circulation cleaning, commonly known as CIP (cleaning-in-place),

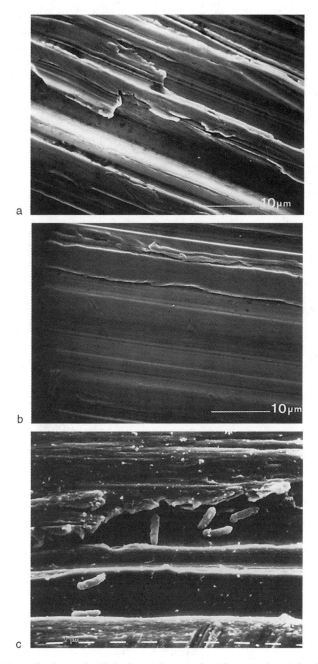

Figure 16.1. Surfaces of polished stainless steel with a number 3 finish (a), with a number 4 finish (b), and with *Pseudomonas* bacteria resting within crevices of an inadequately cleaned surface of 304 stainless steel with a number 4 finish (c). (Courtesy E. A. Zottola, University of Minnesota.)

turbulence of circulating solutions substitutes for the brushing. Turbulence is affected by the rate of flow and the nature of surfaces and is expressed as the *Reynolds number*. The higher the Reynolds number, the greater the turbulence. In pipeline systems a desirable value for circulation cleaning is 30,000, even though turbulent flow begins at about 3,000. Falling films of cleaning solutions on walls of storage tanks are turbulent when the Reynolds number is 200 or higher. The International Dairy Federation has sanctioned the empirical rule that cleaning solutions need to flow at a rate of at least 1.5 m/sec (5 ft/sec) to adequately clean in a CIP system.

Rinsing removes over 90% of soil on most ice cream equipment (Figure 16.2). Furthermore, most of the remaining soil is removed within the first 3 min of circulation cleaning if cleaner and temperature are appropriate and the soil layer is not unusually heavy. However, the rate of removal of soil decreases logarithmically with time, making it necessary to continue circulation for 15–30 min to ensure that tenacious soil is removed.

Cleaning in place of large surfaces such as walls, ceilings, and floors of tanks is done with spraying devices (Figure 16.3) located strategically inside the tanks or placed in them after they have been emptied of their ingredients or products (Figures 16.4 and 16.5). However, in cleaning vessels and tanks there must be a balanced flow rate at the proper pressure and rate of removal. Flooding of the bottom of a vessel may impede cleaning because of low turbu-

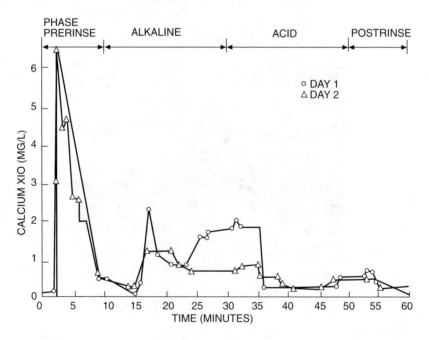

Figure 16.2. Rates of removal of milk soil from surfaces of stainless steel determined by quantities of calcium removed during cleaning in place. [Courtesy Missouri AES and Campbell and Marshall (1975).]

16 CLEANING, SANITIZING AND MICROBIOLOGICAL QUALITY

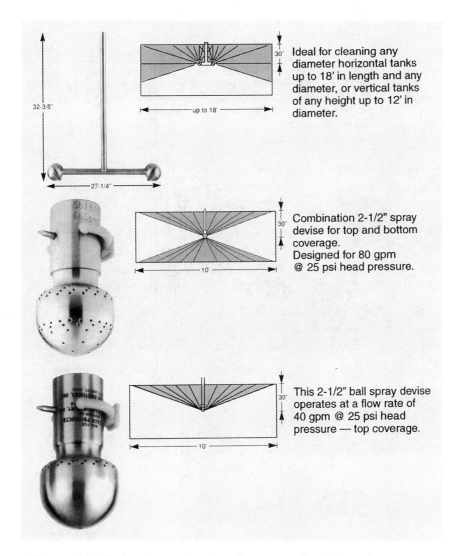

Figure 16.3. Devices for spraying rinse, detergent, and sanitizing solutions on surfaces of equipment. (Courtesy Klenzade Engineering, Beloit, WI.)

lence. Flow control in both lines and tanks is now done with much greater precision because meters and adjustable rate pumps can be tied to programmable logic controllers. Concentrations of detergents can be monitored with conductivity probes and the data used to activate feed pumps when concentrations need to be strengthened.

Cleaning in place permits use of strong cleaners and high temperatures. It virtually eliminates human error in the cleaning operations, especially if sys-

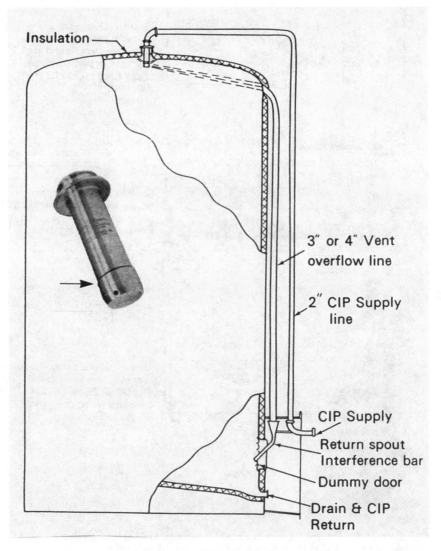

Figure 16.4. Cutaway drawing of a silo storage tank with inset showing a typical permanently installed spray device. Cleaning solutions are distributed to the top of the tank via the annular orifice (arrow). The tank vent line is cleaned by a solution sprayed from the nozzle, in this application through a hole drilled at a 7-degree angle near the bottom. The spray device and the spud through which it is inserted are cleaned by solutions delivered via small holes at the top. (Courtesy Klenzade Engineering, Beloit, WI.)

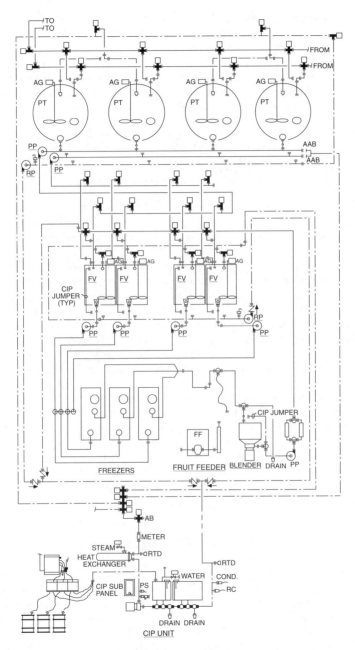

Figure 16.5. Line drawing of a system for CIP cleaning of a medium-size ice cream plant showing CIP makeup and heating systems (bottom), circuits for circulation of solutions with valves and pumps, and equipment to be cleaned: pasteurized mix tanks (PT), flavor vats (FV), freezers, fruit feeder, and blender. (Courtesy Klenzade Engineering, Beloit, WI.)

tems are used that automatically prepare the cleaning solutions and control their concentration, temperature, and time of application. This releases workers for other important work. Importantly, the processes can be monitored effectively and records kept electronically. Computer-based CIP monitoring systems can track on a time basis the critical variables of return temperature, conductivity of the returned solution, flow rate, and more. In monitoring by one major supplier (Floh 1993), the following discoveries were made:

- Each circuit has a distinctive fingerprint when it washes to a preset program; therefore, if any discrepancy is detected, troubleshooting should start.
- The fingerprinting facilitates determination of the quality of cleaning and sanitizing operations with increased certainty.
- Significant savings can be achieved by using the data to optimize cleaning sequences and timing.

Cleaning-out-of-place (COP) is the process of placing components of equipment in a tank in which detergent solution is rapidly circulated to clean the parts.

Postwashing rinsing of equipment is done to remove traces of detergent and suspended soil from surfaces. To minimize mineral deposits, it is recommended to acidify the water to about pH 5. Because chlorine corrodes stainless steel and hypochlorite is unstable in acid conditions, acidified rinse water should be used after washing with chlorinated detergents.

Surfaces should be drained dry after cleaning. Microorganisms that may have escaped removal or death during cleaning cannot grow in the absence of moisture even if organic matter remains on the surface to provide nutrients for growth.

In summary, the four steps that should be taken after use of dairy equipment are: (1) rinse with warm water; (2) wash with a detergent of proven effectiveness, dissolving it in appropriately conditioned water at a temperature commensurate with the soil, equipment, and method of washing (by hand, CIP, or COP); (3) rinse with hot water to promote drying (except that hot water may not be used in some refrigerated equipment); and (4) sanitize immediately before use. Frequent turnover of air in processing rooms is important to promote drying of surfaces.

SANITIZATION OF EQUIPMENT

Sanitization is the process of treating equipment surfaces with chemical or physical agents that will destroy all of the pathogenic microorganisms and most of the spoilage bacteria. Sterilization destroys all microorganisms. Sterilization is seldom necessary in frozen dessert operations, because microbial growth is not possible in the frozen products. However, bacteria may not all be removed during cleaning, postrinse solutions can carry microorganisms to equipment, and some surfaces may be subject to contamination during periods of disuse. Therefore, it is imperative that all product-contact surfaces be treated with sanitizer just prior to use. The commonly used chemical sanitizers include the chlorine-bearing compounds, iodophors, acids, and quaternary ammonium compounds.

Sodium hypochlorite is an effective and inexpensive sanitizing agent. It is readily soluble, unaffected by hard water, and highly efficacious, killing bacterial spores readily. Guidelines for applying sodium hypochlorite found in the Grade A Pasteurized Milk Ordinance call for exposure to at least 50 mg/L (ppm) of the sanitizer for at least 1 min at 24°C (75°F).

Iodine has excellent antimicrobial properties but is sparingly soluble in water, tends to stain, and is volatile at moderate temperatures. Iodophors (literally "iodine carrier" in the Greek) have been developed to overcome most of these deficiencies. An iodophor contains iodine, a nonionic wetting agent, and buffered phosphoric acid. The wetting agent stabilizes the iodine and increases its ability to permeate microbial cells. The acid provides the low pH necessary for quick bactericidal action of the iodine. The color of dilute solutions of iodophor is proportional to the concentration; solutions with a straw color are effective in most applications. Use concentrations range from 12.5–100 mg/L. The activity of iodophors increases with concentration, temperature, and acidity, but instability and corrosiveness also increase. Temperature and pH should be kept below 49°C (120°F) and 5.0, respectively, to ensure that iodine is available to kill microorganisms and that sublimation of iodine does not occur. Organic matter reduces the activity of iodophors but not as much as it affects hypochlorites. Iodophors work well in hard water.

Mixtures of organic acids such as sulfonic acid with anionic or nonionic wetting agents and phosphoric acid provide detergency, solubilization of minerals, and bactericidal action. The lethal effect on microorganisms is provided by the pH of about 2 and the activity of the wetting agent at this low pH.

Quaternary ammonium compounds ("quats") are cationic wetting agents that slowly destroy microorganisms. Another disadvantage is that they are inactivated by hardness components of water and by anionic detergents. However, they are noncorrosive, nonirritating, stable to heat, relatively stable in the presence of organic matter, and active over a wide range of pH. For the latter reasons there are applications in which quats may be the preferred sanitizer. In such cases they should be used at concentrations of at least 200 mg/L at pH 5 or above and for exposure times of at least 30 sec.

Sanitization with heat, a common practice in the fluid milk industry, is not used as much in frozen dessert plants. A major reason is that large changes in temperature can damage freezers and other equipment operated at below-freezing temperatures. In systems containing large numbers of valves, pumps, and gasketed lines, heat from hot water penetrates well into minute areas protected from chemical sanitizers. Where such systems exist in ice cream plants, it may well be desirable to use hot water for sanitization. Exposure to hot water should be for at least 5 min at an outlet temperature of 77°C (170°F). Small pieces of equipment may be sanitized in cabinets with steam or hot air. Exposure to steam at 77°C for 15 min or at 93°C for 5 min is minimal.

SANITARY ENVIRONMENT

Sanitary surroundings and clean personnel are a must if equipment is to be kept in hygienic condition without fail. All rooms, especially toilets and locker rooms, must be kept as clean and sanitary as the area immediately surrounding packaging equipment. The following factors essential in hygienic construction are similar for utensils, equipment, work rooms, buildings, and surroundings:

1. Surfaces, especially those that contact products, must be smooth and free from scratches and grooves. Floors are the only exception. They should be slightly rough to prevent slipping.
2. Surfaces should be sloped to drain and free from depressions.
3. Corners should be rounded with a radius large enough to permit scrubbing with a brush.
4. All surfaces not cleaned in place must be easily accessible or easily disassembled for cleaning. Equipment should have enough clearance above the floor to permit cleaning daily or should be completely sealed to the floor.
5. Materials used in construction must be impervious to moisture and free from odors.
6. Lighting must be adequate for the operation that takes place in the specific area. Locations where the highest amounts of light are required include those where data must be taken or recorded.
7. Ventilation with fresh clean air is essential.
8. Rodents and insects must be kept out of the facility and places for them to be harbored must be eliminated. For example, all dry storage areas should have the shelving set away from the walls adequately for observation and cleaning to take place.
9. Operations must be segregated to minimize chances of pathogenic microorganisms being carried from raw materials to finished products. Persons handling raw milk or cream must not be allowed access to rooms where pasteurized products are exposed unless those persons have first changed their clothes completely and have disinfected themselves.
10. Room air pressures should be maintained successively higher from the mix room to the processing room to the freezing operation to the packaging operation. Thus, the flow of air will be away from the most critical area, the packaging operation.
11. The supply of hot and cold water must be unrestricted, and facilities for disposal of both liquid and solid wastes must be adequate.

Maintaining positive pressures in processing and packaging rooms can be very expensive. One manufacturer of pressurizing systems estimated that the cost for heating and cooling air for one well-controlled 10,000-ft^2 processing room in the midwest United States ranged from \$10,000 to \$14,000 in 1994 (Mans 1994). Specialists in engineering of such equipment may also need to periodically check the air flow and balance in the system.

HYGIENIC PERSONNEL

Hygienic personnel are an absolute must for the frozen dessert manufacturer. No worker should be allowed to perform tasks in the plant who has not been

16 CLEANING, SANITIZING AND MICROBIOLOGICAL QUALITY

adequately taught the necessities of personal hygiene and approved practices within the plant. The importance of these factors can best be conveyed to all employees when top managers regularly demonstrate their endorsement of the firm's adopted practices.

Freedom from chronic contagious diseases should be confirmed yearly by medical examinations. Hygienic practices that should be practiced in the plant include the following:

1. All workers must be clean, including fingernails. Hands must be washed before commencing work and whenever there is a chance that they have been contaminated.
2. Hair restraints must be worn, including coverings for beards.
3. Clean clothes and shoes must be worn, and the shoes should be disinfected in a foot bath any time a person enters the processing or packaging area.
4. Jewelry should not be worn in the processing or packaging environment, and all pens, pencils, and other objects that might fall into a vat or other container of ingredients or products should be placed at or below waist level and in a secure holder.

TESTS OF THE FINISHED PRODUCT

Whether the finished product is soft-serve, packaged hard-serve ice cream, novelties, or mixes to be sold to other manufacturers, it is vital to the selling firm that no product leave the premises that has not met the specifications of the manufacturer. These specifications should be established when the product line is developed. The tests may be dictated by regulations, by quality considerations, or by economic necessity. Specifications are set based on achievable and detrimental levels of important attributes or potential deficiencies in a product. We have already considered numbers of aerobic and coliform bacteria in frozen desserts. Each batch of product should be tested for these microorganisms. In those plants that experience very low counts of coliform bacteria it may be desirable to use a more sensitive test for post-pasteurization contamination. By the addition of 1% glucose to violet red bile lactose agar all of the *Enterobacteriaceae* are able to grow and ferment a carbohydrate. This increases the sensitivity of the coliform type test to include all of the nonsporing, facultatively anaerobic, gram negative rods that are killed by pasteurization. Excellent sanitary conditions are indicated when a very high percentage of samples tested at a 1:1 dilution show no colonies on plates of this medium. This is good evidence that pathogens, such as *Listeria monocytogenes* and *Salmonella,* have not entered the product. Use of this test for enterobacteria may be justified for another reason as well. False coliform tests can appear when sufficient glucose or sucrose are plated with the sample for nonlactose fermenters to produce enough acid to form deep red colonies in the agar. This sometimes happens with mixes high in these sugars when large amounts of sample are plated.

Ingredients that are added to frozen desserts after pasteurization deserve special attention by the microbiological laboratory. Suppliers are likely to specify maximal counts of certain microorganisms in their products. It is advisable to verify that the specifications are being met by random tests. Any firm that uses frozen fruits should be particularly careful to test them for enterobacteria.

Concentrations of the major characterizing ingredients of frozen desserts need to be monitored regularly, especially those on which nutrition labels are based. Of course, tests of fat and total solids should be run on each mix before freezing or direct sale in the liquid form. It is also prudent to check these components again on a statistically valid number of samples of frozen product. Fat content should vary from the target by no more than 0.2% and total solids by no more than 0.5%. Weight control should be within ± 30 g (1 oz) for 0.5-gal packages and within ±90 g (3 oz) for 3-gal containers.

In Chapter 15 it was noted that each type of product should be checked on a regular schedule for sensory qualities. This does not mean that every lot should be checked. In general, as the size of the lots decreases and the number of lots of a given product increases, the need to test every lot for sensory qualities decreases.

Finally, rotation of the product from storage on a first-in, first-out basis is an absolute necessity if product quality is to be high. Inventories should always be kept as low as practicable considering expected demand. It is also practical to consider economic factors such as availability of equipment and labor as well as cost savings available with purchases of supplies in large lots. However, quality should be the over-riding consideration. A policy should be set regarding maximum time of storage of each product, and the responsibility for avoiding longer-term storage should be assigned. Managers must always consider the risks involved in under- and overproduction. If these risks and their potential economic consequences, both good and bad, are faced up front, managers will be more likely to bring smiles to the faces of the Board of Directors than if consideration is not given to the risks during the planning for purchasing, production, and marketing.

SUMMARY

Having for sale the most appealing ice cream or other frozen dessert in terms of appearance, flavor, texture, package, nutrition, and cost is an asset—but only to the manufacturer who has a high degree of assurance that the product is safe to eat and will not be involved in a damage suit by a consumer. Practices in the dairy that determine product safety involve the microbial quality of the ingredients, adequacy of the lethal process(es), especially pasteurization, and protection from recontamination of the pasteurized products by pathogens. It is the latter topic that has had our attention in this chapter.

The emphasis has been on cleaning, sanitization, sanitary and hygienic practices, and cleanliness of the environment. The ice cream manufacturer cannot depend on the freezing process to be lethal to pathogenic microorganisms. Frozen desserts are relatively safe products, but chances for contamination after pasteurization are not insignificant. A program of total quality management is essential for every frozen desserts manufacturer. Although the manufacturer cannot test a product safe, nor can tests of finished product change the characteristics of that product, it is necessary that certain testing be done. The data from those tests serve to protect the firm against claims for variance from the expected and required qualities if those data show the product to be within specifications. They can serve as references to guide in planning future production operations. Some firms err greatly by spending money for tests then not using the data for planning to improve efficiency, quality, safety, economy, and, ultimately, **profit**.

REFERENCES

Campbell, J. R., and R. T. Marshall. 1975. *The Science of Providing Milk for Man.* McGraw-Hill Book Co., New York. Chapter 25.

Floh, R. 1993. Automatic recirculation cleaning in the 90's. *Dairy, Food and Environmental Sanitation* 13:216–219.

Harper, W. J. 1972. Sanitation in dairy food plants. In R. K. Guthrie (ed.) *Food Sanitation,* AVI Publ. Co., Westport, CT.

Mans, J. 1994. Over-pressurized production areas can cost a bundle. *Dairy Foods* June, p. 57.

17
Refrigeration

Refrigeration is the removal of heat from a substance and, therefore, is a process of exchanging heat, i.e., heat transfer. The excess heat in the substance being cooled (refrigerated) is transferred to a cooler substance, which becomes heated. Therefore, "refrigeration" is the reverse of "heating," both occurring simultaneously and being dependent on the same principles and factors of heat exchange. However, the ice cream industry uses the term "refrigeration" to mean cooling to temperatures between 4.4°C and –34°C (40°F and –30°F).

METHODS OF REFRIGERATION

The Ice and Salt Method

Natural ice, in very early times, was harvested from ponds and stored to be used for refrigeration. Since ice melts at 0°C (32°F), lower temperatures could be obtained only by mixing ice and salt. This method, now called "the ice and salt" method, is still used when only a small amount of refrigeration is required. Under usual conditions it is not practical to obtain temperatures below –15°C (0°F) by this method, although theoretically –21°C (-6°F) is possible when using 1 lb of salt to 3 lb of ice.

The lowest temperature obtained by this method depends upon the proportion of ice and salt, the rate at which heat is supplied, the density of the brine, the original temperature of ice and salt before mixing, and the size of both the ice and salt particles. Smaller lumps of ice and more concentrated brine favor rapid cooling and are used when the rate of heat transfer is high.

Mechanical Refrigeration

Mechanical refrigeration is now almost universally used and has the advantages of requiring less labor, being less cumbersome, yielding lower temperatures, and providing more uniform temperature control than the ice and salt method. It is also faster, cleaner, and drier. It is based on the principle that a liquid absorbs heat when it vaporizes, as in the case of water changing into steam; the vapor can be collected, cooled to a liquid state, and used again. The particular liquid, called the refrigerant, to be used in a mechanical refrigeration system depends on many factors, the more important of which are (1) the boiling point of the liquid, (2) pressure characteristics (the pressures under which it can be used), (3) the latent heat of vaporization (amount of heat absorbed when the refrigerant vaporizes), (4) the ease with which a leak can be detected, (5) its corrosive action on metals used in the system, and (6) its toxicity.

TYPES OF REFRIGERANTS COMMONLY USED

Of the various refrigerants, only two have been found sufficiently satisfactory to be widely used in the ice cream industry: ammonia, which has been used a long time, and certain Freons, nontoxic refrigerants used in ice cream cabinets, in small installations (home refrigerators, counter freezers, etc.), and where very low temperatures are not required.

The most important advantages of ammonia are (1) it absorbs a large amount of heat when vaporizing, (2) operating pressures are reasonably convenient, (3) leaks are easily detected, (4) its toxicity is not great in low concentrations though its odor is very pungent, and it has a very pronounced irritating effect on mucous membranes and wet skin, and (5) it usually operates at pressures above atmospheric, thereby keeping foreign gases and liquids out of the system.

Its main disadvantages are (1) it is very corrosive to brass and other copper alloys (since only iron and steel can be used in the system, the equipment is rather bulky), (2) the relatively large volume of gas requires large compressors, (3) mechanical automatic operation is difficult to obtain, (4) maintenance of temperature within a narrow range is difficult (the last two of these disadvantages are due to the large amount of heat absorbed during vaporization but are not significant in many installations), and (5) for some installations the objectionable odor from even minor leaks becomes very important.

Freons 12 and 502 are two of several chlorofluorocarbon (CFC) refrigerants that have been widely used for many years in small ice cream freezers and freezer cabinets. However, these Freons were banned by the 1987 Montreal Protocol, which was agreed to by the leading developed countries of the world. This agreement led to the Clean Air Act in the United States. This Act of the U.S. Congress set limits on the time that chlorofluorocarbon refrigerants can be used in the United States (see p. 195). It permits existing equipment employing CFCs to continue to run on them, but prescribes times when no new equipment will contain them and time limits on when no additional CFCs will be manufactured. The advantages of these refrigerants are (1) they require lower operating pressures than ammonia, (2) their odor is neither objectionable nor toxic, (3) they are not injurious to food materials, (4) they are not corrosive to most metals, although alloys containing magnesium must be avoided, (5) small and compact refrigeration systems can be used, because of lower pressures, (6) they are carriers for oil, thus facilitating the lubrication of compressors, (7) mechanical automatic operation is fairly simple, and (8) uniform temperatures can be maintained.

The main disadvantages of Freon refrigerants are (1) they must be free of moisture, (2) leaks are not easily detected (a special Halide torch that gives a pink flame in the presence of Freon can be used to detect leaks), (3) the system must be tighter than for ammonia to avoid leaks (small leaks are not noticeable), and (4) they are less satisfactory than ammonia for very low temperatures.

A new generation of refrigerants that contain no CFCs is being developed, and one of them, R402B, is on the market. A more promising one, R404A, is being tested. Neither of these refrigerants is suitable for replacement of R12 or R502 without considerable modification of the refrigeration system. Compared with the older CFCs, the new refrigerants cause the systems to run hotter, hence they require larger condensers and more airflow over the condenser to remove sufficient heat. Furthermore, R404A requires redesign of compressors and the use of polyester type oil. R404A is not compatible with the older oil. Because R404A and R402B are most efficient when used at freezing temperatures, another refrigerant, R134A, is recommended for use where temperatures are above freezing.

PRINCIPLES OF MECHANICAL REFRIGERATION

The mechanical refrigeration system has only three essential parts: (1) the compressor, (2) the high-pressure side and (3) the low-pressure side (see Figure 17.1). The compressor consists of one or more cylinders, usually surrounded by a water-jacket for cooling. The cylinders contain pistons similar to those of a gasoline engine, and they are operated by a crankshaft that runs in oil in the crankcase. The purpose of the compressor is to concentrate (compress) the vapor. It takes the vapor from the low-pressure side, in large volume at low pressure and low temperature, and discharges it into the high-pressure side. Thus, the compressor occupies a position dividing the two sides. The high-pressure side extends from the compressor to the expansion valve and includes the condenser and receiver. The hot vapor leaving the compressor passes through the condenser, i.e., coils of pipe cooled by water or air. This cooling in the condenser changes the vapor into a liquid at about room temperature and still under a high pressure, the liquid being collected in a tank (the receiver). This liquid refrigerant passes on to the expansion valve (or, in the case of some ice cream freezers, to the flooded cylinder around the freezing chamber), which is the other component dividing the high and low sides. The low-pressure side extends from the expansion valve to the compressor. The expansion valve is usually an ordinary needle valve permitting fine adjustment and may be operated manually. It derives its name from the fact that the liquid refrigerant passes through the valve and then expands into a vapor. The liquid refrigerant is under high pressure at room temperature before passing through the valve, and under low pressure at low temperature as it leaves the expansion valve to go through the expansion coils or chamber, i.e., coils of pipes or a freezing chamber located where refrigeration is to be produced, and leading back to the compressor. In this way the refrigerant is used repeatedly, being compressed, condensed, and expanded. The refrigerant never wears out, but slight leaks invariably occur, making it necessary to replenish the supply. The refrigeration or cooling is obtained as the liquid refrigerant absorbs heat while vaporizing. The pressure in the expansion coils or chamber determines the lowest temperature obtainable, and this pressure is often called the suction pressure or back

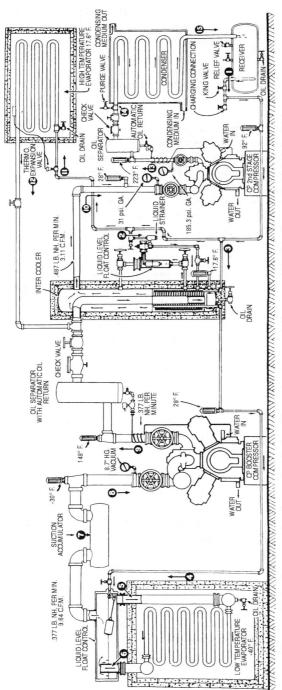

Figure 17.1. An example of a mechanical refrigeration system. (Courtesy APV Ice Cream, Lake Mills, WI.)

pressure of the system. The lower that pressure, the greater the cooling effect. The expansion coils or chamber may be located in the hardening room, in a tank of water or brine, in the ice cream freezer, etc., to give refrigeration in that particular place.

A large amount of heat is absorbed as the liquid changes to a vapor (latent heat), and a smaller amount of heat is absorbed by the vapor when it expands further (sensible heat). This absorbed heat is carried in the vapor to the compressor and on to the condenser, where it is transferred from the hot refrigerant vapor to the cooling water or air around the condenser coils. Sometimes this cooling water is used only once and wasted; in other places it is more economical to reuse this water. In these cases the water is pumped to the top of a cooling tower (usually on the roof) and allowed to trickle down over the tower, being cooled by partial evaporation in the process. This proves very economical where the cost of water is high. In some Freon systems the condenser is air-cooled, usually by a fan blowing air around condenser coils that have fins to facilitate radiation of the heat.

Although the principle involved in the mechanical refrigeration system is rather simple, the construction and installation are too complicated for a brief discussion. Usually it is more economical to obtain the services of a refrigeration engineer to supervise the planning and installation of the system than for the firm to attempt to design and install a system with its own personnel.

OPERATING PRECAUTIONS

Some precautions to observe in operating refrigeration systems are as follows:

1. When opening valves on refrigerant lines, open them slowly.
2. Keep the suction pressure as high as possible. However, pressure must be sufficiently low to give the desired temperature. The pressure on the low side should correspond to an ammonia boiling point 5.6°C (10°F) lower than the temperature of the medium surrounding the expansion coils or chamber for maximum efficiency. A lower back pressure than this reduces the refrigeration capacity. There is a temperature drop of about 5.6°C (10°F) between the medium being cooled and the ammonia due to the wall of the expansion coil. This is the same principle as in a milk cooler. The cooling medium must always be cooler than the temperature to which it is desired to cool the milk. Table 17.1 shows the boiling point of ammonia at different gauge pressures.

    To illustrate, if a minimum temperature of –23.3°C (–10°F) is wanted in the medium being cooled, the pressure on the low-pressure gauge should be that at which ammonia will boil at –28.9°C (–20°F) or 0.53 kg/cm^2 (3.45 lb/in^2). Table 17.1 shows that it is not necessary to carry a vacuum on the low side unless extremely low temperatures are desired. Head pressures (pressures on the high side) are determined by the temperatures of the refrigerant in the condenser. Because liquids transfer heat more efficiently than gasses, lower temperatures can be obtained in the cooled product if equally cold liquid refrigerant, rather than vapor, is in direct contact with the surface of the heat exchanger.

17 REFRIGERATION

Table 17.1. The Relation of Gauge Pressure to Boiling Point of Ammonia and Minimum Temperature Produced In Refrigerated Medium Where Vapors Are Recipients of the Heat

Gauge pressure		Boiling point of ammonia		Minimum temperature	
(lb/in^2)	(kg/cm^2)	(°C)	(°F)	(°C)	(°F)
1.17	0.18	−31.7	−25	−26.0	−15
3.45	0.53	−28.9	−20	−23.3	−10
5.99	0.75	−26.1	−15	−20.6	−5
8.77	1.36	−23.3	−10	−17.8	0
10.93	1.69	−20.6	−5	−15.0	5
15.37	2.38	−17.8	0	−12.2	10
19.17	2.97	−15.0	5	−9.4	15
23.55	3.65	−12.2	10	−6.7	20
27.93	4.33	−9.4	15	−3.9	25
32.95	5.11	−6.7	20	−1.1	30
38.43	5.96	−3.9	25	1.7	35
44.41	6.88	−1.1	30	4.4	40

3. Keep the head pressure as low as possible to save power. The head pressure depends on (a) size of the condenser, (b) temperature of water used for cooling the condenser, (c) amount or volume of water flowing through the condenser, (d) impurities in the refrigerant (mainly oil and air), and (e) cleanliness of the inside and outside of the condenser.
4. Avoid operating the compressor at zero suction pressure or under vacuum, because this favors oil passing by the piston rings and out of the compressor into the condenser.
5. Avoid frost on the compressor. The suction pipe at the compressor should carry frost only up to the compressor when operating most efficiently.
6. Inspect and drain all oil traps regularly. Some oil always passes along with the refrigerant and collects at low spots in the system. Valves at these points permit the oil to be drained out, thereby improving the efficiency of the heat transfer. Worn piston rings favor oil passing into the refrigerant.
7. Keep air out of the system. Air and some other gases do not condense into a liquid, but collect at high spots in the system. Valves at these points permit the removal of the air. Air decreases the efficiency of the condenser, causing excessive head pressures. Usually the air is removed at the high point of the condenser; this operation is called purging.
8. Keep the expansion coils as free from frost as possible. This is especially important in hardening rooms, because frost and ice reduce the rate of heat transfer.

DEFROSTING METHODS

The common methods of defrosting are as follows:

1. Brushing the coils with a stiff or wire bristled brush: This is not very effective as it may leave a thin layer of ice which gradually increases in thickness.

2. Scrubbing with hot water: A wet, messy, disagreeable operation leaving much moisture in the room and warming up the entire hardening room. However, it is nearly always used when the expansion coils are in a separate cabinet through which the hardening room air is circulated.
3. Passing hot liquid refrigerant through the coils: This requires extra valves and pipes in the installation, but does not raise the temperature of the room much. The frost and ice are easily removed from the quickly heated coils before the ice melts.
4. Passing hot refrigerant vapor or gas through the coils: This is similar to the use of hot liquid refrigerant.
5. Using a brine drip or spray over the coils: A trough containing calcium chloride crystals is placed above the coil so that as the crystals absorb moisture the brine drips down over the coil to collect in a vessel at the bottom.

METHODS OF COOLING

Brine

The brine method of cooling (sometimes called the brine system) represents the first application of mechanical refrigeration. The expansion coils of the mechanical refrigeration system are immersed in a large tank of brine (a calcium chloride solution) to cool the brine. Then the brine is pumped through pipes to the freezer (or other place where refrigeration is desired) and back to the brine tank to be cooled again. This method involves additional investment in brine tank, brine solution, pumps, pipes, etc. It is less efficient than a direct expansion system since the heat removed for refrigeration must be transferred to the brine and then to the mechanical refrigeration system. Other disadvantages are the corrosiveness of the brine, the difficulty in obtaining very low temperatures, and the more bulky installation. The most important advantages are that it permits storing up of refrigeration and can be used where ammonia leaks would be dangerous. Although it has been largely replaced in modern factories, it continues to find application in certain operations such as in making ice cream novelties and in making artificial ice.

The care of brine systems is important and may be summarized as follows:

1. Test the brine every month for concentration, alkalinity, and ammonia.
2. Keep the concentration of the brine high enough to give a freezing point at least 5.6°C (10°F) lower than the lowest temperature to which it will be cooled (see Table 17.1). Otherwise, brine will freeze onto the expansion coil, and this ice will act as insulation preventing the heat in the brine from penetrating the expansion coil.
3. Adjust the alkalinity by adding a solution of sodium hydroxide (caustic soda) or of lime until the brine is neutral to litmus or phenolphthalein. If the brine is acid to litmus, it will be too corrosive.
4. Use only one metal, preferably a pure grade of cast iron, in contact with the brine. The use of different metals in a system favors corrosive action.

5. Immerse a bar or strip of zinc in the brine to decrease the corrosion when different metals are used in a system.
6. Add a solution of sodium dichromate and caustic soda to reduce corrosion; however, this will cause irritation of the skin. Care must, therefore, be used in handling the dichromate, as well as the brine containing it. To make the solution, thoroughly dissolve, by stirring, a mixture of 5 lb commercial dichromate and 1.4 lb caustic soda in 1 gal of water. This amount will be sufficient to treat 375 gal of brine the first time. Once a year it will be necessary to add from one-fourth to one-half the original amount.
7. Avoid air coming in contact with the brine since air makes the brine acid and more corrosive. Keep the brine tank covered, and avoid bubbling air through the brine or spraying the brine.
8. Avoid ammonia leaks from the expansion coils, which cause the brine to become more alkaline. They can be detected by boiling a sample of brine in a narrow-necked flask and testing the vapors with red litmus paper. If the red litmus paper turns blue, the steam from the boiling brine contains ammonia.

Direct Expansion

The direct-expansion method of cooling has replaced the brine method in many installations, because it represents increased efficiency and a saving in investment. In this method the brine pipes in the freezer (or other place where refrigeration is desired) are replaced by the expansion coils of the mechanical refrigeration system. These expansion coils may or may not contain much liquid refrigerant in addition to the refrigerant vapor.

Flooded System

The flooded system or method of cooling is a special case of the direct expansion method in which the liquid refrigerant collects in and nearly fills the expansion coils or chamber. The compressor draws off the vapor as the absorbed heat vaporizes the liquids in the operation of this flooded system. The liquid refrigerant under high pressure and at room temperature passes through a valve (usually controlled by a float) to the expansion coils or chamber, where it is a liquid under a lower pressure and lower temperature. As heat is transferred to the liquid refrigerant, evaporation takes place. The vapor from this evaporative process is constantly removed by the compressor, and the liquid level is maintained by the float.

The important advantages of the flooded system are: (1) it is more efficient, because heat is more readily transferred between liquids than between vapors, gas, or liquid to gas; (2) less cooling surface or coil surface is needed; and (3) there is less fluctuation in temperature. The fact that float valves occasionally stick causing liquid refrigerant to enter and damage the compressor is the main disadvantage.

TERMS USED IN REFRIGERATION

1. *BTU:* British thermal unit, or the amount of heat required to raise the temperature of 1 lb of water 1°F at the temperature where water has its greatest density (about 39°F).
2. *Latent heat of evaporation:* The amount of heat (in BTU) required to change 1 lb of liquid into 1 lb of vapor without changing the temperature or pressure.
3. *Latent heat of fusion:* The amount of heat (in BTU) required to change 1 lb of liquid into a solid without changing the temperature or pressure.
4. *Ton of refrigeration:* 288,000 BTU per 24 hr, or the amount of heat required to melt 1 ton of ice per day at 32°F without changing the temperature or pressure. Other convenient ton of refrigeration equivalents are 12,000 BTU/hr and 200 BTU/min.
5. *Ton refrigeration machine*: A compressor or machine that will produce 1 ton of refrigeration during 24 hr of continuous operation under a particular set of conditions. (For example, 5°F suction vapor and 86°F condenser.)

Refrigeration compressors are frequently rated in tons of refrigeration at a certain suction pressure, and this numerical figure is approximately one-half the horsepower rating of the motor driving the compressor. However, many engineers find it more satisfactory to list the size of the compressor in terms of piston diameter and length of piston stroke, since the following factors affect the size of motor and capacity of a compressor:

1. The refrigerant used: Compressors are made for a particular refrigerant.
2. The diameter of the cylinder: This influences the volume of refrigerant vapor that can be handled at one stroke.
3. The length of the piston stroke, which also determines the volume of refrigerant vapor handled at one stroke.
4. The speed, which determines the volume of refrigerant vapor handled per minute, hour, or day.
5. The suction pressure or pressure on the low-pressure side, which determines the weight of refrigerant vapor per volume handled by the cylinders.
6. Volumetric efficiency of machine, i.e., the ratio between the piston displacement and the actual amount of gas delivered.

The first four of these factors are usually constant in a particular installation, but the capacity of the compressor decreases rapidly with a decrease in the suction pressure. Also it should be remembered that the pressure on the high-pressure side has almost no influence on the capacity of the compressor, although it does greatly affect the power consumption of the motor.

18
Sales Outlets

A CONE

A cone, if offered by itself,
might spend a lifetime on the shelf

But, it performs a happy task
Just what? I thought you'd never ask

It takes a product sweet and cold
From hand to mouth, the hand to hold

It makes that product now complete
A thoroughgoing ice cream treat

You eat it here or walk away
In either case, you've made your day

A cone's another claim to fame.
It gives ice cream a good last name.

What does one say about a cone,
Except it cannot stand alone?
     —Arthur Barton, Maryland Cup Corporation

In recent years the merchandising of ice cream has become highly competitive, and profit margins have been greatly reduced. Marketing methods have also changed extensively. Ice cream manufacturers have been required to adapt to these changes to be assured of dependable markets for their products. The supermarket has become a dominant factor in ice cream distribution to the consumer. In contrast to early days in the industry, little bulk ice cream is being sold in drug stores, and more is sold through special confectionery stores and institutional outlets. Several types of markets are available to the ice cream manufacturer: (1) wholesale to specialty confectionery stores, dairy stores, supermarkets, and food service establishments; (2) retail directly to customers through the manufacturer's special ice cream stores, dairy bars, or ice cream parlors; and (3) wholesale to other manufacturers by developing a specialized operation for novelties, special formulations, and mixes or by packaging store brands (private label).

WHOLESALING

If this type of marketing is chosen, the supermarket, convenience store, and dairy store offer good outlets for packaged products. The dairy bar, ice cream parlor, and food service institutions are the major outlets for bulk ice cream. Several major supermarket chains manufacture frozen desserts for their own stores, and some manufacturers establish dairy or confectionery stores to retail to their customers. The industry has become highly concentrated, with single manufacturing plants serving multi-state areas. Especially concentrated are ice cream novelty operations.

RETAILING

Retailing by the manufacturer has been facilitated in recent years by the availability of large shopping centers and by improved transportation and packaging along with the increased competition and reduced profit margins in some areas of wholesale operations. The dairy store, confectionery store, and dairy bar are important retail outlets.

MERCHANDISING

Baumer and Jacobson (1969) submitted an economic and marketing report on frozen desserts which included studies on demand elasticity, consumption patterns, competition among ice cream and related products, and competition from other desserts and snacks. They concluded that ice cream consumption responds readily to price changes. Demand analysis suggested that an industry average price increase of 10% would result in approximately an 8% consumption drop; a 10% increase in the price of private label ice cream might be expected to result in a 17% decline in sales. Thus, the demand for ice cream was seen to be highly elastic.

Consumption patterns for ice cream have remained about constant at about 15 qt per person from 1950 to 1988. At the same time ice milk sales grew from 1 qt per person in 1950 to 5 qt per person in the mid-1960s. Per capita consumption wavered between 5 and 6 qt per capita until the definition of ice cream was changed in 1995 to include the category formerly named ice milk. The product that grew sharply in consumption in the early 1990s was frozen yogurt. It appeared that the drop in ice cream consumption of about 2 qt per person was directly offset by consumption of frozen yogurt. During the 1980s mellorine production dropped so low that the U.S. Department of Agriculture ceased reporting production data. In 1993 the following products commanded the listed percentages of the frozen dessert market: ice cream, 57%; ice milk, 21%; frozen yogurt, 10%; water ices, 4%; sherbets, 3%; other products, 4%.

On a dollar basis, "at-home" sales of frozen desserts made up $5.3 billion while "away-from-home" sales totaled $4.9 billion in 1993. Restaurants and lunchrooms were in a virtual tie with fast-food restaurants with combined

18 SALES OUTLETS

sales of more than 94% of the frozen desserts eaten "away-from-home." Ice cream and frozen custard stands had sales of almost $3 million in 1994.

Profitability within the industry has been published by the International Ice Cream Association as reported by Dunn & Bradstreet (IICA 1994). The data represent the median for establishments included in the survey. *Return on sales* (net profit after taxes as a percent of annual net sales) increased from 2.7% in 1990 to 5.3% in 1993. *Return on assets* (net profit after taxes divided by total assets) rose from 5.8% to 7.0%. *Return on net worth* (net profit after taxes divided by net worth) rose from 11.1% to 14.6%.

DRIVE-IN STORE

The main requirement for the drive-in store (second only to the quality of the store itself, of course) is plenty of parking space for customers' cars. The site chosen should have a pleasing background and a good view, and the parking area should be kept neat and tidy. Since it has been estimated that about 75% of ice cream business is in repeat orders, obviously the drive-in store should be on the edge of a residential area, preferably at a shopping center that has excellent highway access.

The store itself should be attractive, well-lighted, well-arranged, and convenient for service. It should be kept clean and sanitary in every detail.

The manager of such a store (of any food store, in fact) should make it a first rule to become familiar with state and local community health regulations and insist that all employees and persons delivering food products to the establishment comply with such regulations.

THE SODA FOUNTAIN[1]

Soda fountain service is a profitable addition to many types of food outlets. However, this kind of business should be ventured into only if there is sufficient capital to do a good and attractive job of installation, and only if the management understands the soda fountain business and is sufficiently interested to become familiar with the many phases of this kind of merchandising.

If properly equipped and managed, and staffed with optimally trained personnel, the soda fountain can be a great help in selling ice cream. It is imperative that service be given in an accommodating, eye-appealing, and taste-tempting manner. The ice cream dispenser and the soda fountain itself should have a neat, well-cared-for appearance. Everything connected with the fountain and

---

[1]The soda fountain, which was known at one time as "ice cream parlour" is now frequently called "ice cream shop" in many parts of the country. Regardless of what it is called, the soda fountain is characterized by sit-down service of a wide variety of ice cream products.

Figure 18.1. The ice cream store includes a fountain (a), display cases (b), and plenty of comfortable room (b) for customers to make their choices and consume them. (Courtesy Braum's Ice Cream and Dairy Stores, Oklahoma City, OK.)

its environment must suggest wholesomeness, cleanliness, and sanitation (see Figure 18.1).

General Housekeeping

Both the inside and the outside of the fountain need regular and thorough cleaning. For sanitary reasons, the inside must be kept scrupulously clean. A spotless general appearance has definite customer appeal and generally pays off handsomely in additional profits.

Glass, Plastic, or Paper for Serving?

Glass, plastic, or paper cups and serving dishes may be used. Although many people prefer glassware, great care must be taken to prevent breakage and to keep dishes sanitary. With glass there is also the danger that chips of glass may fall into the open ice cream containers, and cleaning dirty dishes is always extra work. When disposable cups and dishes are used, good housekeeping is made easier, serving can be more sanitary, there is no danger of broken glass, and expensive delays for dishwashing are not experienced. Moreover, merchandising can be done most effectively on disposable containers.

PERSONNEL

Soda fountain success depends upon the selection and training of the right type of personnel for giving good service. This applies equally well to dairy bar or dairy store personnel. The men and women behind the counter should be schooled to follow a well-planned routine regarding the best form of greeting and approach to customers in taking the order, in serving the food, in making suggestions for increasing the order, in presenting the customer's bill or check, and in cleaning the counter before the next customer is served. In short, the good dispenser, or salesperson, will cultivate cheerfulness, meet customers with a smile, be dignified and quiet, give quick, accurate, and gracious service, and be dependable and honest.

Personal Hygiene

Persons handling and serving food should be trained to live up to definite and rigid standards of personal hygiene. The public expects and deserves to not be put at risk of contacting a disease from foods. The list of personal care items that managers should require of employees follows: hair—well groomed and restrained, especially if it is long; teeth—clean and recently brushed; hands and fingernails—clean and trim; uniforms—clean, attractive, and neat; and shoes—clean and polished. Soap, water, and single-service towels should be easily accessible.

Keep Hands Clean

"Hands," someone has said, "are the tools of the food worker," essential in many operations in filling orders. The following rules about cleanliness of the hands must be observed by all operators: Keep them clean. Always assume that hands are carrying dangerous microorganisms. Wash them before handling tableware and food. Wash them after each visit to the lavatory; after coughing, sneezing, or combing the hair; and after handling refuse or waste of any kind. Always be sure they are clean before they touch food.

TRAINING FOR FOUNTAIN SERVICE

Almost certainly the profit motive is the reason to own a soda fountain or ice cream store. To continue to build a profitable business, increasing numbers of customers are needed. To attract customers and have them come back for repeat orders, the ice cream must be of high quality. Excellent dairy ingredients and flavoring substances must be used in making it, and an equally excellent quality of syrups and toppings must be used to combine with the ice cream at the fountain. Generally, it is desirable to buy and use only the best and most palatable materials for toppings and syrups, However, when a premium is paid for ingredients, it is imperative that the manager be adept at using them most effectively. Based on this knowledge, the retailer is set to train workers properly. Properly trained workers are absolutely necessary if the proprietor is going see ice cream sales and profits continue to rise. If the staff of a store and fountain is to function well, training for the job is necessary for all employees. The work expected of them should be carefully explained and ably demonstrated to them.

Servers at the fountain, counter, or drive-through window must receive continuous training until they have thoroughly mastered the know-how of what and how to serve. They should be familiar with the best methods of preparing toppings and of blending fruits, nuts, and toppings with simple syrups. They should know how to dip ice cream most economically as well as the most attractive ways in which the variously colored and flavored ice creams should be served.

Employees as Sales Representatives

The men and women employed for fountain service are the principal contact retailers have with the buyers of their products. It is really a waste of money to advertise, build and equip a fine building, and manufacture or buy high-quality ice cream if the employees who represent the establishment create an unfavorable impression on the public. The lesson that should be taught the first day a new person is employed is the lesson of friendliness, courtesy, and service. The customer is KING! No one other than the customer will buy the product.

The Ice Cream Store Manager

Essential characteristics of the manager: honest and conscientious; able to gain the respect of the employees and lead them; efficient; original enough to be able to develop new sales ideas; and, finally, loyal to the proprietor or organization. The manager must understand the principles of food service and business management.

All that has been said about empolyees' training, ability, and qualifications has been said simply to emphasize that the employees really reflect the business and managerial ability of the proprietor. A good manager will make a business prosper. If a business is too large to handle by the owner(s) personally, much will depend on the store or fountain manager chosen to help run the business.

Managers who have thoroughly trained their representatives, i.e., their employees, for their respective jobs, have taken a vital step toward a profitable business.

How to Dip Ice Cream

While it may seem like an easy thing to do, it really takes practice and planning to dip ice cream properly. To lower the ice cream surface evenly in the container, to cut from the highest surface of the ice cream, and to keep the container stationary while dipping the ice cream require skill that comes only with practice. Experienced fountain operators have found that it is best to press the ball of ice cream gently on the side of the cone with the outside of the dipper bowl, pressing just enough to make it stick but not enough to spoil its shape or to break the cone. Employees should learn early at what temperature ice cream will dip or cut easily and still not be so soft as to spoil the texture and cause shrinkage. This feel for dipping comes only by practice (see Figure 18.2).

The dipped package usually refers to carry-out packages filled by the retailer, who dips the hardened ice cream from a bulk package in the cabinet and presses it into a new container (Figure 18.3).

It is common knowledge that the volume of ice cream after dipping is less than the volume before dipping. This decrease in volume is caused by air being compressed or expelled from the ice cream with a corresponding loss in overrun. The retailer sustains what is known as a dipping loss because it is not possible to sell the same volume of ice cream received from the manufacturer. Several factors may affect dipping losses, among which are overrun, dipping temperature, and composition. These factors were studied in detail by Bierman (1926a,b); his findings still offer valuable information.

The percentage dipping loss increases as the overrun of the ice cream increases (Table 18.1). Hand-packed quarts that had the most uniformity of weight were dipped from ice cream containing 80–100% overrun. The dipping temperature also influences the number of servings that can be dipped from a container of ice cream (Table 18.2). A dealer who recognizes and controls these variables will be in a better position to establish proper dispensing procedures.

It appears from the above results that if dipping losses are to be kept at a minimum, ice cream of average composition (15% sugar) should be dipped at 8°F or lower. Proper dipping temperature is also affected by the composition, especially the sugar content of the product. Dipping temperature should be varied inversely 1–1.5°F for each 1% difference in sugar content; to make dipping of ice cream of normal composition practical at 8°F or lower, the ice cream should contain approximately 90% overrun.

Composition has only a slight effect on dipping losses when compared to the effects of overrun and dipping temperature. Dipping losses in high-fat ice cream are slightly less than in average or lowfat ice cream. When the NMS content is high, dipping losses tend to be greater than when the product contains low

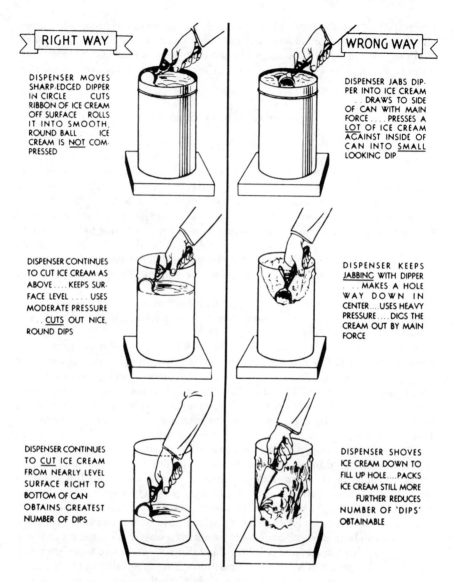

Figure 18.2. How to dip ice cream. (Courtesy of C. P. Gundlach & Co.)

NMS, provided dipping is done at one temperature. This is caused by the increased sugar (lactose) content introduced in the mix when the NMS content is increased. Ice cream with a high sugar content has a greater dipping loss than ice cream containing a lower percentage of sugar unless the dipping temperature is varied to give equal hardness to the ice cream; then the dipping loss will be approximately the same regardless of the sugar content.

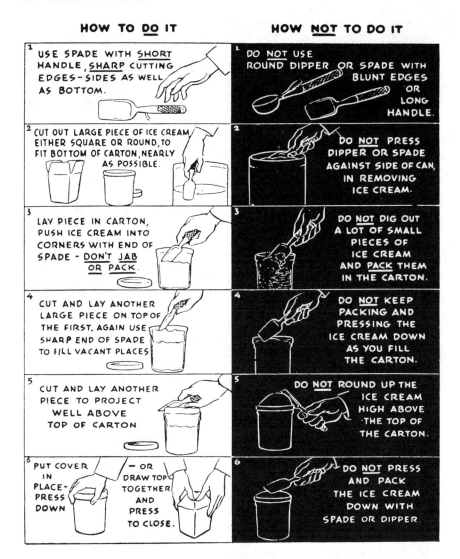

Figure 18.3. How to dip ice cream into carry-out packages. (Courtesy of C. P. Gundlach & Co.)

Dipping studies indicate that two factors influence the weight of ice cream dipped: (1) the resistance offered by the ice cream, which prevents the expulsion of the air; and (2) the amount of force applied to push the dipper into the ice cream.

Some customers like to purchase ice cream in the hand-packed form. In the smaller operation this permits sale of all of the bulk products without having

Table 18.1. How Overrun Affects Dipping Losses and Weight Per Quart[a]

Average overrun (%)	Dealer dips from 20-qt containers (qt)	Average dipping loss (%)	Weight per quart (oz)	
			Before dipping	After dipping
60.3	16.96	15.18	22.95	27.05
80.6	15.37	23.14	20.37	26.52
100.8	14.07	29.66	18.30	26.25
118.6	13.45	32.76	16.81	25.14
Average	14.96	25.18	19.61	26.24

[a]From Bierman (1926a).

Table 18.2. How Dipping Temperature Affects Dipping Losses and Weight Per Quart[a,b]

Dipping temperature (°F)	Dealer dips from 20-qt containers (qt)	Average dipping loss (%)	Weight per quart after dipping (oz)
3–8	15.33	23.35	25.72
9–16	14.65	26.75	26.82
17–20	15.86	20.70	24.46

[a]Compiled from Bierman (1926b).
[b]Average overrun, 90%; weight per quart before dipping, 19.34 oz.

them prepackaged. Thus, it cuts down on freezer space and number of different packages required. However, the advantages of hand packing ice cream are outweighed by the following disadvantages:

1. An increased chance for unintentional contamination during the dispensing operation
2. A wide variance in the amount of ice cream placed in containers of equal size unless each package is filled by weight. [However, selling by weight may lead to large variations in container fill. The main factors influencing variation in quantity are (a) temperatures of ice cream, (b) differences in technique among persons doing the dipping, and (c) variable composition and air content of the ice cream.]
3. The inevitable delay in serving customers
4. The necessarily higher temperature of the ice cream when it leaves the retail store. (It has to be softer for dipping than it would be if prepackaged; therefore, many more of the ice crystals have melted by the time it reaches the destination. Therefore, on freezing, larger ice crystals will be formed. This reduces the quality of the product.)
5. The inability of the retailer to control operating costs. (Since it is impossible for the retailer completely to fill equal-sized containers with equal weights of dipped ice cream, it is impossible to operate on a narrow margin.)

Numbers of dipper portions per 72-oz gallon are given in Table 18.3.

Table 18.3. Average Number of Dips Per Gallon When Weight Per Gallon is 72 oz (78% Overrun)

Dipper size (no.)	Number of dips	Size of dip (oz)
10	19	3¾
12	23	3⅛
16	29	2½
20	38	1⅞
24	44	1⅝
30	60	1³⁄₁₆

BASIC SODA FOUNTAIN PREPARATIONS

Ice cream can be served in countless ways in tempting flavors with a combination of different toppings, syrups, whipped cream, nuts, candies, or other thoughtful inclusions. It has become the favorite fun food, and few if any foods enjoy greater popularity with consumers of all ages.

Because of these attributes, ice cream provides great potentials for merchandising at the fountain. Basic ice cream items of fountain service are the cone, the dish of ice cream with or without fruit, the sundae, the soda, the milk shake, the banana split, the parfait, and novelty items. The art of preparing and serving these items constitutes an important phase of ice cream merchandising.

Dish of Ice Cream with Fruit
1. Place two dippers of ice cream in dish.
2. Surround with 1 oz sliced fruit of preference.

Sundae
1. Place ½ oz crushed, sliced, or whole fruit into a dish.
2. Add two dippers of ice cream.
3. Surround ice cream with crushed, sliced, or whole fruit.
4. Add nuts.
5. Top with whipped cream and a single item of fruit.

Soda
1. Place 1½ oz syrup into 14-oz cup or glass.
2. Stir spoon of ice cream or 1½ oz coffee cream into syrup.
3. Fill glass three-quarters full with carbonated water.
4. Float two dippers of ice cream into mix.
5. Mix gently.
6. Top with whipped cream.

Milk Shake
1. Place two dippers of ice cream into chilled cup.
2. Add 1½ oz syrup.
3. Add 6 oz milk.
4. Mix thoroughly and rapidly.
5. Pour into serving container.

Parfait
1. Place 1 tablespoon (tbs) of crushed fruit in parfait glass.
2. Add one dipper of ice cream.
3. Cover with 1 tbs crushed fruit.
4. Add one dipper ice cream.
5. Cover with 2 tbs crushed fruit.
6. Add one dipper ice cream.
7. Cover with 1 tbs crushed fruit.
8. Top with whipped cream.
9. Decorate with single piece of fruit or nut.

Banana Split
1. Split banana in half, lengthwise, with peel on. Place one half on each side of dish, flat side down, then remove peel.
2. Place three dippers of ice cream on banana halves. Vanilla in center and other flavors on each side.
3. Cover each dipper of ice cream with different topping.
4. Garnish the top and between dippers with whipped cream.
5. Add single piece of fruit.
6. Place slices of banana or fruit around center of item.

The person, time, and place of creation of the banana split has been a matter of curiosity among many in the retail ice cream business. An article in the January 1984 issue of *The Sundae School Newsletter* published by the National Ice Cream and Yogurt Retailers Association (NICRYA) sheds light on the matter. A newspaper clipping claims that "It happened in 1904 when a young David Strickler was learning the drug business in the Tassell Pharmacy (in Latrobe, PA) which later became Strickler's." Mr. Strickler's daughter added that "There were no dishes for such a concoction, so he drew up his own and had the people in Grapeville (Pennsylvania) design it." An article in the October, 1906 issue of *The Soda Fountain* magazine indicated that the banana split first came into public notice at the 1905 convention of the National Association of Retail Druggists. The article noted that "One of the features of the Boston convention was hospitality which was offered by the manufacturers and supply houses, yet among all the beverages dispensed here, none was more novel with the ladies than the banana split."

SPECIALS FOR SEASONS AND MEMORABLE OCCASIONS

One of the best ways to promote sales is by featuring specials and by having a "season-minded" approach. Many special days in every season will suggest to the imaginative soda fountain dispenser new ideas for taste-tempting formulas and decorative schemes. The intelligent manager must, of course, not only make these specials but see that they are well advertised so the public will know about them well in advance of the special days they are featured.

REFERENCES

Baumer, E. F., and R. E. Jacobson. 1969. Economic and marketing report on frozen desserts. IAICM, Washington, D.C.
Bierman, H. R. 1926a. How the composition of ice cream affects dipping. *Ice Cream Trade J.* 23(2):45.
Bierman, H. R. 1926b. Effect of temperature on dipping. *Ice Cream Rev.* 10(8):126.
IICA. 1994. The Latest Scoop. International Ice Cream Association. Washington, D.C.

19
Formulas and Industry Standards

Basic mix formulas may have variations and modifications as dictated by the ingredients available, consumer buying preferences, costs, and finished product quality expected.

The recipes given are for 10 gal (approximately 92 lb) of flavored mixes unless otherwise indicated. The numbers in all of the formulas are percentages unless otherwise specified. The percentage of stabilizer-emulsifier will vary with the blend, so manufacturers' recommendations should be followed.

PLAIN ICE CREAM

Formulas

Fat	10.0	12.0	14.0	16.00	18.0
NMS	11.5	11.0	10.0	8.50	7.0
Sugar	15.0	15.0	15.0	17.00	18.0
Stabilizer-emulsifier	0.3	0.3	0.3	0.25	0.2
Total Solids	36.8	38.3	39.3	41.75	43.2

Variations

Vanilla. To 10 gal plain mix add 6–12 oz flavor and 8 ml yellow color. The amount of flavor needed varies with strength of vanilla, composition of the mix, and personal preference.

Coffee. To 10 gal plain mix add 1 qt coffee extract (a strong coffee extracted from 1 lb of ground coffee may be used in place of the coffee extract); 5–7 oz of 50% burnt sugar color may be added.

Maple. To 10 gal plain mix add 3 oz pure maple extract and 2 oz burnt sugar coloring. If maple sugar is used, a special mix is prepared with 10–12% sucrose and 3–6% maple sugar.

Caramel. To 10 gal plain mix add enough caramel flavor to impart a satisfactory caramel taste and color. Alternatively use 3 qt caramel syrup per 9.25 gal of plain mix (approximately 4 oz per gallon of ice cream). A caramel syrup can be made from 5 parts sugar and 3 parts 20% cream boiled to a light brown color.

Mint. To 10 gal plain mix add 4–8 oz pure mint extract, and color to light green.

19 FORMULAS AND INDUSTRY STANDARDS

Butterscotch. To 10 gal plain mix add 1 gal butterscotch syrup and color yellow.

Bisque ice cream. To 9.5 gal plain mix add 4–8 lb macaroons, sponge cake, ladyfingers, grapenuts, cookie dough, brownie pieces, or similar products broken into small pieces to the ice cream in the batch freezer or with a fruit feeder. The ice cream carries the name of the product used (such as grapenut, sponge cake, or macaroon). Advanced techniques permit using these "bakery mix-ins" along with variegates of appropriate sauces such as raspberry, chocolate, and caramel.

Tortoni, biscuit. To 10 gal 14–16% fat ice cream mix add 1 pt dark rum or rum flavoring. Freeze at overrun of 50% or lower. Place in small fluted cups, and sprinkle with macaroon crumbs.

CANDY OR CONFECTION ICE CREAM

Numerous candy and confection combinations for various flavors occur on the market under various names. These contain 5–8% candy or confection. In general, these mix-ins are added to partially frozen ice cream in the batch freezer or to continuously frozen product with an ingredient feeder.

Peppermint stick. To 9.5 gal plain mix add 4–5 lb crushed peppermint candy.

Butter crunch. To 9.5 gal plain mix add 6 lb ground butter crunch candy.

Chocolate chip. To 9.5 gal plain mix add 2 qt milk chocolate chips, or add milk chocolate syrup to the partially frozen ice cream in the batch freezer.

Peanut brittle. To 9.5 gal plain mix add 4–6 lb crushed peanut brittle.

Marshmallow. To 9.5 gal plain mix add 4–6 lb marshmallows.

Chopped chocolate. To 9.5 gal plain mix add 4–5 lb chopped milk chocolate.

Licorice. To 10 gal plain mix add 4 oz licorice paste and 1.3 oz oil of anise. Color black.

Molasses taffy. To 9.5 gal plain mix add 4–6 lb molasses taffy.

Toffee. To 9.5 gal plain mix add 4–6 lb of broken toffee candy.

Mint chip. To 9.5 gal plain mix add 4–6 lb green or red or mixture of mint chips.

English toffee. To 9.5 gal plain mix add 4–6 lb crushed English toffee (consisting of 2 lb butter, 5 lb sugar, 1.5 lb nutmeats, and ½ tsp soda). Heat butter and sugar to 320°F. Remove from heat and add soda, stirring thoroughly. Mix in nutmeats and spread to cool.

Ginger. To 9.5 gal plain mix add 8 lb preserved chopped ginger root, or add a no. 10 can of ginger root flavoring.

CHOCOLATE ICE CREAM

Formulas

Fat	8.0	10.0	12.00	14.00	16.0
NMS	12.5	12.0	11.00	10.00	8.0
Sugar	16.0	16.0	16.00	17.00	19.0
Cocoa	2.7	3.0	3.00	3.00	3.5
Stabilizer-emulsifier	0.3	0.3	0.25	0.25	0.2
Total Solids	39.5	41.3	42.25	44.25	46.7

The quantity of cocoa needed to flavor chocolate ice cream ranges from 2.7–3.0 lb; of cocoa-liquor blend, 3.5–4.5 lb; and of chocolate liquor alone, 4.5–5.5 lb per 10 gal of mix.

Variations

Chocolate. To 9 gal plain mix add chocolate syrup made from 3 lb cocoa or 4 lb chocolate liquor, 3 lb sugar, and 4–6 qt water. Mix cocoa and sugar together and add enough water to make a paste. Heat in a steam-jacketed kettle. As the syrup thickens, add the water gradually, constantly stirring. Heat to 80°C (175°F), draw off the syrup, cool, and use. Just enough water is needed to prevent an excessively thick syrup. The syrup can be made up in quantities and stored for a few days in a cold storage room at 2–4°C (35–40°F).

Chocolate mint flake. To 9.5 gal chocolate mix add 4–6 lb mint flakes.

Chocolate malt. To 9 gal plain mix add half the chocolate syrup used for chocolate ice cream and 1 qt of malt syrup. If malt syrup is not available, use 1–2 lb of malted milk.

Chocolate almond. To 9.5 gal chocolate mix 4–6 lb broken almonds.

Chocolate marshmallow. To 9.5 gal chocolate mix add 4–6 lb marshmallow.

Mocha. To 7.5 gal plain mix add 2.5 gal chocolate mix and 8–12 oz coffee syrup or enough coffee extract to give a mild coffee flavor.

German chocolate, Swiss chocolate, or other variations. These may be prepared by using chocolate flavoring or other flavorings typical of the product desired.

FRUIT ICE CREAMS

Basic mixes presented for plain ice creams may be flavored with 10–15% fruit, depending on the kind of fruit and the flavor intensity expected. For frozen fruit a fruit to sugar ratio within the range of 2.5:1 to 4:1 is recommended, depending on the type of fruit.

Fruit suppliers have the best information about the use of their particular fruits and flavorings. They should be consulted for recommendations. The manufacturer can then personalize the flavor by making minor adjustments or using a unique mix formulation. The following general recommendations should be adjusted to suit the fruit preparation.

Variations

Strawberry. To 8 gal mix add 2 gal berries, aseptically processed. A mixture of puree and "solid fruit" is recommended. Add red color to adjust color to desired intensity of pink.

Peach. To 8 gal plain mix add 2 gal peaches, By adding a portion of the peaches as puree, intensities of both the flavor and color are enhanced. Color to light yellow.

Cherry. To 8.5–9.0 gal mix add 4–6 qt cherries. Use maraschino cherries or chose a sour variety and add cherry concentrate to fortify the flavor. Color to light red.

Apricot. To 8 gal plain mix add 2 gal apricots. Color to light yellow.

Pineapple. To 9 gal plain mix add 1 gal crushed pineapple.

Apple. To 8 gal plain mix add 2 gal sliced apples. Frozen apples should be packed with a fruit to sugar ratio of 7:1. Fortify apple flavor with about 320 ml of apple essence.

Banana. To 9 gal plain mix add 1 gal crushed bananas. If fresh bananas are used they must be fully ripe. Add a small amount of citric acid or lemon juice to prevent discoloration of the fresh fruit and to enhance flavor. A high fruit to sugar ratio gives best results.

Orange. To 9 gal plain mix add 4 qt orange juice and 1 qt lemon juice. Add 3 lb sugar to the juice. Color light orange.

Orange-pineapple. To 9 gal plain mix add 1 gal crushed pineapple and enough orange concentrate to bring out the orange flavor. Color light orange.

Lemon. To 9 gal plain mix add 4 qt fresh or frozen lemon juice plus 1 qt orange juice and 3 lb sugar. Color lemon yellow.

Raspberry. To 9 gal plain mix add 1 gal red or black raspberry puree with a 1:1 sugar ratio. Raspberry extract is usually needed to fortify the flavor.

Fig-nut. To 9 gal plain mix add 4 qt canned figs and 2 lb chopped nuts. Color light tan.

Grape. To 9 gal plain mix add 1 gal concentrated grape juice.

Blueberry. To 8 gal plain mix add 2 gal blueberry puree.

Cherry-vanilla. To 8 gal plain mix add 2 gal nonbleeding cherries injected with an ingredient feeder.

Date. To 8 gal plain mix add 2 gal crushed dates.

FROZEN YOGURT

Frozen yogurt is a food that is prepared by freezing while stirring a pasteurized mix containing milk fat, NMS, sweetener, stabilizer, and yogurt. It may contain any of numerous flavoring agents, but it is most often flavored with fruits. There is no federal standard for frozen yogurt at this writing. Where there are state regulations, it is commonly required that the yogurt ingredient must have been cultured with a mixture of *Lactobacillus bulgaricus* and *Streptococcus thermophilus* bacteria after the milk has been pasteurized. If pasteurization has been done after culturing, it is accepted that the label will declare so. Usually a very high heat treatment, e.g., 85°C (185°F) for 15 min, is given the milk before it is inoculated with the yogurt culture. The common industry practice is to consider titratable acidity of the finished yogurt mix of 0.30% as establishing a theoretical minimum amount of yogurt to be added to the mix. The amount added by most manufacturers ranges from 10–20% of the total weight of the mix.

In general manufacturers attempt to limit the amount of the acetaldehyde flavor in the frozen yogurt, believing that most customers do not prefer that flavor which characterizes plain yogurt. Yogurt definitely has a somewhat acidic flavor as compared with lowfat ice cream containing the same amount of fat.

The apparent reason that frozen yogurt has been preferred over the similarly comprised and prepared ice milk product (now named ice cream with an appropriate descriptor) is that the yogurt bacteria are thought by many people to assist in digestion of lactose and to have other health-promoting properties. It remains to be seen what effect the new labeling regulations that permit the use of modifiers to the name ice cream will have on the consumers choice of those products vs. frozen yogurt.

Frozen yogurt products are all relatively low in fat content. The label of this class of products varies with the fat content, with the same names applying as in ice cream products of the same fat content: < 0.5 g total fat per 4 fl oz serving is **nonfat frozen yogurt**; 0.5–3.0 g per serving is **lowfat frozen yogurt**; and above 3 g per serving is labeled **frozen yogurt**. It is not expected that the industry will market reduced fat frozen yogurt, because the highest fat content of frozen yogurt is in the vicinity of 4%. Thus, it is obvious that a reduced fat product would contain about 3% fat. If the finished product weighs 70 g per 4 fl oz serving, at 3% fat a serving would contain only 2.1 g of fat and would qualify for the lowfat label.

Formulas

Strawberry yogurt:

Plain mix (5% fat)	3 gal
Lowfat yogurt	1 gal
Yogurt standardizer (commercial)	1 gal
Strawberry puree	2.25 qt
Strawberries (solid pak)	2 qt
Yogurt acid (lactic acid)	1.5 oz
Citric acid (liquid)	1 oz
Fresh strawberry flavor	0.25 oz

Peach yogurt:

Plain mix (5% fat)	3 gal
Lowfat yogurt	1 gal
Yogurt standardizer	1 gal
Nectarine puree	1 pt
Peach puree	1 pt
Peach cubes	2 qt
Peach flavor extract	3 oz
Yogurt acid (lactic acid)	1 oz
Citric acid	1 oz

NUT ICE CREAMS

A basic plain mix containing 10–14% fat can be flavored with 3–6 % nutmeats. Butter and caramel flavors are excellent companions with nuts.

Variations

Burnt almond. To 9.5 gal plain mix add 4–5 lb burnt or roasted almonds. Almond flavor may also be added, as well as some burnt sugar or caramel color.

Pistachio. To 9.5 gal plain mix add 4 lb chopped pistachio nutmeats, pistachio extract to taste, and color light green. Frequently English walnuts or pecan nutmeats are used instead of pistachio nuts, in which case the pistachio flavor is secured from the extract.

Butter pecan. To 9.5 gal plain mix add 3 lb butter crunch candy plus 2 lb chopped pecans. Butter pecan ice cream may also be made by using 9 gal plain mix, adding 10 lb butter pecans, half chopped and half whole, or all ground to a coarse size.

Maple walnut. To 9.5 gal plain mix add 2–3 lb chopped walnut meats. Add 3 oz pure maple extract.

Maple pecan. To 9.5 gal plain mix add 2–4 lb chopped pecans. Add 3 oz pure maple extract.

Black walnut. To 9.5 gal plain mix add 4–5 lb broken black walnut meats.

Caramel nut. To 9.5 gal plain mix add 3 lb crushed nuts, 2 oz burnt sugar coloring, and caramel flavoring.

Pineapple nut. To 9 gal plain mix add 1 gal pineapple and 3 lb nuts.

Pecan crunch. To 9.5 gal plain mix add 6 lb ground pecan crunch candy.

Banana nut. To 9.25 gal plain mix add 0.75 gal crushed bananas and 2 lb chopped nuts.

Chocolate almond. To 9.5 gal chocolate mix add 4 lb whole and broken almonds.

Almond toffee. To 9.5 gal plain mix add 4 lb broken toffee candy and 2 lb broken almonds.

Coconut pineapple. To 9 gal plain mix add 3 qt crushed pineapple and 4 lb ground coconut.

Peanut. To 9.5 gal plain mix add 4–6 lb peanuts, crushed or peanut butter.

PUDDINGS

Variations

Nesselrode pudding. To 8–8.5 gal of 14–16% fat mix, add 6–8 qt special commercial fruit mixture suitable for nesselrode pudding. This should be added after the mix is partly frozen by use of a fruit feeder. A nesselrode mixture may be prepared from the following chopped fruits: 1 qt crushed pineapple, 1 qt candied cherries, 1 qt maraschino cherries, 1 qt raisins, 12 oz candied orange peel, 2 lb each walnuts, almonds, and pecans, chopped or coarsely ground. Color light orange. Standard plain mix is often used instead of the richer mix in making puddings.

Tutti-frutti. To 9 gal plain mix add 4–8 qt of a mixture of several fruits of a suitable mixture for tutti-frutti. Color to a light pink. Usually a prepared mixture consists of the following fruits, chopped into small pieces: 2 qt red cherries, 2 qt green pineapple, 2 lb raisins, 2 qt crushed pineapple, 2 lb nuts. Add red coloring.

Frozen pudding. Add nutmeats and rum, or rum flavor, to tutti-frutti ice cream.

English plum pudding. To 8 gal parfait mix (16% fat) add 3 lb chocolate syrup and the following fruits and nuts after the mix is partially frozen: 2 lb figs, 2 lb dates, 8 lb mixed candied fruits, 2 lb walnuts, 8 oz vanilla extract, 2 lb pecans, 6 tsp cinnamon, 1 tsp each ginger, allspice, and cloves. Fruits and nuts should be chopped and may be mixed with the spices before they are added to the partly frozen mix, or the spices may be added to the mix before freezing begins.

Plum pudding. To 8 gal plain mix add 8 lb egg yolks, 6 lb sugar, 4 qt orange juice, 4 lb pecans, 4 lb walnuts, and 8 oz vanilla.

Date pudding. To 9 gal plain mix add 6 lb dates, pitted and ground, plus 2 lb chopped walnuts.

PARFAIT

To 2 gal of ice cream mix containing 16% fat add 2.5 lb dried egg yolk or 8 lb whole egg and cook to a custard (about 41°C for 30 min), and cool to 38°C. Add this custard to 7 more gal of the same ice cream mix. The resulting 10 gal of parfait mix can then be flavored as an ordinary ice cream mix.

MOUSSE

For a small quantity, whip 1 gal of whipping cream to a stiff consistency, and gently stir in 2 lb sugar, 1–2 drops desired color, and 0.5–1 oz vanilla. Dates, nuts, and drained fruit chopped into small pieces can be stirred in also. Place into molds and harden without further stirring.

For a larger quantity, use 6 gal heavy whipping cream (40%), 2 gal milk, 14 lb sugar, 6 oz stabilizer, and 6 oz vanilla. To disperse and dissolve the stabilizer, mix it with about 2 lb of sugar and stir it into the milk. heat the milk to the temperature necessary to dissolve the stabilizer, then combine with the remaining ingredients. This larger mix is prepared, chilled to about 0°C, placed in a batch freezer, drawn at a low overrun, and hardened before serving.

FRAPPÉ

Frappés are made in the same manner as ices except they are served in a soft condition. To make 5 gal of ice, combine 9 lb sugar, 4.5 lb corn syrup solids, 3 oz pectin, 1 gal fruit juice, and sufficient water (about 23 lb) to make 5 gal. Serve soft-frozen. If necessary to hold for a limited time, maintain temperature at −7°C (20°F).

SORBETS

A sorbet is similar in composition to an ice. It has a relatively high sugar (30%) and fruit and fruit juice (30–50%) content, generally contains egg white (2.6% solids) and pectin or gum stabilizer (0.4%), and has an overrun of 20% or less. Citric acid may be added to enhance flavor. The remainder is water. Exotic flavors are often used in sorbets.

PUNCH

Punch is an ice in which fruit juices have been reinforced with an alcoholic beverage. Often rum flavoring (nonalcoholic) is used instead of liquor. To about 7 gal of water add 20 lb sugar, 1 qt lemon juice, 1 qt grape juice, and rum flavoring as desired. If the product is to be served frozen, add 5 oz stabilizer.

GRANITE

Granite is made from the same ingredients as ice (Chapter 13), but it is frozen much harder and with little whipping or stirring during freezing. The product is coarser than an ice.

SHERBETS

See Chapter 13, Sherbets and Ices.

SOUFFLÉ

This product is made from sherbet mix to which whole egg is added. Freezing is done with high enough overrun to give a fluffy product. Use 7.5 gal base sherbet mix, 8 lb whole eggs, and 2 gal pureed fruit. By substituting the fruit used as flavoring, many other soufflés can be made.

LACTO

Lacto is made from sherbet mix that is composed from cultured sour milk, buttermilk, or other fermented milk product. For grape lacto use 3 gal cultured buttermilk, 9 lb sugar, 1.5 lb eggs with yolks separated from whites, 3 cups lemon juice, and 1 qt grape juice. Dissolve the sugar in the milk and add the fruit juices and beaten egg yolks. Whip egg whites and add at the **batch freezer**. Other flavors can be made by substituting raspberry, cherry, or orange juice for the grape juice.

FRUIT SALAD

This product consists of large pieces of mixed fruits in combination with whipped cream or ice cream. About 1 gal of a mixture of pineapple (sliced or cubed), red cherries, apricots, peaches, pears, and/or pitted prunes is generally folded into ⅔ gal whipped cream that has been flavored to taste with mayonnaise. This mixture is allowed to set to stiffen somewhat before being placed into molds to harden. About ⅔ gal of softened ice cream can be substituted for the whipped cream. Addition to the chilled fruit mixture of a 3-oz package of gelatin (unflavored or flavored as preferred) dissolved in 1 cup hot water will enhance the body of the fruit salad.

FANCY MOLDED ICE CREAM

An aufait is a molded ice cream and fruit combination of two or more layers of ice cream with a layer of fruit between the layers of ice cream. To make a strawberry aufait, layer a strawberry mold half full of slightly softened ice cream, harden, cover with a thin layer of pectinized strawberries (if the layer is too thick, it cannot be cut), and finish filling the mold with strawberry or vanilla ice cream. Many combinations of types of fruit and numbers of layers are possible. Also, it is possible to gently stir the pectinized fruit into the ice cream as it is discharged from the freezer. The fruit should be viscous enough that it makes a more or less continuous line through the ice cream showing a marbled appearance in the finished product.

Spumoni

Press vanilla ice cream into about ¼ the depth of a spumoni cup and force it up the walls of the cup. Add chocolate ice cream to half fill the cup. Finish filling with a mixture of fruit and whipped cream and harden. The fruit-whipped cream mixture is prepared by whipping 1 gal whipping cream, then gently stirring in 1–1.5 gal of drained fruit to which has been added 1 lb confectioners sugar.

FROSTED MALTED

To 3.5 gal lowfat ice cream mix add 1 qt malt syrup (or 1–2 lb malted milk powder) and chocolate syrup made of 0.5 lb cocoa (or 1 lb chocolate liquor), 0.75 lb sugar, and 1–1.5 qt water. Freeze and serve from the freezer.

SPECIALS

Egg nog. To 9 gal plain mix add 1 gal egg nog base (can be a commercial preparation). An egg nog base may be prepared by using 15 lb whole fresh eggs, 7.5 lb fresh egg yolks, 4 lb dry eggs, 8 lb sugar, and 45 lb 20% cream. This mixture is heated to 71°C (160°F) for 20–30 min and homogenized at about 2,000 lb/in^2. Egg nog ice cream may also be prepared by using 10 gal plain mix, 1.5 lb dried egg yolks, 3 oz vanilla extract, 2 oz lemon extract, 1 pt rum, 3 oz egg color, and spices (as desired).

Rum raisin. To 9 gal plain mix add 25 lb soft seedless raisins and 4–6 oz Roman punch or rum flavor and yellow color. If the raisins are not soft, they should be soaked in water and drained before being added to the ice cream.

Date walnut. To 9 gal mix add 15 lb soft pitted dates, 3–4 lb English walnuts, and color as desired. The soft pitted dates and English walnuts pieces may be blended with a small amount of sugar and added to the mix.

Sweet potato. To 7 gal plain mix add 28 lb sweet potato puree (7:1 sugar pack), 2 pt lemon chips, 2 oz vanilla extract, and 4 oz Prussian orange color.

Avocado. Grind finely and add 1 lb sugar to each 4 lb fruit. Mix well and add 16 lb of mixture to 90 lb of mix.

Honey. Approximately half the sugar on a cane sugar basis can be replaced by honey to obtain a honey flavor. A suitable mix contains 10% milkfat, 11% NMS, 8% sugar, 9% honey (72.4% solids), and 0.3% stabilizer. If more honey is used, freezing and hardening difficulties may be experienced. There is the possibility of using honey in combination with low- or medium-DE corn syrup solids, because they do not depress the freezing point as much as other sweeteners. This combination would allow the use of a greater percentage of honey without freezing difficulties. Under these conditions, the following formula should give a highly flavored and adequately sweetened product: 10% milkfat, 1% NMS, 8% low-DE CSS, 12% honey, and 0.25% stabilizer. The milder flavored and lighter colored honeys are preferred for ice cream. Sweet clover, alfalfa, or clover honey are desirable.

Persimmon. To 8 gal plain mix add 2 gal persimmon pulp (4:1 sugar pack), 1 pt grated orange rind, and 2 oz vanilla. The persimmon should be prepared by separating the peel and seed from the pulp. Pulp should be medium coarse. The fruit gives a golden yellow product that is smooth in texture and rich in flavor.

Watermelon. To 10 gal plain mix add 10–12 oz watermelon flavor. Color pink.

Rhubarb. To 8.5 gal plain mix, add 1.5 gal cooked, crushed rhubarb (3:1 sugar pack), 2 lb broken pecans, 2 pt orange chips (or 1 pt preserved orange peel, or orange rind), 2 oz vanilla, and ½ oz mace. Color pink.

Pumpkin. To 8 gal plain mix add 2 gal cooked pumpkin puree, 2 pt orange chips (or 1 pt preserved orange peel or orange rind), 2 oz vanilla, 0.5 oz cinnamon, and 0.25 oz nutmeg.

HOMEMADE ICE CREAM

Each of the following basic formulas will produce about 1 gal of ice cream when frozen in a salt and ice freezer and when an extract type flavor is used. Generally, 1 tablespoon of vanilla extract is needed to flavor these mixes. (The gelatin must be dissolved in 1/2 cup cold water.)

Basic Formulas

(a)	Quantity	(b)	Quantity
Light cream	1.75 qt	Cream (30–40%)	1.3 qt
Evaporated milk	0.5 pt	Milk	1.5 pt
Sugar	2 cups	Sugar	2 cups
Gelatin	1 tsp	Gelatin	1 tsp
Salt	Pinch		

(c)	Quantity	(d)	Quantity
Cream (30–40%)	1.5 qt	Cream (30–40%)	1.5 qt
Milk	1.5 qt	Evaporated milk	15 oz
Sw. cond. milk	15 oz	Sugar	2 cups
Sugar	1 cup	Gelatin	1 tsp
Gelatin	1 tsp		

(e) Cooked base	Quantity	(f) Diabetic	Quantity
Cream (30–40%)	1.5 qt	Cream (30–40%)	1 qt
Milk	1 pt	Milk	1 qt
Sugar	2 cups	Saccharin	4 grains
Eggs	6	Gelatin	1 tsp

Other flavors as desired may be developed by using these formulas and recipes as a guide.

Berry Ice Cream

Fruit mixture:

Flavored gelatin (same flavor as the fruit) dissolve in 1 cup water	6 oz
Berries (preferably frozen with sugar)	20 oz

Mix berries with solubilized gelatin but do not allow the gel to set.

Ice cream base mix:

Eggs, beaten	4
Sweetened condensed milk	2 cups
Sugar	1.33 cups
Half and half	2 qt
Instant vanilla pudding mix	6 oz

Blend ingredients of base mix well before adding the fruit mixture. Makes enough mix for a 6-qt freezer.

LOW- OR REDUCED-LACTOSE ICE CREAM

A significant number of consumers suffer from some degree of inability to completely digest lactose. These lactose malabsorbers sometimes experience discomfort in the lower bowel when lactose that escapes being absorbed in the small intestine is fermented into acid and gas in the colon. This can cause gas pains, and, in severe cases, diarrhea. Most of these lactose malabsorbers can eat a serving of ice cream without symptoms of lactose malabsorption, especially if the ice cream is consumed with a meal. In this case the stomach empties over a long period of time, giving the essential intestinal enzyme the time needed to hydrolyze most of the lactose consumed. The solution for those persons who are highly affected is to consume frozen desserts in which there is a low amount of lactose. Such products can be made by either ultrafiltering and/or diafiltering the milk to remove much of the lactose or by prehydrolyzing the lactose. (In diafiltration, after about one-half of the volume is removed by ultrafiltration, water is added to the retentate and filtration is continued until the volume is again reduced to about 50% of the initial volume.) Of course a combination of these treatments is also possible. The author has made high-quality reduced lactose goat milk ice cream from goat milk concentrated by ultrafiltration. Geilman and Schmidt (1992) developed formulas for ice cream made with ultrafiltered and diafiltered cow's milk. They found, as had others (Hofi 1989; Masters and Kosikowski 1986:Tong et al. 1989) that these products were harder and melted more slowly than those produced with equivalent NMS, probably because the ultrafiltered retentate retained virtually all of the proteins but lost two-thirds of the lactose. The traditional mix contained 4% protein and 6.6% lactose, whereas mixes made with ultrafiltrate contained 9% protein and 2.2% lactose. The type of sugar added in replacement of lactose caused significant differences in hardness of the finished products.

The removal of lactose by ultrafiltration reduces the concentration of dissolved solids in the product and raises the freezing point. On the contrary, hydrolysis of the lactose produces two molecules for every lactose molecule hydrolyzed, and, therefore, lowers the freezing point. By removing 50% of the lactose by ultrafiltration and then hydrolyzing the remaining portion of lactose, the freezing point can be kept at the level it would have been had neither treatment been done.

It remains to be seen whether the public will become conscious enough of the need for lactose-reduced ice cream for a market to be developed. It definitely raises the cost of production to use either of the suggested methods of reducing lactose content. Furthermore, lactose is the least costly of the milk-derived components in ice cream.

ICE CREAMS OF LOWERED FAT CONTENT

The following ice cream formulas have been collected from various firms and institutions in an effort to show the variety of components and the extent of their recommended use in frozen desserts containing relatively low amounts of fat. Sources of formulas are shown under each heading. The mention of proprietary product is in no way an endorsement of the product or the formula in which it appears. Some of the formulas have been modified slightly to fit the format of the section.

Fat-Free, Sugar-Free Ice Cream		Fat-Free, Sugar-Free Frozen Yogurt	
Ingredients	Percentage	Ingredients	Percentage
Hoechst Food Ingredients		*Hoechst Food Ingredients*	
Water	67.75	Water	67.65
NMS	13.0	NMS	13.60
Polydextrose	8.0	Polydextrose	8.10
Sorbitol, crys.	5.0	Maltodextrin (10 DE)	5.80
Maltodextrin, 10 DE	5.0	Fat mimetic[b]	3.50
Stabilizer[a]	1.2	Stab/emulsifier	0.70
Acesulfame K	0.023	Milkfat	0.20

Aspartame	0.023	Acesulfame K	0.27
		Aspartame	0.18
Total solids	**32.25**	**Total solids**	**32.35**

[a]Mixture of microcrystalline cellulose, mono/diglycerides, cellulose gum and carrageenan, standardized with whey powder.

[b]A whey-based fat replacer such as Dairy Lo™.

Fat Free Ice Cream

Ingredients	Percentage	Ingredients	Percentage
SPI Polyols (Hard pack)		*Kerry Ingredients (Hard pack)*	
Water	60.8	Water	60.7
Maltitol, crys.	10.4	NMS	10.9
NMS	12.5	Fat replacer[a]	5.0
Sugar	12.0	Sucrose	8.2
Corn syrup solids, 35 DE	4.0	High-fructose corn syrup	4.5
		Corn syrup solids, 36 DE	9.5
Stab/emulsifier	0.3	Malodextrin	1.2
Total solids	**39.2**	**Total solids**	**39.3**

[a]Prolo 11, a whey-based fat replacer that contains stabilizer.

Ingredients	Percentage	Ingredients	Percentage
Grain Processing Corp (Soft-serve)		*Grain Processing Corp (Hard pack)*	
Water	69.71	Water	63.81
NMS	11.0	NMS	11.7
Sugar	15.2	Sugar	14.2
Maltodextrin (5 DE)	3.0	Maltodextrin (10 DE)	7.2
Milkfat	0.4	Maltodextrin (5 DE)	2.0
Cellulose gel	0.5	Milkfat	0.4
Locust bean gum	0.06	Cellulose gel	0.5
CMC	0.05	Locust bean gum	0.06
Mono/diglycerides	0.03	CMC	0.05
Xanthan gum	0.02	Mono/diglycerides	0.03
Carrageenan	0.02	Xanthan gum	0.02
Polysorbate 80	0.01	Carrageenan	0.02
		Polysorbate 80	0.01
Total solids	**30.29**	**Total solids**	**36.19**

Ingredients	Percentage	Ingredients	Percentage
Penn State (Soft-serve)		*Danisco Ingredients USA (Hard serve)*	
Water	68.15	Water	62.2
NMS	12.0	NMS	13.0
Sugar	13.0	Sugar	11.5
Corn syrup solids, 36 DE	6.0	Corn sugar	8.5
		Maltodextrin, 5 DE	2.0
Milkfat	0.5	Polydextrose	2.0
Stabilizer	0.35	Stabilizer[a]	0.8
Total solids	**31.85**	**Total solids**	**37.8**

[a]Consists of cellulose gel, mono/diglycerides, cellulose gum, locust bean gum, and carrageenan.

Bunge Foods		Bunge Foods	
Water	66.8	Water	64.4
NMS	12.0	NMS	11.2
Sucrose	12.0	Sucrose	9.7
Corn syrup solids, 36 DE	8.0	Corn syrup solids, 36 DE	9.7
Stab/emulsifier[a]	1.2	Maltodextrin, 10 DE	2.0
		Whey protein concentrate	1.5
		Egg white solids	0.5
		Stab/emulsifier[a]	1.0
Total solids	**33.2**	**Total solids**	**35.6**

[a]Consists of a blend of cellulose gel, mono/diglycerides, locust bean gum, cellulose gum, carrageenan, and dextrose for standardization.

Light, Sugar Free		Low Fat	
SPI Polyols		Grain Processing Corp (Soft-serve)	
Water	70.1	Water	68.7
NMS	13.0	NMS	10.0
Sorbitol	11.5	Sugar	13.0
Milkfat	4.0	Maltodextrin, 5 DE	5.0
Cellulose gel	1.0	Milkfat	2.6
Stab/emulsifier	0.4	Cellulose gel/gum	0.51
Total solids	**29.9**	Locust bean gum	0.06
		CMC	0.05
		Mono/disaccharides	0.03
		Xanthan gum	0.02
		Carrageenan	0.02
		Polysorbate 80	0.01
		Total solids	**31.30**

REFERENCES

Geilman, W. G., and D. E. Schmidt. 1992. Physical characteristics of frozen desserts made from ultrafiltered milk and various carbohydrates. *J. Dairy Sci.* 75:2670–2675.

Hofi, M. A. 1989. The use of ultrafiltration in ice cream making. *Egypt J. Dairy Sci.* 17:27.

Masters, A. R., and F. V. Kosikowski. 1986. Effect of protein and solids content on low lactose ice cream from ultrafiltered milk. *J. Dairy Sci.* 69 (Suppl. 1):78. (Abstr.)

Tong, P. S., L. A. Jensen, and L. Harris. 1989. Characteristics of frozen desserts containing retentate from ultrafiltration of skim milk II. Some physical properties. *J. Dairy Sci.* 72(Suppl.):129.(Abstr.)

APPENDIX A
Historical Chronology of Ice Cream Industry

The development of the ice cream industry can be most quickly told by listing the approximate dates of some important events.

1700 Ice cream probably came to America with the English colonists. A letter written in 1744 by a guest of Governor Bladen of Maryland described having been served ice cream.

1774 First public recorded mention of ice cream in America was made by Philip Lenzi, a caterer, announcing in a New York newspaper that he was prepared to supply various confections including ice cream.

1777–1800 Early advertisement of ice cream by Philip Lenzi, *New York Gazette Mercury*, May 19, 1777, and November 24, 1777; by J. Corree in 1779 and 1781 in the *Gazette*; by Joseph Crowe in the *New York Post Bay*, June 8, 1786; by A. Pryor on May 18, 1789. Mr. Hall was selling ice cream in New York in 1785 and Mr. Bosio established a retail business in Germantown, Pennsylvania in 1800.

1789 Mrs. Alexander Hamilton, wife of the Secretary of the Treasury, served ice cream at a dinner attended by George Washington.

1812 Ice cream was served in the White House at the second inaugural ball by Mrs. Dolly Madison, wife of the fourth President.

1846 The first hand-cranked freezer was invented by Nancy Johnson, but she failed to patent it.

1848 Patents were granted to a Mr. Young on a revolving household type of hand freezer with dasher.

1851 The father of the wholesale ice cream industry of America, Jacob Fussell, a Baltimore milk dealer, began to manufacture ice cream in Baltimore. He established plants in Washington, D.C., in 1856; and in New York in 1864.

1856 Patent granted Gail Borden in August 1856 for the process of condensing milk. The first condensed milk factory was established in Wolcottville, Connecticut.

1858 Ice cream plant opened in St. Louis by Perry Brazelton, who learned the business from Jacob Fussell.

1864 The Horton Ice Cream Company was started in New York.

APPENDIX A

1879 Ice cream soda was introduced at Centennial Exposition, Philadelphia.

1892 The Pennsylvania State College established the first course in ice cream making.

1895 Pasteurizing machines were introduced.

1892–1906 Investigation and development of the dry milk industry in America. One of the first dry milk plants was established by Merrell Soule Company at Arcade, New York, in May 1906. The first spray process plant was built in Ferndale, California, in 1911.

1896–1904 Italo Marchiony, an Italian emigrant, produced the first ice cream cone in 1896 in New York. The patent issued on his special mold in December 1903. In 1904 E.A. Hamwi, a Syrian waffle concessionaire, rolled waffles into the shape of a cone for the benefit of an ice cream vendor in an adjoining booth at the World's Fair in St. Louis, Missouri.

1899 The homogenizer was invented by August Gaulin in France and was in use within two years. The U.S. patent was dated April 11, 1904.

1900 The Association of Ice Cream Manufacturers was formed; later the name was changed to International Association of Ice Cream Manufacturers (IAICM); and again to International Ice Cream Association (IICA).

1902 The horizontal circulating brine freezer was invented.

1904 The *Ice Cream Trade Journal* was made the official organ of IAICM.

1910–1912 First State Agricultural Experiment Station Bulletins concerning ice cream were published, including 1) R. W. Washburn, 1910. "Principles and Practices of Ice Cream Making," Vermont State Bull. *155*. 2) M. Mortensen, 1911. "Classification of Ice Cream and Related Frozen Products—Score Cards for Judging Ice Cream," Iowa State Bull, *123*. 3) B. W. Hammer, 1912: "Bacteria in Ice Cream," Iowa State Bull. *134*.

1911 The homogenizing process was applied to condensed or evaporated milk.

1913 The direct expansion freezer was introduced. The continuous freezing process was patented.

1915–1919 Textbooks on ice cream were published in the United States: 1) J. H. Frandsen and E. A. Markham, 1915. *The Manufacture of Ice Creams and Ices.* 2) W. W. Fisk and H. B. Ellenberger, 1917. *The Ice Cream Laboratory Guide.* 3) W. W. Fisk, 1919: *The Book of Ice Cream.*

1921 The Eskimo Pie was patented by C. Nelson, Waukon, Iowa. This was the first of the coated ice cream and novelty sticks.

1922 Development of direct expansion refrigeration adapted to freezers.

1925 Dry ice (solid CO_2) was used to facilitate delivery of ice cream.

1926 The counter freezer for soft ice cream appeared.

1928 The Vogt continuous freezer was developed by Henry Vogt of Louisville, Kentucky.

1929–1935 Development and acceptance of continuous freezers. The Vogt instant freezer was first introduced by Cherry Burrell and installed commercially in 1929. The Creamery Package continuous freezer was introduced in 1935.

1940–1945 Development of low-temperature storage units for the home.

1946 Carry-home packages marketed through chain grocery stores gained popularity. Soft ice cream and drive-in stores appeared.
1950 Appearance of vegetable fat products in the ice cream industry.
1942–1953 FDA hearings on federal standards for ice cream.
1951 Ice cream centennial held in Baltimore, June 15.
1953 High-temperature–short-time pasteurization of ice cream mix (175°F, 25 sec) approved by U.S. Public Health Service, February 13.
1960 Definitions and Standards for Frozen Desserts approved by FDA of the U.S. Dept. of Health, Education, and Welfare.
1965–1970 Introduction and development of highly automated, high volume processing equipment.
1974–1981 Revision of definitions and standards of identity for frozen desserts.
1983 Ice cream standards and regulations revised.
1983 Ice Cream for America Day.
1984 July, National Ice Cream Month.
1993 Nutrition Labeling Education Act passed by Congress.
1994 Food labeling regulations issued.
1995 Federal standard for ice milk nullified in favor of using descriptors in labeling ice creams not meeting the Standard of Identity for 10% milkfat in ice cream.

Appendix B
Miscellaneous Tables

HOW TO USE TABLE A.1

Example I. Required to make 50 gal of 30° Baumé sugar syrup.
Formula. Read directly from Table A.1 the Baumé degree and the amount of sugar and water needed to make 1 gal of that degree syrup. Note also the weight per gallon of syrup. Multiply each of these amounts by the number of gallons to be made:

$$5.80 \times 50 = 290 \text{ lb sugar}$$
$$4.70 \times 50 = 235 \text{ lb water}$$
$$10.50 \times 50 = 525 \text{ lb, 50 gal 30° Baumé syrup}$$

Example II. To find the amount of water required to dilute 1 gal of 34° Baumé syrup to 30° Baumé syrup.
Solution. Use the formula

$$\frac{(\text{lb water per gal dilute syrup}) \times (\text{lb sugar gal original syrup})}{\text{lb sugar per gal dilute syrup}} - \text{lb water per gal original syrup}$$
$$= \text{lb water required}$$

From Table A.1 substitute into the formula:

$$\frac{4.70 \times 6.85}{5.80} - 4.03 = 1.52 \text{ lb water}$$

For any larger quantity multiply 1.52 by the number of gallons to be diluted.

Example III. To find the amount of sugar required to thicken 1 gal of 30° Baumé syrup to 34° Baumé.
Solution. Use the formula.

331

$$\frac{\text{(lb sugar per gal thickened syrup)} \times \text{(lb water gal original syrup)}}{\text{lb water per gal thickened syrup}} - \text{lb sugar per gal original syrup}$$
$$= \text{lb sugar required}$$

From Table A.1 substitute into the formula:

$$\frac{6.85 \times 4.70}{4.03} - 5.80 = 2.19 \text{ lb sugar}$$

For any larger quantity multiply 2.19 lb by the number of gallons to be thickened.

Table A.1. How to Make Sugar Solutions and Determine Their Concentration Using the Baumé and Brix Hydrometers[a]

Degrees Baumé at 68°F[b]	Degrees Brix at 68°F	Weight in air at 68°F (lb/gal)	Weight of sugar (lb/gal)	Weight of water (lb/gal)
30.0	55.2	10.50	5.80	4.70
31.0	57.1	10.59	6.05	4.54
32.0	59.1	10.68	6.31	4.37
33.0	61.0	10.78	6.58	4.20
34.0	63.0	10.88	6.85	4.03
35.0	64.9	10.97	7.12	3.85
36.0	66.9	11.07	7.41	3.66
37.0	68.9	11.18	7.70	3.48
38.0	70.9	11.28	8.00	3.28
39.0	72.9	11.39	8.30	3.09
40.0	74.9	11.49	8.61	2.88

[a]This table is based on information given in U.S. Bureau of Statistics Circular No. 375, Table I.
[b]The relationship of Baumé degrees (B) to specific gravity (sp.gr.) is given by B = 145 − (145/sp.gr.), and sp.gr. = 145/(145 − B).

HOW TO USE TABLE A.2

Example IV. Find Baumé of syrup wanted in leftmost column. Read across to column beneath Baumé of syrup to be diluted. The figure given is the amount of water, in fluid ounces, to be added to 1 gal of syrup.

Table A.2. Fountain Syrup Dilution per Gallon of Syrup[a]

Syrup wanted (Baumé degrees)	Syrup to be diluted (Baumé degrees)			
	31	32	33	34
30	6	11.5	17.5	23.5
31	—	6	11.5	17.5
32	—	—	5.5	11.0
33	—	—	—	5.5

[a]Numbers indicate amount (fluid ounces) of water to be added.

APPENDIX B

WEIGHT, MEASURE AND FOOD SOLIDS IN VARIOUS ICE CREAMS

All foods contain water, and everything not water is called a "solid."

In producing ice cream having certain characteristics (for example, weighing so much per gallon, having certain volume per unit of weight, or certain food solids per gallon, and having at all times 12% fat), the relationships in Tables A.3 and A.4 will be helpful.

A mix of 31–48% TS, having the weights per gallon as shown, then made to take on 100% overrun, or to have 1.6 lb of food solids per gallon, or to weigh 4.5 lb per finished gallon, will give the figures shown in Table A.4.

Table A.3. Weight of Finished Ice Cream, with Various Weights of Mix and Various Overruns (12% fat)

Overrun (%)	Mix weight (lb/gal)				
	8.75	9.00	9.25	9.50	9.75
50	5.80	6.00	6.17	6.33	6.50
60	5.47	5.63	5.78	5.94	6.09
70	5.15	5.29	5.44	5.59	5.74
75	5.00	5.14	5.29	5.43	5.57
80	4.86	5.00	5.14	5.28	5.42
85	4.73	4.86	5.00	5.14	5.27
90	4.60	4.74	4.85	5.00	5.13
95	4.48	4.63	4.74	4.97	5.00
100	4.38	4.50	4.62	4.75	4.87
105	4.27	4.39	4.51	4.63	4.76
110	4.17	4.29	4.40	4.52	4.64
115	4.07	4.19	4.30	4.41	4.53
120	3.98	4.09	4.20	4.32	4.43

Table A.4. Relationships of Mix Factors (12% fat)

	Mix		Yield at 100% overrun		Food solids weight 1.6 lb/gal		Ice cream weight 4.25 lb/gal	
TS (%)	Weight (lb/gal)	TS weight (lb/gal)	Weight frozen (lb)	Weight of food solids (lb)	Yield (%)	Weight frozen (lb)	Yield (%)	Weight of food solids (lb/gal)
31	8.95	2.95	4.49	1.62	84.4	4.85	110.6	1.40
36	8.98	3.23	4.49	1.71	102.0	4.45	111.0	1.53
38	9.00	3.42	4.50	1.85	114.0	4.21	111.8	1.62
40	9.03	3.61	4.52	1.88	126.0	4.01	112.6	1.70
42	9.07	3.81	4.53	1.90	138.1	3.81	113.4	1.79
46	9.14	4.20	4.57	2.10	162.5	3.48	115.0	1.96
48	9.16	4.40	4.58	2.20	175.0	3.33	115.4	2.04

Table A.5. Composition, Relations, and Densities of Milks and Creams

Fat (%)	NMS (%)	TS (%)	Ratio of fat to NMS	Fat, percentage of solids	Specific gravity at 68°F	Weight lb/gal
Milk						
3.0	8.33	11.33	1:2.77	25.20	1.034	8.61
3.1	8.40	11.50	1:2.71	26.95		
3.2	8.46	11.66	1:2.64	27.47		
3.3	8.52	11.82	1:2.58	27.93		
3.4	8.55	11.95	1:2.52	28.41		
3.5	8.60	12.10	1:2.46	28.90	1.033	8.60
3.6	8.65	12.25	1:2.40	29.40		
3.7	8.69	12.39	1:2.35	29.85		
3.8	8.72	12.52	1:2.30	30.30		
3.9	8.76	12.66	1:2.25	30.77		
4.0	8.79	12.79	1:2.20	31.25	1.032	8.59
4.2	8.86	13.06	1:2.11	32.15		
4.4	8.92	13.32	1:2.03	33.00		
4.6	8.98	13.58	1:1.59	33.90	1.032	8.58
4.8	9.04	13.84	1:1.88	34.72		
5.0	9.10	14.10	1:1.82	35.46	1.031	8.58
Cream						
18.0	7.31	25.31	1:0.41	71.11	1.015	8.48
20.0	7.13	27.13	1:0.36	73.71	1.013	8.43
25.0	6.68	31.68	1:0.27	78.91	1.008	8.37
30.0	6.24	36.24	1:0.21	82.78	1.004	8.36
35.0	5.79	40.79	1:0.16	85.81	1.000	8.32
40.0	5.35	45.35	1:0.13	88.20	0.995	8.28
45.0	4.90	49.90	1:0.09	90.11	0.985	8.22

Table A.6. Relationship of Baumé to TS in Sweetened Condensed Skim Milk[a]

Baumé at 120°F (degrees)	Sucrose (%)	NMS (%)	TS (%)
37.4	45.63	27	72.63
37.6	45.00	28	73.00
37.8	44.38	29	73.38
38.0	43.75	30	73.75
38.2	43.13	31	74.13
38.4	42.50	32	74.50

[a] Assumes a sucrose:water ratio of approximately 62%. While the Baumé reading for TS is approximately right, too much value should not be given this test for there may be considerable variation due to varying proportions of serum solids and sucrose, the amount of fat content, etc.

APPENDIX B

Table A.7. The Freezing Point Relationships of Calcium Chloride Brine

Calcium chloride anhydrous (%)	Calcium chloride hydrous CaCl–6H$_2$O (%)	Specific gravity 18/4 C	Weight (lb/gal)	Degrees Baumé	Degrees Salometer	Freezing point (°F)
5	9.90	1.0409	8.67	5.7	22	28.9
10	19.80	1.0847	9.04	11.3	44	22.8
11	21.78	1.0937	9.12	12.4	48	21.6
12	23.76	1.1029	9.19	13.5	52	20.1
13	25.74	1.1121	9.28	14.6	58	18.3
14	27.72	1.1214	9.35	15.7	62	16.7
15	29.70	1.1307	9.42	16.8	68	14.7
16	31.68	1.1402	9.50	17.8	72	12.9
17	33.66	1.1497	9.58	18.9	76	10.8
18	35.64	1.1594	9.67	19.9	80	8.4
19	37.62	1.1692	9.75	21.0	84	5.5
20	39.60	1.1791	9.83	22.0	88	2.7
21	41.58	1.1890	9.91	23.1	92	−0.6
22	43.56	1.1990	10.00	24.1	96	−4.4
23	45.54	1.2090	10.08	25.1	100	−8.3
24	47.52	1.2192	10.16	26.0	104	−13.2
25	49.50	1.2294	10.24	27.1	108	−18.8
26	51.48	1.2398	10.34	28.1	112	−25.1
27	53.46	1.2503	10.42	29.0	116	−32.8
28	55.44	1.2610	10.51	30.0	120	−42.2
29	57.42	1.2718	10.60	31.0	124	−54.4
29.8	58.80	1.2804	10.67	31.8	128	−67.0

Table A.8. Cost of Bulk Ice Cream by Scoops (cents/serving)

Price ($/gal)	Size of scoop (average servings/gallon)						
	30 (59)	24 (51)	20 (41)	16 (34)	12 (26)	10 (25)	8 (23)
2.00	3.39	3.92	4.88	5.88	7.69	8.00	8.70
2.50	4.24	4.90	6.10	7.35	9.62	10.00	10.87
3.00	5.08	5.88	7.32	8.82	11.54	12.00	13.04
3.50	5.94	6.86	8.54	10.30	13.46	14.00	15.22
4.00	6.78	7.84	9.76	11.76	15.38	16.00	17.40
4.50	7.62	8.82	10.98	13.24	17.30	18.00	19.56
5.00	8.48	9.80	12.20	14.70	19.24	20.00	21.74
5.50	9.32	10.78	13.41	16.18	21.15	22.00	23.91
6.00	10.17	11.76	14.63	17.65	23.08	24.00	26.09

Table A.9. Seasonal Sales Expectancy—National Average

Month	Percentage of annual sales	Month	Percentage of annual sales
January	3.42	July	16.58
February	3.80	August	14.27
March	5.33	September	10.04
April	7.17	October	6.17
May	10.90	November	4.19
June	14.50	December	3.63

Appendix C
Frozen Desserts Plant Inspection Form

The FDA has recommended good manufacturing practices in manufacturing, processing, packaging, and holding human food. It includes general provisions for definitions and personnel; building and facilities involving plants and grounds, sanitation facilities and controls, and sanitation operations; equipment and procedures; and production and process controls. The following is a typical frozen desserts plant inspection form.

FROZEN DESSERT PLANT INSPECTION FORM

INGREDIENTS PURCHASED DAILY   GALLONS MANUFACTURED DAILY

Mix Sources: _____ Gals. ___   Ice Cream _____
Milk & Cream: Sources: _____ Gals. ___ Mix _____ Other _____
Other Milk Products: _____   Total Gallons _____

Name _____ Location _____

Sir or Madam: An inspection of your plant has been made today, and you are notified of the defects with a cross (X). Violation of the same item on two successive inspections may call for suspension of your license.

(1) *Floors*—Smooth finish; no pools; wall joints and floor surface impervious; trapped drains; no sewage backflow; clean .... ( )

(2) *Walls and ceilings*—Smooth; washable; light-colored finish; good repair; clean ( )

(3) *Doors and windows*—All outer openings effectively protected against entry of flies and rodents; outer doors self-closing; screen doors open outward .... ( )

(4) *Lighting and ventilation*—Adequate light in all rooms; well ventilated to preclude odors and condensation; filtered air with pressure systems .... ( )

(5) *Miscellaneous protection from contamination*—Tanks and vats covered; ports protected; no woven-wire strainers; no straining of pasteurized mix except through perforated metal; no drip from mezzanine or overhead pipes; rooms of sufficient size; ingredients not unloaded directly into processing rooms; pasteurized product not in contact with equipment used for raw or lower grade products unless sanitized; no plant operations in living quarters.. ( )

(6) *Toilet facilities*—Comply with plumbing code; good repair; clean; outside ventilation; no direct opening; self-closing doors; free of flies; washing sign; effluent into public sewer or as approved by the state .... ( )

(7) *Water supply*—Constructed and operated in compliance with state standards;

APPENDIX C

no cross-connections between safe and unsafe water; complies with bacteriological and chemical standards .... ( )

(8) *Hand-washing facilities*—Adequate, convenient; hot and cold water, soap, sanitary towels; hands washed after toilet; good repair; clean; improper facilities not used .... ( )

(9) *Sanitary piping*—Easily cleanable size, shape and length; smooth, non-toxic, uncorroded surfaces; sanitary fittings; interior surfaces accessible for inspection .... ( )

(10) *Construction and repair of containers and equipment*—Easily cleanable, non-toxic, smooth, corrosion-resistant surfaces; no open seams; good repair; accessible for inspection; self-draining; pressure-tight seats on submerged thermometers .... ( )

(11) *Plant cleanliness*—Neat; clean; no evidence of insects and rodents; trash and garbage properly handled; approved pesticides, safely used .... ( )

(12a) *Cleaning of containers and equipment*—Multi-use containers thoroughly cleaned after each usage; equipment cleaned each day .... ( )

(12b) *Bactericidal treatment of containers and equipment*—Containers treated after each cleaning in steam cabinet 170°F. for 15 min. or 200°F. for 5 min.; or hot-air cabinet 180°F. for 20 min.; or steam jet 1 min.; or immersed in standard chlorine or 170°F. water for 2 min.; or automatic washers (residual count not over 1 per cc of capacity); assembled equipment treated daily immediately before run, with flow of 200°F. steam or 170°F. water at outlets for 5 min. or standard chlorine flow for 2 min.; supplementary treatment for surfaces not thus reached .... ( )

(13) *Storage of containers*—Stored to assure drainage and protected from contamination .... ( )

(14) *Handling of containers and equipment*—No handling of surfaces to which ingredients or products are exposed .... ( )

(15) *Storage and handling of single-service containers*—Received, stored and handled in a sanitary manner; paperboard containers not reused .... ( )

(16) *Pasteurization*—

(16a) *Thermometers:* Comply with code specifications .... ( )

(16b) *Time and temperature controls: Batch*—Entire mix excluding fruits, nuts, flavorings, etc. pasteurized; no raw products bypass around pasteurizers .... ( )

Adequate agitation throughout holding period; agitator sufficiently submerged; indicating and recording thermometers on each vat throughout pasteurization; recorder reads no higher than indicator; thermometer bulbs submerged. ( )

Inlet and outlet valves and connections in compliance with specifications ( )

Charts show 155°F. for 30 minutes, plus emptying time if cooling begun after outlet valve opened (also plus filling time when required); no milk added after holding begun .... ( )

Air heating—Air in vats and pockets heated to at least 5°F. above mix temperature during heating and kept at 160°F. or higher during holding, with approved device; approved trap on steam line; approved air thermometer; bulb at least 1 inch above mix.. ( )

Vat and pocket covers and cover ports—No drainage from top of cover into vat, open or closed; ports surrounded by raised edges; pipes, thermometers, etc. through cover have aprons unless joint watertight; covers kept closed .... ( )

Preheating holders—Holders not used as heaters are preheated to pasteurization temperature just before run, also when empty after shutdown exceeding holding period, unless outlet has flow-diversion valve... ( )

(16c) *Time and temperature controls: HTST*—Flow diversion valve complies with requirements; divert flow line self-draining; proper assembly and operation; product held at minimum pasteurization temperature .... ( )

Recorder-controller complies with requirements; recording thermometer no higher than indicating thermometer; cut-in and cut-out temperature at or above 175°F.; sensor located properly .... ( )

Holding tube complies with requirements; no short-circuiting; proper slope ( )

Satisfactory means to prevent adulteration with added water .... ( )

Pasteurized mix in regenerator automatically under greater pressure than raw mix at all times .... ( )

(16d) *Charts*—Used only 1 day, preserved 3 months; complies with requirements ( )

(17) *Cooling*—All fluid milk products cooled to ≤45°F. on receipt unless to be pasteurized within 2 hours; pasteurized mix cooled to ≤ 45°F. and held there until frozen; header gap on surface coolers not less than ¼ inch or thickness of header at gap; condensation and leakage from cooler supports and headers, unless completely enclosed in covers, directed away from tubes and milk trough; recirculated water and refrigerant of required sanitary quality; cooler covered or in separate room; cooler shields tight fitting .... ( )

(18) *Handling of mix*—If not frozen where pasteurized, mix transported in sealed containers; protected against contamination, no dipping, kept covered .... ( )

(19) *Packaging*—If not approved automatic equipment: no contact surfaces handled, packages adequately covered immediately after filling; packages handled in sanitary manner by trained persons .... ( )

(20) *Overflow or spillage*—Discarded .... ( )

(21) *Returns*—No opened containers of mix or frozen desserts returned except for inspection .... ( )

(22) *Personnel, cleanliness and health*—Clean outer garment, washable for inside employees; hands clean; no person with infected wound or lesion; no use of tobacco in processing areas .... ( )

(23) *Vehicles*—Clean; covered; no contaminating substances transported .... ( )

(24) *Bacterial standards of pasteurized mix or frozen desserts*—Not to exceed 20,000 standard plate count or 10 coliform per gram in three out of last five consecutive official samples .... ( )

(25) *Ingredients*—Clean, fresh wholesome flavor, odor, and appearance; stored above floor, kept covered, properly handled; milk products meet bacterial standards; ingredients added after mix pasteurized are of approved quality .... ( )

Remarks:

Date	Inspector

Index*

Absorption, description of, 41
Acesulfame, K, 63
Acid flavor in ice cream, 259
Acidity of,
  ice cream mixes, 38
  sherbets and ices, 236
Acids as ice and sherbet ingredients, 236
Aerating unit, 193
Agar (agar-agar), 29
Agglomerates, 37, 72
Aging of ice cream mix, 158
Air cells in ice cream, 168–170
  "fluffy" texture from, 318
Airmass flow controller, 191
Algebraic method for milk and cream standardization, 105
Algin and alginates 29, 72
Almond toffee ice cream mix formula, 319
American Dairy Products Institute, 146
American Diabetes Association, 63
American Dietetics Association, 63
Ammonia,
  boiling point as related to gauge pressure, 297
  as refrigerant, 215, 293
Apple ice cream,
  fruit and sugar for, 95–96
Apricot ice cream,
  fruit amount and preparation for, 95
  mix formula for, 317
APV Crepaco, 181, 183, 185
Arithmetical method of mix calculation, 107

Aspartame, 63
Asset, most important to ice cream manufacturers, 276
Aufait ice cream,
  definition of, 20
  preparation of, 253

Bacterial counts, 277
Balanced mix, description of, 33
Ball Electronics Systems Division, 219
Banana ice cream,
  fruit amount and preparation for, 95
  nut type, mix formula for, 319
  split, recipe for, 312
Baskin-Robbins, 4
Batch freezers, 4, 195–198
  container filling from, 198
  flavoring in, 160
  operation of, 197–198
Batch pasteurization of ice cream mix, 140, 147–148
Baumé readings,
  of corn and sucrose syrups, 68
  of sugar solutions, 332
  of sweetened condensed skim milk, *334*
Berry ice cream,
  mix formula for, 324
Biscuit tortoni mix formula, 315
Bisque ice cream mix formula, 315
Black walnut ice cream mix formula, 319
Blackberry ice cream,
  fruit amount and preparation for, 95
Bleeding defect in sherbets and ices, 240

*Page numbers in italic refer to tables and illustrations.

339

INDEX

Blenders,
  high shear type, 142–145
Blueberry ice cream,
  fruit amount and preparation for, 95–96
  mix formula for, 317
Body of ice cream, 262
BonBons, 214
Braum's Ice Cream and Dairy Stores, 304
Brine,
  cooling method using, 298
  freezing point of, 335
British Thermal Unit (BTU),
  definition of, 300
Brix readings,
  of sucrose syrups, 68, 332
Brown sugar, as an ice cream sweetener, 62
Bulk ice cream,
  cost by scoop, 335
  packaging of, 205
  sales outlets for, 301–312
Bulking agents, 59
Bulky-flavor ice cream, 1
Butter,
  as ice cream ingredient, 50, 52, 54
Butter crunch ice cream mix formula, 315
Butter oil (anhydrous milkfat),
  as an ice cream ingredient, 52, 54
Butter pecan ice cream mix formula, 318
Buttermilk, cry, 56
Butterscotch ice cream mix formula, 315
"Buttery" texture defect in ice cream, 263
Butyric acid, 14

Cacao bean, cultivation and processing of, 87
Cakes, ice cream, 253–255
Calcium,
  in ice cream, 15
  salts, as ice cream ingredients, 32
Calcium chloride brine,
  freezing point, relations of 335
Calculations for ice cream mixes, 104–138
  arithmetical method, 107
  available ingredients/Desired composition table, 114
  complex mixes, 117
  computer use in, 127
  formulas for, 110
  importance of, 104
  mathematical processes used, 104
  mix decisions, 113
  Pearson square, 106
  simple mixes, 113
  standardizing milk and cream, 106
  serum point method, 108
  using leftovers, 124
  using proof sheets, 108
Calories in ice cream, 10–13, 173, 204
  and related products, 10
Caproic acid, 14
Caramel ice cream mix formula, 314
Caramel nut ice cream mix formula, 319
Carbohydrates in ice cream, 10–12
Carboxymethylcellulose (CMC; cellulose gum), 29, 72–73
Carob, *see* Locust beam gum
Carrageenan, 29, 66
Casein, 49
Caseinates, 32, 57
Cellulose as stabilizers,
  gel (microcrystalline), 66
  gum (carboxymethylcellulose), 73
Cherry ice cream,
  fruit amount and preparation for, 95
  recipe and procedure for, 97
Chlorine,
  in sanitizing solutions, 287
Chlorofluorocarbons (CFCs) as refrigerants, 195, 213
Chocolate, *see also* Cocoa, 87–91
  coatings, 250–252
  confections, 91
  ice cream, *see* Chocolate ice cream, 90, 315
  liquor, 88
  sweet milk type, 88
  syrup, 90
  with vegetable fat, 251
Cholesterol in ice cream and related products, 14
Chocolate chip ice cream formula, 315
Churning of ice cream, 37, 43
  influence of emulsifiers, 78
  influence of mineral salts, 39
CIP (cleaned-in-place) system in ice cream manufacture, 151
  cleaning solutions for, 279–280
  of continuous freezers, 195
  important factors in, 280–286
Citrates as ice cream ingredients, 32

INDEX

Citric acid as an ice and sherbet ingredient, 236–239
Citrus fruit in sherbets, 2
Classification of ice cream and related products, 18
Clean Air Act, 293
Cleaning,
  equipment, 278–285
  hands, 305
  soft-serve freezers, 226
  spray type devices used in, 283–285
Cocoa, *see also* chocolate
  beans, 87
  breakfast type, 89
  butter, 88, 251
  color, 89
  Crillos, 87
  defatted, 89
  Dutch process, 88
  fat, 88
  Foresteros, 87
  natural process, 88
  nibs, 88
  sweet, 251
Coconut pineapple ice cream mix formula, 319
Code of Federal Regulations, 3, 215
Codex Alimentarius, 90
Coffee ice cream mix formula, 314
Cold elite, 224
Coliform bacteria, 277, 289
Colloids, 34–35
Coloring materials,
  Certified, 101
  FD&C, 101
Combining ingredients, 139
Community Right to Know Act, 215
Composition,
  cream, *334*
  ice cream, 2, 10, *22–23, 46*
  milk, 45, *334*
Computers, 180–187
Concentrated milks, 56
Cones, ice cream, 4, 127, 208, 256, 301
Constituents in ice cream, role of, 26, 34
Consumption of ice cream, 6, 302
Contact plate freezers, 214
Containers, frozen dessert, 200, 305
Continuous ice cream freezers, 177–195
Continuous phase, 166
Continuous process system, 142
Cooked flavor, 259

Corn sugar, 14
Corn sweeteners, 29, 61
Corn syrups, 61
  physical constants, *69*
Cost of ice cream per scoop, *335*
Council on Scientific Affairs, 63
Cream,
  frozen, 49
"Crumbly" body defect, 262
Cryohydric point, 165
Cryoscope for freezing point, 42
Cups for ice cream, 208
Curdy melt defect, 268

Daily value, 203
Dairy and Food Industry Supply Association, 4, 186, 248
Dairy Queen, 5
Dashers, 182–184
Date walnut ice cream, 322
Demulsification of fat, 37
Defects. *See also* individual defects,
  in ice cream, 258–269
    body, 262
    flavor, 258–262
    melting quality, 268
    texture, 263
  in sherbet, 239
Defrosting methods, 297
Density of mixes, 38
Detergents, 278–280
  acids, 280
  alkalis, 279
  chelating agents, 280
  functions, 279
  phosphates, 279
  surfactants, 280
Dextrose, 61
Dextrose equivalents (DE), 29, 174
Digestibility of ice cream, 16
Diglycerides, 75
Dipping frozen desserts, 208, 307–311
Dips per gallon, *311*
Direct expansion refrigeration, 299
Discriptors of ice cream, 202
Dispensing machine for soft-serve, 230
Disperse phase, 166
Dispersion of stabilizers, 143
Drive-in stores, 303
Dry ice, shipping with, 220
Dry milk, 56
Dryness of body of ice cream, 37

Du Nouy apparatus, 40
Dutch process cocoa, 88

Egg nog, 322
Egg yolk solids, 29, 79, 222
Electron micrographs of internal structure of ice cream, 169–172
Emery Thompson, 196
Emergency facility response plan, 215
Employees, 306
Emulsifiers, 30, 43, 71, 75
Emulsions,
    fat-in-serum, 31
    stability, 36
Energy value of ice cream, 10, 15
English plum pudding, 319
English toffee, 315
Enterobacteriacerae, 289
Environmental impact, 201
Enzymes, 50
Eskimo pie, 4
Ether extraction for fat content, 164
Evaluating frozen desserts for qualities, 269–274
Extruded ice cream novelties, 242, 244–248

Fancy molded novelties, 321
Fat, 47–49
    analogs, 64
    crystallized, 170
    extender, 64
    fractionated, 55
    globules, 47, 75–76, 159, 167, 169
    mimetic, 64
    sparer, 64
    substitute, 64
Fat free ice creams, 325–327
Fatty acids, 14, 47
    omega–3 type, 54
Federal Standards, 1, 3, *24,* 203
Fig-nut ice cream,
    mix formula for, 317
Filler, 191
    rotary, 209
    parallel lanes, 210, 256
Filling from a batch freezer, 198
Fill weight, 211
Flavor, 33
    categories, 20–21, 82
    combinations, 98–100
    groups, *81*

Flavor defects,
    acid (sour), 259
    categories, 258
    cooked, 259
    egg yolk, 260
    lacks freshness, 259
    lipolyzed, 260
    old ingredient, 260
    oxidized, 259
    salty, 260
    unnatural, 261
    whey, 260
Flavor for nonfat and lowfat ice cream, 102
Flavoring mixes, 160
Flavorings, 81–90
    complex, 97
    liquer, 83
    natural, 82
    synthetic, 83
Flow diversion, 150
Flow meter, 142, 178, 190, 193
Foamy melt, 269
Food solids in ice cream, 333
Forming machine for bulk containers, 206
Formulas for frozen desserts. *See also,*
    specific flavors, 315–327
    complex flavors, 98
    fat-free, 325–327
    frozen yogurt, 318
    homemade, 323–324
    ices, 249
    light, 327
    low fat, 327
    low-lactose, 324
    reduced-lactose, 324
    sherbets, 238
    soft-serve, 230–231
    sugar-free, 325–327
Frappè, 20, 320
Freezing point, 42, *43,* 59, 167, *173*
    method of predicting, 175
Freezing process, 164–194
Freezing systems,
    Batch,
        Coldelite, 224
        Emery Thompson, 196
        SaniServe, 223
        Taylor, 228
    Continuous,
        APV Crepaco, 181, 183, 185

INDEX **343**

Gram Equipment Company of
  America, 192
Tetra Laval Hoyer, 189
Waukesha Cherry Burrell, 179–180
Programming elements, 188
Freezing time,
  ice cream, 177
  ices and sherbets, 239
Freezers,
  Batch, 177, 195–198
    cleaning, 194
    construction, 177
    operation, 197
  soft-serve, 177, 222–229
  Continuous, 177–195
    heat treatment type, 228
    programmable, 190
    schematic drawing, 183
    shut down of, 194
French vanilla ice cream, 2
Frequency inverter, 178
Freons, 195, 293
Frosted malted, 321
Frozen,
  confections, 2
  cream, 49
  custard, 2
  yogurt, 2, 317–318
Fructose, 61
Fruits,
  amounts, 95
  aseptic, 91–93
  candied, 96
  dried, 96
  fresh, 94
  frozen, 94
  glaced, 96
  ice creams, 91, 316
  pack, 94
  preparation, 95
  processed, 91
  salad, 321
Fudge bars, 249

B-D-galactosidage, 14, 15
Gelatin, 29
Gelatin cube, 20
Gelato frozen dessert, 232
Ginger ice cream,
  mix formula for, 315
Glacier-freezing tunnel, 248
Globules of milk fat, 49, 76, 167, 169

Glucose (dextrose), 58–60
Goat milk ice cream, 2
Good Humor ice cream bar, 4
Gram Equipment Company of America,
  frozen dessert equipment, 245–248
Granite, 20
Grape ice cream,
  fruits and sweetener for, 95
  mix formula for, 317
Gravity flow storage/retrieval racks, 216
Guar gum, 29, 66, 72–73
"Gummy" body defect, 262
Gums in ice cream. *See* individual gums

HACCP (Hazard Analysis Critical Control Point), 277
Handling frozen desserts, 216
Hands, cleanliness, 305
Hardening of frozen desserts, 211–215
  in cabinets, 213
  in cold cells, 213
  on contact plates, 214
  facilities for, 212
  rate of, 212
  storage capacity for, 215
  time of, 212
  in tunnels, 213
Hazardous waste, 215
HAZWOPER regulation, 215
Heat absorbed during freezing, 175
Heat Treatment Freezer for soft-serve products, 228
Heat, types of,
  latent heat, 165, 176
  specific heat, 173, 176
High-temperature short-time pasteurization, of frozen dessert mixes, 147–151
History of ice cream industry, 3, 328–330
Homemade ice cream formulas, 323–324
Homogenization, 151–158
  equipment, 152–154
  pressures, 158
Homogenized mixes, 42, 156
Honey in ice cream, 63
  mix formulas and preparation, 323
Housekeeping, 304
How to dip ice cream, 307
Hydration of solid ingredients, 35
  effects of minerals on, 36
Hydrocolloids in frozen desserts, 64, 71

Hygienic standards,
  of construction, 288
  of personnel, 288, 305
  practices thereof, 289
Hypochlorites as sanitizers, 287

IAMFES, a professional society, 186
Ice cream,
  bars, 250
  bulky flavored, 19
  cakes, 253, 255
  clinics, 272
  dipping of, 307
  federal standards, 1, 3, 24, 203
  fruit flavored, 19
  improvers, 75
  light type, 19, 25
  lowfat type, 19, 25
  pies, 253
  plain, 19
  reduced fat type, 19, 25
  store, *304*
Ice and salt method of freezing, 292
Ice crystals in frozen desserts, 171, 263–267
  effect of milk fat on, 27
Ices,
  formula for, 249
  preparation of, 236
Ingredient feeder, 161–162, 191
Ingredients of frozen desserts,
  composition, *52*
  density, *52*
  optional, 32
  specifications for, 289
  weight, *52*
Inspection form for frozen dessert facilities, 336–338
Interfacial tension, 41
Internal structure of ice cream, 168, 263–267
International Dairy Federation, 200
International Ice Cream Association, 248
Inventory,
  control, 216
  rotation of products, 290
Invert sugar, 14
Iodophors as sanitizers, 287
I-Scream bar, 4
Iso malt as bulking agent and sweetener, 67

Kilocalories, 10
Klondike bar, 214

Labeling of frozen desserts, 201
Labels for nutrition facts, 12–13, 201–205
Lactose (B-D-galactosidase), 14, 50
Lacto (ice cream product), 321
Lactobacillus bulgaricus, yogurt culture, 317
Lactose,
  in dry whey, 58
  in ice cream, 14
  in milk, 50
  in nonfat dry milk (NDM), 58
  low-lactose milk solids, 32, 57
  malabsorption, 14
  sweetness of, 59
Laminar flow in pasteurizers, 150
Latent heat,
  of evaporation, 300
  of fusion, 165, 176, 300
Laxation, 67
Lecithin, 79
Lemon ice cream formula, 317
Licorice ice cream formula, 315
Life cycle analysis, 201
Light (lite) ice cream, 203
Linear programming, 127
Lipolyzed flavor, 260
Liquefying dry ingredients, 142
Liquid nitrogen as a refrigerant, 212
Listeria monocytogenes, 160, 276, 289
Locust bean gum (carob), 73
Lowfat ice cream, 203
Low-lactose milk solids, 32, 57

Magnetic flow meter, 193
Malted milk, 232
Manager, of ice cream store, 306
Maltitol, a sugar alcohol, 67
Mannitol, a sugar alcohol, 67
Maple ice cream formula, 314
Maple sugar, 62
Marshmallow ice cream formula, 315
Mass flow meter, 142, 178, 190
Mathematical calculations of mixes, 104
McSundae, 5
Mechanical refrigeration, 294
Mellorine, 2, 19, 232
Melting quality of ice cream, 267–269
Merchandising, 302
Micelles, of casein, 76, 170
Microbial gums, 72

INDEX

Microcrystalline cellulose (cellulose gel), 73
Microparticulation, 65
Microprocessor, 142
Microscopic appearance of mix, 156
Migration of packaging materials, 201
Milk fat, 13, 27, 47, 264
    fractions, 55
    globules, 49, 76, 166–170
    sources of, 51
    sugar blended with, 54
Milk shake, 2, 311
    base, 232
    mixes, 231
Mineral salts, 58
    in ice cream, 15, 49
Minimum weight, 202
Mint chip ice cream formula, 315
Mint ice cream formula, 314
Mixes,
    acidity, 38
    complex, 104
    density, *38*
    processing, 139–163
    simple, 104
Mixers, high shear, 142
Molasses toffee ice cream formula, 315
Molded frozen dessert products, 241
Molding method, 243
Monoacylglycerols, 14
Monoglycerides, 75
Montreal Protocol of 1987, 293
Mousse, 20, 320
Multi-flavored ice creams, 97–100
    from Bunge Foods, 98
    from Consolidated Flavors, 99
    from Creative Flavors, 99
    from David Michael and Co., 99
    from Fantasy Blanke Baer, 99
    from Fruitcrown Products Corp., 99
    from Geurnsey Dell, 99
    from Pecan Deluxe Candy Co., 99–100
    from Star Kay White, 100
    from Virginia Dare, 100

National Ice Cream and Yogurt Retailer's Association (NICYRA), 312
Nisselrode pudding, 319
Nitrogen, liquid, 212
Nonfat dry milk (NDM), 56, 58
Nonfat ice cream, 203
Nonfat milk solids (NMS), 16, 27, 55–58

Norse Dairy Systems cone and cup filler, 256
Northfield Freezing Systems spiral freezer, 214
Novelties, 214, 241–257
    equipment, principles and guidelines, 248
    extruded types, 242, 244–248
    molded types, 243, 248
    rotary machine for, 248
    styles of, 250
Nucleation, 172
Nut ice cream formulas, 318–319
Nutrition Facts labeling, 203, *204*
Nutritional Labeling and Education Act of 1990, 201
Nuts for frozen desserts, 100
    recommended amounts, 101, 318–319

Occupational Safety and Health Administration (OSHA), 215
Off-flavors in ice cream, 258–261
Optional ingredients, 32
Orange ice cream formula, 317
Orange-pineapple ice cream formula, 317
Oven method for total solids, 164
Overrun in frozen desserts, 2, 32, 165, 178–182, 266
    and dipping losses, *310*
    control of, 187
    typical thereof, *208*
    and weight, 333
Overwrap, 208

Packages and packaging, 200–220
    appearance, 267
    bulk ice cream type, 206
    defects of, 267
    for direct sale to consumers, 207–210
    forming machine for, 206
    for mixes, 159
    "Life Cycle Analysis" of packaging systems, 201
    optimum lot size, 210
    overwrap of, 208
    size shares, 208
    target weight, 211
    Technical Guide of the International Dairy Federation (IDF), 200
Palatability, 16
Parfait, 312, 320

Particle size,
  in milk, *48*
  in ice cream, *48*
Parvine, 2
Pasteurization, 147–151
  batch (Low-temperature long-time), 148
  continuous
    High-temperature short-time (HTST), 148
    Higher-heat shorter-time (HHST), 147–151
    Ultra High temperature (UHT), 147–148
Pasteurized milk ordinance (Grade A PMO), 287
Peach ice cream, 96–97
  formula, 316
Peanut brittle ice cream formula, 315
Pearson square method,
  for standardization of milk and ice cream, 106
Pectins for stabilizing sherbets, 72
Peppermint ice cream formula, 315
Personnel in ice cream stores, 305
Personal hygiene in ice cream stores, 305
Phases of the physical state of frozen desserts,
  continuous phase, 166
  disperse phase, 166
Phosphates as casein solubilizers, 32
Phosphorus,
  of the human body, 15
  in ice cream, 16
Phospholipids,
  in milk fat, 14
  as stabilizers, 153
Physical structure,
  of homogenized mixes, 156
  of ice cream, 2, 166–172
Pineapple ice cream formula, 317
Pistachio ice cream formula, 318
Plain ice cream formula, 314
Plain mixes, formulas for calculating, 110
Plant exudates as stabilizers, 72
Plant inspection form, 336–8
Plastic cream, 54
Polydextrose as an ice cream bulking agent, 66
Polyols as sugar replacers in ice cream, 67

"Poly" types of emulsifiers, 41–43
  Spans, 77
  Tweens, 41–42, 77
Polymers,
  list of for packages, 201
Polyoxyethylene sorbitan monoleate, 41, 77
Popsicle, 4
Positive displacement pump, 160, 162, 178
Powder funnel, 146
Pre-emulsifying device (aerator), 182
Premium ice cream, 25
Processing mixes, 139–163
  flow chart, 140
Product code, 217
Production statistics for frozen desserts, 6, 8
Profitability within the ice cream industry, 303
Programmable controller (PC), 181, 187
Programmable logic controller (PLC), 180, 283
Proof sheet, *111*
Protective clothing and masks, 215
Protein,
  in ice creams, 13
  in market ice cream of representative types, 205
Puddings,
  description of, 19
  formulas for,
    Date, 320
    English plum, 319
    Frozen, 319
    Nesselrode, 319
    Plum, 320
    Tutti-frutti, 319
Punch, formula for, 321

Quality control, 216, 220, 276–291
Quaternary ammonium compounds, as sanitizers, 287

Rancid (lipolyzed) off-flavor in ice cream, 260
Rasberry ice cream,
  formula and procedure for, 97
  fruit amount and preparation for, 317
Reduced fat ice cream, 203
Reference amounts of nutrients, 202
Refrigeration, 292–300

INDEX 347

brine method, 298
calculating requirements for, 175
compressors, 300
condenser, 294
direct expansion, 299
flooded system, 299
high pressure side, 294
ice and salt method, 292
mechanical, principles of, 294, *295*
operating precautions, 296
receiver for refrigerant, 294
tons of, 300
Retail sales, 301–302
Reynolds number, 282
Riboflavin in ice cream, 16
Risk management, 216
Rotation of inventory, 290

Safety, 290
Sales,
  by season, 335
  representatives, 306
Salmonellae, 160, 277, 289
Salt, 101
Sandiness defect, and nonfat milk solids, 28
SaniServe, freezer supplier, 223, 225
Sanitary environment, 287
Sanitary Standard, 3A, 148, 154, 207
  for freezers, 186
Sanitizers and sanitization, 286–289
  hypochlorites, 287
  inspection, 336–338
  iodophors, 287
  quaternary ammonium compounds, 287
  of soft-serve freezers, 226
  sulfonic acid, 287
Sawvel, T.D. Company,
  bulk filler by, 207
Score card, 271
Scoring frozen desserts,
  guide, 270–273
  methods, 270–274
Seaweed extracts as stabilizers, 72
Seed gums as stabilizers, 72
Sensible heat, 165
Serum point method, 108
Serving size, 202, 205
Sherbets,
  characteristics, 235
  composition, 234

  defects, 239
  description, 2
  formulas, 237
  fruits, 2
  ingredients, 235
  preparation, 237
  stabilizers, 236
  sweeteners, 235
Shipping with dry ice, 220
Shrinkage defect, 43, 264
Silo tank, 284
Simple mixes, calculation, 113–117
Skim milk, 55
Soda, ice cream,
  date introduced, 4
  method of preparation, 4, 311
Soda fountain, 303
Soda fountain preparations, 311–312
Sodium,
  caseinate, as an ice cream ingredient, 57
  hypochlorite, as a sanitizer, 287
  in frozen desserts, 101
Soft-serve frozen desserts, 222–233
  custard, 222
  dispensers of, 230
  freezers for, 222–226, 228
  ice cream, 222, 229–231
  mixes, 229
  overrun, 231
  temperature, 231
  yogurt, 231
Solubilization of stabilizers, 143
Solutions, true, 34
Sorbets, 320
Sorbitol, a sugar replacer, 66–67
Soufflé, 20, 321
Spans, as emulsifiers, 77
Specials, 241, 312, 322
Specific heat, 173, 176
Specifications, of products, 289
Spices, 101
Spumoni, 20, 255, 257
Stabilizers, 29, 71–75
  agar, 72
  alginates, 72
  blends, 79
  carboxymethylcellulose (CMC) (also cellulosegum), 73
  carrageenans (Irish moss), 73
  cellulose gel (microcrystalline cellulose), 73
  for sherbets, 236

functions of, 30
guar gum, 74
locust bean gum (also carbo bean gum), 74
Stainless steel, 280
Standardizing mixes, 106
Standard of Identity, Federal, 18
Statistical quality control, 211
Stiffness of ice cream, 178–182
Storage/retrieval systems, 216, *218*
Storage tank, 284
Stores,
  ice cream, 303–310
  manager of, 306
Strawberry,
  ice cream, 96
  fruit for, 95, 316
  puree, 96
  solid pack, 96
Sucralose, 64
Sucrose, 14, 60, *172*
  equivalent sweetness, 174
  freezing point of solutions, *173*
Sugar,
  alcohols, 97
  brown, 62
  corn, 14, 62
  maple, 62
Sugar-free, 21, 100
Sundae, 3, 11
Supercritical fluid extraction, 88
Superpremium ice cream, 25
Surface tension,
  of mixes, 40
Sweet potato ice cream, formula for, 323
Sweetened condensed milk, 57
Sweeteners, 28, 58–64
  defects of, 261
  nonnutritive types, 63
  syrups, 262

Target weight of packaged products, 211
Taylor, freezer manufacturer, 228
Temperature monitor, 218, *219*
Tempering frozen desserts, 269
Tests for,
  milkfat, 290
  total solids, 164, 290
Texture defects,
  buttery, 263
  coarse, 264
  fluffy, 266–267

  sandy, 28, 267
Thompson, Emery
  batch freezer, 4, 196
Titratable acidity, 51
Toffee ice cream formula, 315
Tortini ice cream formula, 315
Total solids,
  in ice cream, 30, 290
  oven test method, 164
Training of ice cream store employees, 306–309
Tunnels, hardening,
  Tray type, 213
  Spiral type, 213
Tweens, 41, 77, 80

Ultrafiltration, 15, 203
Unnatural flavor, 261
U.S. Department of Health and Human Services, 201
U.S. Food and Drug Administration, 248
U.S. Occupational Safety and Health Administration, 215
U.S. Recommended Daily Allowance, 203

Vanilla, 83–87
  Bourbon, 85
  concentrated, 85
  extract, 85
  flavor notes, 86
  fold, 85
  imitation, 86
  Indonesian, 85
  Mexican, 85
  paste, 85
  powder, 85
  Tahitian, 85
  unnatural flavor, 261
Vanilla ice cream formula, 314
Vanillin, 85
Variable speed drive, 178
Vegetable fat frozen dessert, 232
Viscosity,
  apparent, 40
  of mixes, 39, 177
  true, 40
Vitamins, 16, 51, 203
Vogt freezer, 179–180

Walnut ice cream, formulas, 95, 319
Water,
  hardness, 278

in ice cream, 31
quality, 278
state of in frozen desserts, 166
Water ice, 2, 20, 235
Wattage required, 178
Waukesha Cherry-Burrell, 161–162, 179–180
Weak body defect, 263
Weight of frozen desserts,
   per gallon, 202
   per serving, 202
Whey, 58
Whey protein concentrate, 78

Whey proteins, 49
Whey protein isolate, 78
Whey in a off defect, 269
Whipping rate, 42
Wholesale trade, 301–302
Woodson, Inc, 217

Xylitol for sugar-free frozen desserts, 67

Yield of product, 333
Yogurt, frozen,
   formulae for, 318